MATHS IN PERSPECTIVE 2

Mechanics

Barbara Young
Tarporley County High School, Cheshire

D0419803

Edward Arnold
A division of Hodder & Stoughton
LONDON   MELBOURNE   AUCKLAND

ACKNOWLEDGEMENTS

Thanks are due to my colleagues for their help and encouragement and, in particular, to J.D. Hatton for his invaluable advice and ideas, not to mention his help with trialling the material in this text. Thanks are also due to the many students who have helped with comments, suggestions and error spotting. But most of all thanks are due to Jane Attwater, Matthew Burton, Alex Degnen, David Forrester, Sandy Griffin, Graham Holden, Mark Jones, Jonathan Kendall, Peter Ravenhill, Chris Shorrock, Ben Stephenson, David Ward and Julian Williams who have borne the brunt of the experimentation needed to produce this book and whose helpful and constructive criticisms have produced many improvements.

And, finally, thanks to my husband, without whose practical help and encouragement none of these texts would have been produced.

The publishers would like to thank the following examining boards for their permission to include questions from past examination papers: The Associated Examining Board; Joint Matriculation Board and University of London School Examinations Board.

© 1989 Barbara Young

First published in Great Britain 1989

British Library Cataloguing in Publication Data

Young, Barbara
  Mechanics.
  1. Mechanics. —For schools
  I. Title   II. Series
  531

ISBN 0 7131 7822 1

Phototypesetting by Thomson Press (India) Limited, New Delhi.
Printed and bound in Great Britain for the educational publishing division of Hodder and Stoughton, Mill Road, Dunton Green, Sevenoaks, Kent by Richard Clay Ltd, Bungay, Suffolk.

CONTENTS

Unit 4 FORCES AND MOMENTS: RIGID BODY EQUILIBRIUM

Unit 5 WORK, ENERGY, POWER, IMPULSE, MOMENTUM

Unit 6 CENTRES OF GRAVITY

Unit 11 VECTOR APPLICATIONS TO MECHANICS: RELATIVE VELOCITY

Unit 12 FURTHER CASES OF RIGID BODY EQUILIBRIUM: SYSTEMS OF FORCES

INTRODUCTION

This is an A level course in mechanics—not an A level text book. It was developed to meet the following objectives.

- To provide a course that would cater for A level classes that include both average and very able students.
- To enable the students to spend as much time as possible *doing* mathematical problems, rather than writing up dictated notes.
- To make each piece of work skill-centred rather than just topic-centred. To make each piece of work include plenty of rote practice of the skills involved but also include work to stretch abler students.
- To develop a 'rolling' course where techniques and topics are met over and over again rather than just in one chapter. In particular to repeat techniques that students find difficult or tend to forget easily.
- To include examples that teach mathematical 'tricks of the trade' as well as basic techniques.
- To introduce students to A level questions as early in the course as possible.
- To include A level questions as part of the course material as often as possible.
- To provide material for abler students (or for any keen students) to do extra work on their own.
- To provide a comprehensive set of reference/revision notes.

To meet these objectives:
- The course consists of a series of sections of work. The time allowed for each section is one (60–70 minute) lesson plus an equal amount of private study time.
- This allocated time includes time to introduce and/or explain the theory involved, to go over examples and to do the work associated with the section.
- Each piece of work includes plenty of practice in the skills and techniques that are to be developed at that point and also includes more difficult questions for abler students. Students do as much as they have time for, but the exercises are so planned that even slow students should have tried all the basic techniques in the time allowed.
- The teacher provides all the theory and explanations required in each case. The only theory provided in the text is that needed for reference and revision.
- The examples in the text cover not only basic skills but also many 'tricks of a mathematicians's trade'.
- Certain techniques that students find difficult or tend to forget easily are repeated and revised at regular intervals throughout the course.
- At the end of each unit of work (comprising 3–12 sections) one or two miscellaneous exercises are provided (leading possibly to a test on the section).
- These miscellaneous exercises are of two distinct types. At the beginning of the course each section has a Miscellaneous Exercise A consisting of simple questions similar to those covered in the work just done. There is also a second exercise—B consisting of A level questions.
- **Most Important** This second type of miscellaneous exercise is composed of A level questions, but not only on the topic just done. Each question *includes* a skill developed in the section just covered but covers also topics and skills met earlier in the course.

- The sections leading to the second type of miscellaneous exercise are labelled P.Q. + to indicate that they require more time than the other sections. Sufficient questions are included to provide *both* practice on A level questions during the course *and* revision of the course prior to the A level exams.

Notes on teaching the course

If a teacher introduces and explains any necessary background and theory, it is quite feasible for students to work through the examples in many of the sections themselves. However, the teaching value of the examples of greatly increased by either the teacher going through the examples with the students, or the students tackling the examples as a group (with guidance), without reference to the model answers.

Also, whilst care has been taken to include one example for each technique required, often students need to go through several examples of the same kind before the idea is assimilated.

Students' Guide

All A level Mechanics syllabuses contain a common core. Thus most of the topics covered in this course are included in all Mechanics syllabuses. The table below lists the sections NOT included in each of the syllabuses listed.

The A level questions included in the Miscellaneous Exercises NB, at the end of most units, each have a bracket after the question which tells the student
  (i) the A level board: A = AEB, J = JMB and L = London
  (ii) the section(s) which contain the relevant theory used in the question.
This last piece of information has two purposes:
  first—to tell the student *not* to attempt this question if any one of the sections listed is not in his/her syllabus;
second—to instruct the student where to look, within the course, for help with the question.

Thus it would help the student if he/she used a piece of card as a bookmark upon which were listed the sections not in his/her syllabus, taken from the table below.

Table of Non-Core Sections

An X indicates that the section is NOT in the syllabus.
An F indicates that the section is in the Further Mechanics syllabus.

	J	L	A	S	O	C	O&C	W	NI
1.7							X		
2.3, 2.9									X
2.10	X		X	X	X	X			X
3.5									X
4.7				X			X		
5.9					X	X			
6.3, 6.4, 6.5						X			
7.6, 7.7		F	F	X	X	X			X
7.8, 7.9		F	F	X	X	X	X		X
8.6				X	X	X			X
8.7		F				X	X		
9.3		F				X			
9.4		F				X			
9.5		F				X	X		
10.1, 10.2						X			
10.3						X	X		
10.4		F		X		X			
10.5		F		X		X			
10.6		F		X		X	X		
10.7		F	F	·		X			
11.1					X	X	X		
11.2					X	X	X		
11.3	F	·			X	X	X		
11.4	F								
11.5	F								
11.6	F								
11.7	F				X		X		X
11.8	F				X		X		X
11.9	F					X	X		
12.1	X	X							
12.2		X					X		
12.3		X		X			X		
12.4		X					X	X	X
12.5		X			X	X			

J = JMB; this course covers the whole of the JMB Mechanics Syllabus (MI)
L = London; this course covers all the mechanics in the Mathematics Syllabus B371.
A = AEB; this course covers the whole of the Applied Mathematics Common Paper for use as 602/1 and 636/2.
S = SUJB; this course covers all the mechanics in AMI.
O = Oxford; this course covers all the mechanics in 9850/2.
C = Cambridge; this course covers the Particle Mechanics in 9205.

O & C = Oxford and Cambridge; this course covers the mechanics in 9650 with the exception of moments of inertia and rigid body rotational dynamics.

W = WJEC; this course covers all the mechanics syllabus with the exception of second order differential equations.

NI = Northern Ireland; this course covers all the mechanics in the Mathematics syllabus with the exception of second order differential equations, the method of dimensions and the law of gravitation.

Note for students who wish to tackle extra questions on their own to try and improve their grade potential.
The more A level questions you do, the better you will get at doing them. So, at the end of each unit there is a large selection of A level questions which can be tackled at any time after the work in that unit has been completed. But, before you attempt any question, check, by looking at the end of the question and at the above table, to see whether it includes any sections that are not on your syllabus.

A DYNAMIC INTRODUCTION

1.1 Vectors

A **scalar** quantity has magnitude only.
  E.g. mass, temperature, time, distance, speed.
A **vector** quantity has both magnitude and direction.
  E.g. force, displacement, velocity.

Representation of a vector

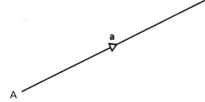

A vector quantity can be represented by a directed line segment.
$\overrightarrow{AB}$ represents the vector with magnitude and direction given by the line segment AB.

Vectors can also be written as single letters:
$\overrightarrow{AB} = \underset{\sim}{a}$ (handwritten) or $\overrightarrow{AB} = \mathbf{a}$ (bold type when printed)
Note that $\overrightarrow{BA} = -\mathbf{a}$.

Modulus of a Vector

The **modulus** of a vector is its magnitude.
  E.g. the modulus of a velocity is its speed.
The modulus of $\overrightarrow{AB}$ is written as $|\overrightarrow{AB}|$ or AB.
The modulus of $\mathbf{a}$ is written as $|\mathbf{a}|$ or a.

Unit Vectors

Unit vectors are vectors with unit magnitude.
$\hat{\mathbf{a}}$ is a unit vector in the direction of the vector $\mathbf{a}$.
$\mathbf{i}, \mathbf{j}, \mathbf{k}$ are unit vectors parallel to the axes Ox, Oy, Oz.

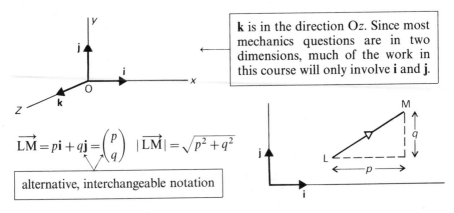

$\mathbf{k}$ is in the direction Oz. Since most mechanics questions are in two dimensions, much of the work in this course will only involve $\mathbf{i}$ and $\mathbf{j}$.

$$\overrightarrow{LM} = p\mathbf{i} + q\mathbf{j} = \begin{pmatrix} p \\ q \end{pmatrix} \quad |\overrightarrow{LM}| = \sqrt{p^2 + q^2}$$

alternative, interchangeable notation

Parallel Vectors

$$\binom{ka}{kb} = k\binom{a}{b}$$

$\binom{ka}{kb}$ is parallel to $\binom{a}{b}$ and k times its length.

Example 1   Find the vector parallel to $3\mathbf{i} + 2\mathbf{j}$ and three times its length.

Vector required $= 3(3\mathbf{i} + 2\mathbf{j}) = \underline{9\mathbf{i} + 6\mathbf{j}}$

Position Vectors

$\overrightarrow{OP}$ = position vector of P with respect to O.
$\overrightarrow{PQ}$ = position vector of Q with respect to P.

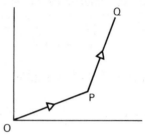

Addition and Subtraction of Vectors

1   If **x** is represented by  and **y** by

then **x** + **y** is represented by

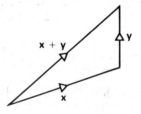

and **x** − **y** is represented by

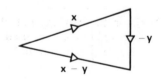

2   If $\mathbf{x} = a\mathbf{i} + b\mathbf{j}$ and $\mathbf{y} = p\mathbf{i} + q\mathbf{j}$

then $\mathbf{x} + \mathbf{y} = (a + p)\mathbf{i} + (b + q)\mathbf{j}$
and $\mathbf{x} - \mathbf{y} = (a - p)\mathbf{i} + (b - q)\mathbf{j}$

3 If $\mathbf{x} = \begin{pmatrix} a \\ b \end{pmatrix}$ and $\mathbf{y} = \begin{pmatrix} p \\ q \end{pmatrix}$

Then $\qquad\qquad\qquad\qquad \mathbf{x} + \mathbf{y} = \begin{pmatrix} a+p \\ b+q \end{pmatrix}$

and $\qquad\qquad\qquad\qquad \mathbf{x} - \mathbf{y} = \begin{pmatrix} a-p \\ b-q \end{pmatrix}$

The **resultant** of two vectors is the sum of the vectors.

with respect to

Example 2 The position vector of P w.r.t. O is $\begin{pmatrix} 3 \\ 2 \end{pmatrix}$. The position vector of Q w.r.t. P is $\begin{pmatrix} 3 \\ 5 \end{pmatrix}$. Find the position vector of Q w.r.t. O.

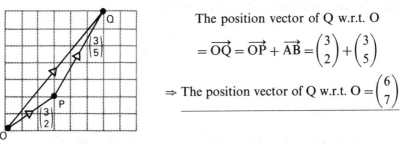

The position vector of Q w.r.t. O

$$= \overrightarrow{OQ} = \overrightarrow{OP} + \overrightarrow{AB} = \begin{pmatrix} 3 \\ 2 \end{pmatrix} + \begin{pmatrix} 3 \\ 5 \end{pmatrix}$$

$\Rightarrow$ The position vector of Q w.r.t. O $= \begin{pmatrix} 6 \\ 7 \end{pmatrix}$

Example 3 If $\mathbf{a} = 3\mathbf{i} + \mathbf{j}$ and $\mathbf{b} = 2\mathbf{i} - 5\mathbf{j}$, find
(a) the resultant of $\mathbf{a}$ and $\mathbf{b}$ (b) $\mathbf{a} - 2\mathbf{b}$.

(a) $\mathbf{a} + \mathbf{b} = \begin{pmatrix} 3 \\ 1 \end{pmatrix} + \begin{pmatrix} 2 \\ -5 \end{pmatrix} = \begin{pmatrix} 5 \\ -4 \end{pmatrix} \Rightarrow$ resultant $= \underline{5\mathbf{i} - 4\mathbf{j}}$

(b) $\mathbf{a} - 2\mathbf{b} = \begin{pmatrix} 3 \\ 1 \end{pmatrix} - 2\begin{pmatrix} 2 \\ -5 \end{pmatrix} = \begin{pmatrix} 3 \\ 1 \end{pmatrix} - \begin{pmatrix} 4 \\ -10 \end{pmatrix} = \begin{pmatrix} -1 \\ 11 \end{pmatrix}$

$\Rightarrow \mathbf{a} - 2\mathbf{b} = \underline{-\mathbf{i} + 11\mathbf{j}}$

> **Note:** In the working out of vector questions, either the $\mathbf{i}$, $\mathbf{j}$ notation or column vectors may be used, but the answers should be given in the same notation as the question.

Example 4 Find the magnitude and direction of the vector $3\mathbf{i} - 2\mathbf{j}$.

Let $\mathbf{r} = 3\mathbf{i} - 2\mathbf{j} \Rightarrow$

$\Rightarrow \quad |\mathbf{r}| = \sqrt{3^2 + 2^2} = \sqrt{13}$

$\tan\theta = \dfrac{2}{3} \Rightarrow \theta = 33.7°$

$\therefore \underline{3\mathbf{i} - 2\mathbf{j}}$ has magnitude $\sqrt{13}$ and is at 33.7° below the positive x-axis.

Example 5   A vector has magnitude 4 units and a bearing of 330°. If **i** is a unit vector due east and **j** is a unit vector due north, express this vector in **i**, **j** notation

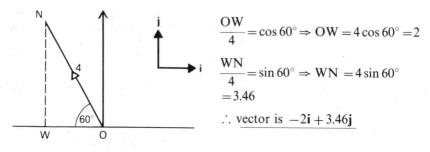

$$\frac{OW}{4} = \cos 60° \Rightarrow OW = 4\cos 60° = 2$$

$$\frac{WN}{4} = \sin 60° \Rightarrow WN = 4\sin 60°$$
$$= 3.46$$

∴ vector is $-2\mathbf{i} + 3.46\mathbf{j}$

Example 6   Find the vector which has magnitude 25 units and is parallel to the vector $6\mathbf{i} + 8\mathbf{j}$.

Let the required vector be $k(6\mathbf{i} + 8\mathbf{j}) = 6k\mathbf{i} + 8k\mathbf{j}$
It has magnitude $= \sqrt{[(6k)^2 + (8k)^2]} = \sqrt{[36k^2 + 64k^2]} = 10k$
$$\therefore 10k = 25 \Rightarrow k = 5/2$$
$$\Rightarrow \text{vector is } \frac{5}{2}(6\mathbf{i} + 8\mathbf{j}) = 15\mathbf{i} + 20\mathbf{j}$$

- Exercise 1.1   See page 15

1.2 Distance, Displacement, Speed and Velocity

Velocity is a vector quantity.
Speed is a scalar quantity.
   E.g. If a car has a velocity 90 km/h due west, then its speed in 90 km/h.
Displacement is a vector quantity, that is, distance moved in a particular direction.
Distance is a scalar quantity.

Motion in a Straight Line

In straight line motion, direction is restricted to forwards–backwards, up–down etc. Here the direction of the velocity or displacement can be distinguished using + or − signs.

E.g.   if   ⟶ is denoted by 3 m/s
                      3 m/s

          then ⟵ is denoted by -3 m/s.
                  3 m/s

Constant Velocity and Constant Speed in Straight-Line Motion

The letter v is used to denote speed and velocity.
The letter s is used to denote distance and displacement.
The letter t is used to denote time.

For **constant** speed or velocity $\boxed{s = vt}$

> When using this relationship, the units must be consistent. If the speed is in m/s then the distance must be in metres and the time in seconds.

Example 7   Find the distance travelled in 10 minutes by a body moving at a constant speed of 20 km/h.

$v = 20\,\text{km/h}$

$t = 10\,\text{min} = \frac{1}{6}h$ ←———

> The speed is given in km/h so the time unit must be in hours.

$s = vt \Rightarrow s = 20 \times \frac{1}{6}\,\text{km} = \underline{3\frac{1}{3}\,\text{km}}$

Changing Units of Speed

Example 8   Express 20 km/h in m/s.

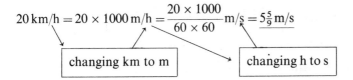

$20\,\text{km/h} = 20 \times 1000\,\text{m/h} = \dfrac{20 \times 1000}{60 \times 60}\,\text{m/s} = \underline{5\frac{5}{9}\,\text{m/s}}$

| changing km to m | | changing h to s |

Example 9   Find the time taken to travel 50 m at 20 km/h.

> You could convert 50 m to fraction of a km and then convert the very small fraction of an hour obtained to minutes or seconds, but it is simpler to change 20 km/h to m/s.

$$20\,\text{km/h} = 5\frac{5}{9}\,\text{m/s} = \frac{50}{9}\,m/s \quad \text{(from Example 6)}$$

$$s = vt \quad \text{rearranges} \Rightarrow t = \frac{s}{v}$$

$$\therefore \text{time} = \frac{50}{50/9}\,\text{s} \qquad = \underline{9\,\text{s}}$$

Average Speed

When speed or velocity is not constant

$$\text{average speed} = \frac{\text{total distance travelled}}{\text{time taken}}$$

$$\text{average velocity} = \frac{\text{total displacement}}{\text{time taken}}$$

Example 10   A, B, C are three points on a straight road, with AB $= 500$ m, BC $= 120$ m and B lies between A and C. A woman jogs from A to B at 3.5 m/s, runs from B to C at 5 m/s and then jogs from C to B at 3.5 m/s. Find (a)   her average speed and (b)   her average velocity.

First we need to find the times for the three points of the run.

Using $t = \dfrac{s}{v}$ for A to B $\Rightarrow t_1 = \dfrac{500}{3.5} = 142.8$ s

Using $t = \dfrac{s}{v}$ for B$\rightarrow$C $\Rightarrow t_2 = \dfrac{120}{5} = 24$ s $\quad \Rightarrow$ total time $= 201.1$ s

Using $t = \dfrac{s}{v}$ for C$\rightarrow$B $\Rightarrow t_3 = \dfrac{120}{3.5} = 34.3$ s

(a)   average speed $= \dfrac{\text{distance}}{\text{time}} = \dfrac{500 + 120 + 120}{201.1} = \underline{3.68 \text{ m/s}}$

(b)   average velocity $= \dfrac{\text{displacement}}{\text{time}} = \dfrac{500 + 120 + (-120)}{201.1} = \underline{2.49 \text{ m/s}}$

Graphical Representation

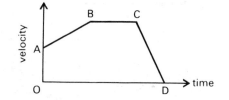

Acceleration $=$ gradient of the velocity–time graph.
Distance travelled $=$ area under the graph.
AB represents motion under constant acceleration.
BC represents motion with constant velocity.
CD represents motion under constant retardation.

Example 11   A car accelerates from rest at 2 m/s² until it reaches 20 m/s. It runs at this speed for 30 seconds. Sketch a velocity–time graph and find how far it has travelled in this period.

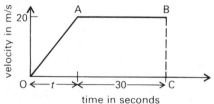

First we need to find the value of t.

Acceleration $=$ gradient $\Rightarrow 2 = \dfrac{20}{t} \Rightarrow t = 10$ s

Distance travelled = area of trapezium OABC
$$= \tfrac{1}{2}(30 + 40) \times 20$$
$$= 700\,\text{m}$$

> Area of trapezium $= \frac{1}{2}(a + b)h$ where a and b are the lengths of the parallel sides and h is the distance between them.

● Exercise 1.2   See page 16

1.3 Motion under Constant Acceleration

Let

> $u = $ initial velocity
> $v = $ final velocity
> $a = $ acceleration
> $t = $ time
> $s = $ displacement

$$\text{Acceleration} = \frac{\text{change in velocity}}{\text{time}}$$

$\Rightarrow \qquad a = \dfrac{v - u}{t} \qquad \Rightarrow v = u + at \qquad (1)$

$$\text{Average velocity} = \frac{\text{displacement}}{\text{time}}$$

$\Rightarrow \qquad \dfrac{u + v}{2} = \dfrac{s}{t} \qquad \Rightarrow s = \dfrac{(u + v)t}{2} \qquad (2)$

Substituting for v from (1) into (2)

$\Rightarrow \qquad s = \dfrac{(u + u + at)t}{2} \qquad \Rightarrow s = ut + \tfrac{1}{2}at^2 \qquad (3)$

Substituting for t from (1) into (2)

$\Rightarrow \qquad s = \dfrac{(u + v)}{2} \times \dfrac{(v - u)}{a} \Rightarrow v^2 = u^2 + 2as \qquad (4)$

These **constant acceleration formulae** are very important and may be quoted.

> $v = u + at$
> $s = \dfrac{(u + v)t}{2}$
> $s = ut + \tfrac{1}{2}at^2$
> $v^2 = u^2 + 2as$

When the direction of motion remains constant, then the distance travelled is the same as the displacement and the speed is the same as the velocity.

Example 12   A car is travelling at 20 m/s and accelerates at $\tfrac{1}{2}$ m/s² for 30 seconds.

(a) How fast will it be going at the end of the 30 seconds?
(b) How far will it have travelled in that time?

(a) Given $u = 20 \, \text{m/s}$
$v = ?$
$a = \frac{1}{2} \text{m/s}^2$
$t = 30 \, \text{s}$
$s = \text{---}$

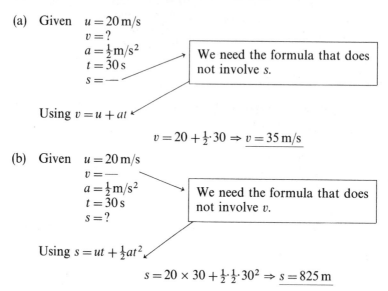

We need the formula that does not involve s.

Using $v = u + at$

$$v = 20 + \tfrac{1}{2} \cdot 30 \Rightarrow v = 35 \, \text{m/s}$$

(b) Given $u = 20 \, \text{m/s}$
$v = \text{---}$
$a = \frac{1}{2} \text{m/s}^2$
$t = 30 \, \text{s}$
$s = ?$

We need the formula that does not involve v.

Using $s = ut + \tfrac{1}{2}at^2$

$$s = 20 \times 30 + \tfrac{1}{2} \cdot \tfrac{1}{2} \cdot 30^2 \Rightarrow s = 825 \, \text{m}$$

Example 13   A toboggan, starting from rest, travels 28 m in the first 8 seconds. How fast is it going at the end of that time?

Given $u = 0$
$v = ?$
$a = \text{---}$
$t = 8 \, \text{s}$
$s = 28 \, \text{m}$

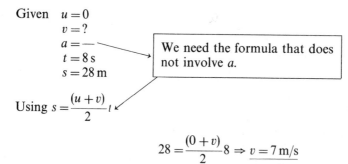

We need the formula that does not involve a.

Using $s = \dfrac{(u + v)}{2} t$

$$28 = \frac{(0 + v)}{2} 8 \Rightarrow v = 7 \, \text{m/s}$$

acceleration is an increase in velocity in a given time.
retardation is a decrease in velocity in a given time—that is, a negative acceleration.

Example 14   The brakes of a train car produce a retardation of $2 \, \text{m/s}^2$. If the train is travelling at 90 km/h, how far from the station must the brakes be applied, if the train is to stop at the station?

Given $u = 90 \, \text{km/h} = \dfrac{90 \times 1000}{60 \times 60} \, \text{m/s} = 25 \, \text{m/s}$

$v = 0$
$a = -2 \, \text{m/s}^2$ Retardation is negative acceleration.
$t = \text{---}$
$s = ?$

Using $v^2 = u^2 + 2as$

$$0 = 25^2 - 2 \cdot 2 \cdot s$$
$$\Rightarrow s = 156.25\,\text{m}$$

$\therefore$ brakes must be applied $\underline{156.25\,\text{m}\ \text{before the station}}$

● **Exercise 1.3**   See page 18 .

1.4 Vertical Motion Under Gravity

> Acceleration due to gravity $= g \approx 9.8\,\text{m/s}^2$

Example 15   A book falls off a shelf 1.6 m above the floor. How long will it take to fall, and with what speed will it hit the floor?
Given $u = 0\downarrow$
$\quad\quad v = ?$
$\quad\quad a = 9.8\,\text{m/s}^2\downarrow$
$\quad\quad t = ?$
$\quad\quad s = 1.6\,\text{m}\downarrow$

Using $s = ut + \frac{1}{2}at^2$

$$1.6 = 0 + \tfrac{1}{2} \times 9.8t^2$$

$$\Rightarrow t^2 = \frac{1.6}{4.9} \Rightarrow t \approx 0.57\,\text{s}.$$

$\therefore$ the book will take $\underline{\text{approximately } 0.57\,\text{s to fall}}$

Using $v^2 = u^2 + 2as$

$$v^2 = 0 + 2 \times 9.8 \times 1.6$$
$$\Rightarrow v = 5.6$$

$\therefore$ it will hit the floor with a speed of $\underline{5.6\,\text{m/s}}$

Example 16   A sailor throws a parcel, with an initial speed of 11 m/s, vertically up to a sailor on a higher deck. The second sailor fails to catch it and it falls to the bottom of a hold, 12 m below the thrower. How long does the flight take?
(Use $g = 10\,\text{m/s}^2$)

Given $u = 11\,\text{m/s}\uparrow$
$\quad\quad v = -\!-$
$\quad\quad a = 10\,\text{m/s}^2\downarrow = -10\,\text{m/s}^2\uparrow$
$\quad\quad t = ?$
$\quad\quad s = 12\,\text{m}\downarrow \quad = -12\,\text{m}\uparrow$

> Before using any of the formulae, you must ensure that the arrows of the vectors are all pointing the same way.

Using $s = ut + \frac{1}{2}at^2$

$$-12 = 11t - \tfrac{1}{2}(10)t^2 \Rightarrow 5t^2 - 11t - 12 = 0$$
$$\Rightarrow (5t + 4)(t - 3) = 0$$
$$\Rightarrow t = 3 \text{ or } -\tfrac{4}{5}$$

But, $t > 0$, therefore time of flight is 3 seconds

Example 17   A stone, dropping vertically, falls past two points, A and then B, a distance of 2.45 metres apart. It travels the distance AB in 0.5 seconds. If the stone fell from rest, find how far above A it started its fall.

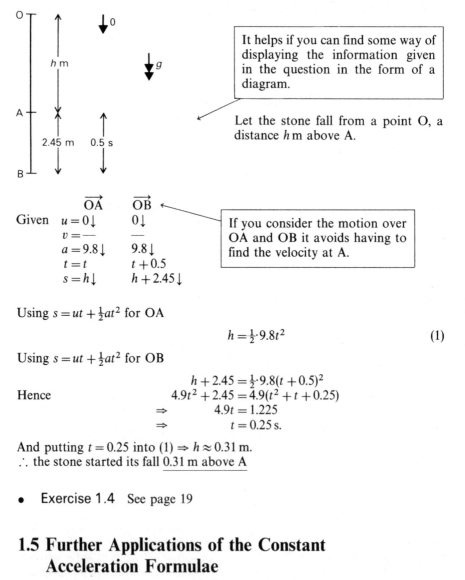

It helps if you can find some way of displaying the information given in the question in the form of a diagram.

Let the stone fall from a point O, a distance h m above A.

Given $\overrightarrow{OA}$ $\overrightarrow{OB}$

$u = 0\downarrow$ $0\downarrow$
$v = -$ $-$
$a = 9.8\downarrow$ $9.8\downarrow$
$t = t$ $t + 0.5$
$s = h\downarrow$ $h + 2.45\downarrow$

If you consider the motion over OA and OB it avoids having to find the velocity at A.

Using $s = ut + \frac{1}{2}at^2$ for OA

$$h = \tfrac{1}{2}\cdot 9.8t^2 \qquad\qquad (1)$$

Using $s = ut + \frac{1}{2}at^2$ for OB

Hence
$$h + 2.45 = \tfrac{1}{2}\cdot 9.8(t + 0.5)^2$$
$$4.9t^2 + 2.45 = 4.9(t^2 + t + 0.25)$$
$$\Rightarrow\qquad 4.9t = 1.225$$
$$\Rightarrow\qquad t = 0.25\,\text{s}.$$

And putting $t = 0.25$ into (1) $\Rightarrow h \approx 0.31$ m.
$\therefore$ the stone started its fall 0.31 m above A

● Exercise 1.4   See page 19

1.5 Further Applications of the Constant Acceleration Formulae

Example 18   A train approaching a station covers two consecutive half kilometres in 16 s and 20 s respectively. If the retardation is uniform, find how much further the train goes before stopping.

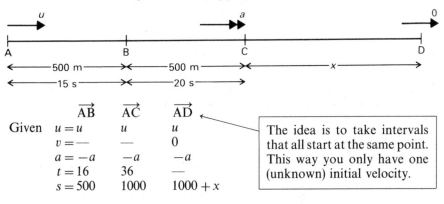

Given

	$\overrightarrow{AB}$	$\overrightarrow{AC}$	$\overrightarrow{AD}$
$u =$	u	u	u
$v =$	—	—	0
$a =$	$-a$	$-a$	$-a$
$t =$	16	36	—
$s =$	500	1000	$1000 + x$

> The idea is to take intervals that all start at the same point. This way you only have one (unknown) initial velocity.

Using $s = ut + \tfrac{1}{2}at^2$ for $\overrightarrow{AB}$

$$\Rightarrow \quad 500 = 16u - \tfrac{1}{2}a256$$
$$\Rightarrow \quad 125 = 4u - 32a \tag{1}$$

Using $s = ut + \tfrac{1}{2}at^2$ for $\overrightarrow{AC}$

$$\Rightarrow 1000 = 36u - \tfrac{1}{2}a1296$$
$$\Rightarrow 1000 = 36u - 648a \tag{2}$$

Solving (1) and (2) simultaneously $\Rightarrow a \approx 0.347$ and $u \approx 34.03$ (3)

Using $v^2 = u^2 + 2as$ for $\overrightarrow{AD}$

$$\Rightarrow 0 = u^2 - 2a(1000 + x)$$

$$\Rightarrow x = \frac{u^2}{2a} - 1000$$

and (3)

$$\Rightarrow x = \frac{(34.03)^2}{0.347 \times 2} - 1000 = 669 \text{ m}.$$

$\therefore$ train travels 669 m further

Example 19   A cyclist, travelling along a straight road with uniform acceleration, 3a passes a certain point with velocity 3u. Three seconds later, another cyclist, moving with constant acceleration 4a, passes the same point with velocity u. The second cyclist overtakes the first when their velocities are 8.1 m/s and 9.3 m/s. Find the values of u and a, and the distance travelled before the second cyclist passes the first.

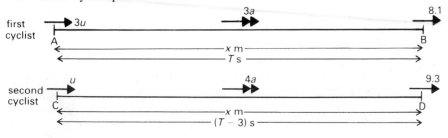

$$\overrightarrow{AB} \qquad \overrightarrow{CD}$$

$\overrightarrow{AB}$	$\overrightarrow{CD}$
$u = 3u$	u
$v = 8.1$	9.3
$a = 3a$	$4a$
$t = T$	$T - 3$
$s = x$	x

> The cyclists meet when they are at the **same place** at the **same time**.

Using $v = u + at$ for $\overrightarrow{AB}$

$$\Rightarrow \quad 8.1 = 3u + 3aT \Rightarrow 2.7 = u + aT \tag{1}$$

Using $v = u + at$ for $\overrightarrow{CD}$

$$\Rightarrow \quad 9.3 = u + 4a(T - 3) \tag{2}$$

Using $v^2 = u^2 + 2as$ for $\overrightarrow{AB}$

$$\Rightarrow 8.1^2 = 9u^2 + 6ax \tag{3}$$

Using $v^2 = u^2 + 2as$ for $\overrightarrow{CD}$

$$\Rightarrow 9.3^2 = u^2 + 8ax \tag{4}$$

$$4 \times (3) - 3 \times (4) \Rightarrow 4 \times (8.1)^2 - 3 \times (9.3)^2 = 33u^2$$

$$\Rightarrow u^2 = 0.09$$
$$\Rightarrow u = 0.3 \, \text{m/s} \tag{5}$$

Substituting for T from (1) into (2)

$$\Rightarrow \quad 9.3 = u + 4(2.7 - u) - 12a$$
$$\Rightarrow 12a = 1.5 - 3u$$
and (5)
$$\Rightarrow 12a = 0.6$$
$$\Rightarrow \quad a = 0.05 \, \text{m/s}^2$$

And, finally, (3) $\Rightarrow x = \dfrac{8.1^2 - 9u^2}{6a} = \dfrac{8.1^2 - 0.81}{0.3}$

$$\Rightarrow x = 216 \, \text{m}$$

● **Exercise 1.5**   See page 20

1.6 Using the Constant Acceleration Formulae and the Velocity–Time Graph

Example 20   A train starts from rest, accelerates uniformly for 500 m, then covers 1.5 km at a constant speed and, finally, is brought to rest under uniform retardation in 250 m. If the time for the whole journey is 5 minutes, find the acceleration and retardation.

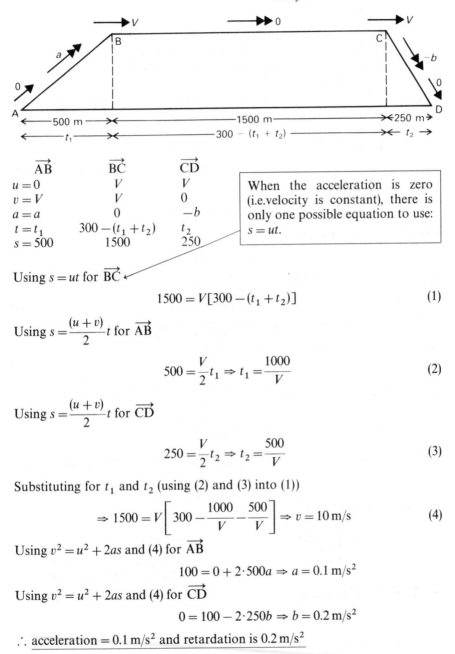

$\overrightarrow{AB}$	$\overrightarrow{BC}$	$\overrightarrow{CD}$
$u = 0$	V	V
$v = V$	V	0
$a = a$	0	$-b$
$t = t_1$	$300 - (t_1 + t_2)$	t_2
$s = 500$	1500	250

When the acceleration is zero (i.e. velocity is constant), there is only one possible equation to use: $s = ut$.

Using $s = ut$ for $\overrightarrow{BC}$

$$1500 = V[300 - (t_1 + t_2)] \tag{1}$$

Using $s = \dfrac{(u+v)}{2}t$ for $\overrightarrow{AB}$

$$500 = \frac{V}{2}t_1 \Rightarrow t_1 = \frac{1000}{V} \tag{2}$$

Using $s = \dfrac{(u+v)}{2}t$ for $\overrightarrow{CD}$

$$250 = \frac{V}{2}t_2 \Rightarrow t_2 = \frac{500}{V} \tag{3}$$

Substituting for t_1 and t_2 (using (2) and (3) into (1))

$$\Rightarrow 1500 = V\left[300 - \frac{1000}{V} - \frac{500}{V} \right] \Rightarrow v = 10\,\text{m/s} \tag{4}$$

Using $v^2 = u^2 + 2as$ and (4) for $\overrightarrow{AB}$

$$100 = 0 + 2 \cdot 500a \Rightarrow a = 0.1\,\text{m/s}^2$$

Using $v^2 = u^2 + 2as$ and (4) for $\overrightarrow{CD}$

$$0 = 100 - 2 \cdot 250b \Rightarrow b = 0.2\,\text{m/s}^2$$

$\therefore$ acceleration $= 0.1\,\text{m/s}^2$ and retardation is $0.2\,\text{m/s}^2$

● **Exercise 1.6**   See page 20

1.7 Motion under Constant Velocity: Revision of Unit

If velocity is constant

$$\boxed{s = vt}$$

Example 21   If a body moves with a velocity $(8\mathbf{i} + 3\mathbf{j})$ m/s for 2 seconds, find the vector giving its displacement in that time.

$$\mathbf{s} = (8\mathbf{i} + 3\mathbf{j}) \times 2 = \underline{16\mathbf{i} + 6\mathbf{j}}$$

Example 22   A particle has an initial position vector $(5\mathbf{i} + 9\mathbf{j})$ m. If it moves with a velocity $(2\mathbf{i} - \mathbf{j})$ m/s, find its position vector after 4 s.

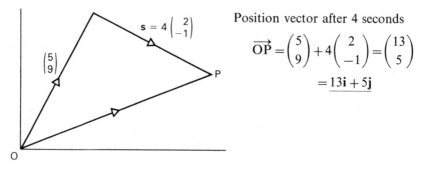

Position vector after 4 seconds

$$\overrightarrow{OP} = \binom{5}{9} + 4\binom{2}{-1} = \binom{13}{5}$$

$$= \underline{13\mathbf{i} + 5\mathbf{j}}$$

Example 23   A particle, initially with position vector $(2\mathbf{i} + \mathbf{j})$ m, is moving with velocity $3\mathbf{j}$ m/s. A second particle, initially with position vector $(7\mathbf{i} + 11\mathbf{j})$ m, is moving with velocity $(-\mathbf{i} + \mathbf{j})$ m/s. Show that the particles collide and find the time of collision.

Position vector of first particle after t seconds $\quad = \binom{2}{1} + t\binom{0}{3}$

Position vector of second particle after t seconds $= \binom{7}{11} + t\binom{-1}{1}$

Assume that the particles collide—that is, they have the same position vector for one particular value of t.

$$\Rightarrow \quad \binom{2}{1} + t\binom{0}{3} = \binom{7}{11} + t\binom{-1}{1}$$

$$\Rightarrow t\left[\binom{0}{3} - \binom{-1}{1}\right] = \binom{7}{11} - \binom{2}{1}$$

$$\Rightarrow t\binom{1}{2} = \binom{5}{10} \leftarrow$$

If no value of t exists satisfying this equation then the particles will not collide.

$$\Rightarrow \quad t = 5$$

$\therefore$ the particles collide after 5 s

• Miscellaneous Exercise 1A   See page 21

1.8 + A Level Questions

• Miscellaneous Exercise 1B   See page 22

Unit 1 Exercises

Exercise 1.1

(Give answers correct to 2 decimal places or 0.1°)
1   Express each of the following vectors in the form $a\mathbf{i} + b\mathbf{j}$, given that the squares in the grid are of unit length:

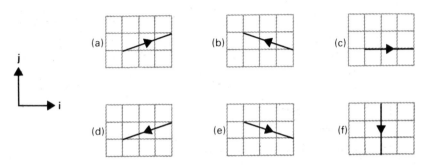

2   Express each of the following vectors as column vectors (i.e. in the form $\begin{pmatrix} p \\ q \end{pmatrix}$), given that the squares in the grid are of unit length:

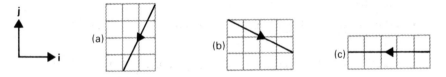

3   If $\mathbf{a} = 3\mathbf{i} + 4\mathbf{j}$, $\mathbf{b} = -7\mathbf{i} + 24\mathbf{j}$, $\mathbf{c} = \mathbf{i} - 2\mathbf{j}$ and $\mathbf{d} = 3\mathbf{j}$ find:

(a)   the modulus of $\mathbf{a}$ (written as $|\mathbf{a}|$)
(b) $|\mathbf{b}|$
(c)   the resultant of $\mathbf{a}$ and $\mathbf{c}$
(d)   the resultant of $\mathbf{c}$ and $\mathbf{d}$
(e) $|\mathbf{c}|$
(f) $|\mathbf{d}|$
(g)   a vector parallel to $\mathbf{c}$ and twice as long as $\mathbf{c}$
(h)   a vector parallel to $\mathbf{a}$ and half its magnitude
(i) $\mathbf{c} - \mathbf{d}$
(j) $\mathbf{a} + 2\mathbf{d} - 3\mathbf{c}$
(k) $|\mathbf{c} + \mathbf{d}|$.

4   Express each of the following vectors in the $\mathbf{i}, \mathbf{j}$ form given that $\mathbf{i}$ is a unit vector due east and $\mathbf{j}$ is a unit vector due north:

(a) $\mathbf{p}$ has magnitude 6 and a bearing of 060°
(b) $\mathbf{q}$ has magnitude 10 and a bearing of 045°
(c) $\mathbf{r}$ has magnitude 3 and a bearing of 180°
(d) $\mathbf{s}$ has magnitude 4 and a bearing of 135°
(e) $\mathbf{t}$ has magnitude 8 and a bearing of 210°
(f) $\mathbf{u}$ has magnitude 7 and a bearing of 270°.

5 Find the magnitude and direction of each of the following vectors, giving the direction from the positive x-axis:

   (a) $5\mathbf{i} + 12\mathbf{j}$ (b) $\mathbf{i} + \mathbf{j}$
   (c) $-\mathbf{i} + \mathbf{j}$ (d) $2\mathbf{j} - 3\mathbf{j}$.

6 Find the vector which has magnitude 10 and is parallel to $24\mathbf{i} + 7\mathbf{j}$.

7 Find the vector which has magnitude 6 and is parallel to $-3\mathbf{i} + 4\mathbf{j}$.

8 If $\overrightarrow{AB} = \begin{pmatrix} 3 \\ 5 \end{pmatrix}$ $\overrightarrow{CD} = \begin{pmatrix} 1 \\ -1 \end{pmatrix}$ $\overrightarrow{EF} = \begin{pmatrix} 0 \\ 3 \end{pmatrix}$ find:

   (a) $|\overrightarrow{AB}|$ (b) $|\overrightarrow{CD}|$ (c) $\overrightarrow{AB} + \overrightarrow{CD}$
   (d) $\overrightarrow{AB} - \overrightarrow{EF}$ (e) $|\overrightarrow{AB} - \overrightarrow{EF}|$ (f) $|\overrightarrow{AB}| - |\overrightarrow{EF}|$.

9 The position vector of B w.r.t A is $\begin{pmatrix} 3 \\ 1 \end{pmatrix}$. The position vector of C w.r.t.
   A is $\begin{pmatrix} -2 \\ 2 \end{pmatrix}$. The position vector of D w.r.t. A is $\begin{pmatrix} 0 \\ -3 \end{pmatrix}$.
   Draw a diagram showing the relative positions of the points A, B, C, D. Find:

   (a) the position vector of B w.r.t. D
   (b) the position vector of C w.r.t. D
   (c) the position vector of B w.r.t. C
   (d) the position vector of C w.r.t. B.

*10 If $\mathbf{p} = -3\mathbf{i} + 4\mathbf{j}$, $\mathbf{q} = 5\mathbf{i} - 12\mathbf{j}$, $\mathbf{r} = -\mathbf{i} + \mathbf{j}$ find:

   (a) a vector parallel to $\mathbf{q}$ with the same modulus as $\mathbf{p}$
   (b) a vector parallel to $\mathbf{p}$ with the same modulus as $\mathbf{q}$
   (c) a vector parallel to $\mathbf{q}$ with the same modulus as $\mathbf{r}$
   (d) a vector parallel to $\mathbf{r}$ with the same modulus as $\mathbf{p}$.

Exercise 1.2

1 Express 72 km/h in m/s.

2 Express 35 km/h in m/s.

3 Express 65 m/s in km/h.

4 Express 10 m/s in km/h.

5 Express 3 km/min in m/s.

6 A cyclist travelling at a constant speed covers 50 m in 8 seconds. Find his speed.

7 A train travelling at a constant speed covers 5 km in 2 minutes. Find its speed.

8 Find the distance travelled in 3 minutes by a body moving with a constant speed of 6.2 m/s.

9 Find the distance travelled in 3 minutes by a body moving with a constant speed of 6.2 km/h.

10 Find the time taken to travel 1 km by a body travelling at a constant speed of 50 m/s.

11 Find the time taken to travel 700 m by a car travelling at a constant speed of 50 km/h.

12 The speed of sound is 340 m/s. If it takes 4 seconds for the sound of thunder to reach me, how far away was the thunderclap?

13 The speed of light is 3×10^8 m/s. The mean distance from the sun to the earth is 1.5×10^8 km. How long does it take light from the sun to reach the earth?

14 Linford Christie won the European Games 100 m gold medal in 1986 with a time of 10.15 seconds. What was his average speed in: (a) m/s (b) km/h?

15 If an athlete runs the 1500 m with an average speed of 23 km/h, what will his time be?

16 A girl walks 200 m due north in 90 s, then 50 m due south in 30 s. Find: (a) her average speed (b) her average velocity.

17 A boy cycles from Tarvin to Clotton, a distance of 5 km, at a speed of 24 km/h. At Clotton he has a puncture and he then walks from Clotton to Tarporley, a distance of 3 km, at a speed of 5 km/h. Calculate:

  (a)   the time taken to travel from Tarvin to Clotton
  (b)   the time taken to travel from Clotton to Tarporley
  (c)   the average speed for the whole journey
  (d)   the average speed for the journey if he had stopped at Clotton for ten minutes.

18 A heavy goods train starts from rest and accelerates uniformly for 4 minutes, by which time it has achieved a speed of 36 km/h. It runs at this speed for 5 minutes and then decelerates uniformly, coming to rest in 2 minutes. Sketch the velocity–time graph and find the total distance travelled.

*19 A man joins the M1 in London at 10.00 am. He stops for petrol at the Newport Pagnell Service Area, 83 km up the motorway at 10.40 am. It takes him 10 minutes. He drives on and later realises that he has left his Visa card at the petrol station. He turns round at junction 18, 49 km further on, at 11.20 am. He arrives back at Newport Pagnell at 11.40 am.

  (a)   What is his average speed from the beginning of the motorway to Newport Pagnell?

(b)   What was his average speed from Newport Pagnell to junction 18?

(c)   What was his average speed from junction 18 back to Newport Pagnell?

(d)   What was his average speed for the whole journey?

(e)   What was his average velocity for the whole journey?

*20   A train starts from rest and accelerates at $0.9 \, \text{m/s}^2$ for 20 seconds. It then travels for 100 seconds at a constant speed and finally undergoes a uniform retardation which brings it to rest in 30 seconds. Sketch the velocity–time graph and find the retardation and the total distance travelled.

Exercise 1.3

Note: Any quadratic equation in this exercise can be factorised.
Questions 1–10 involve the motion of a body under uniform acceleration in a straight line from A to B.

1   Initial velocity $= 2 \, \text{m/s}$, acceleration $= 4 \, \text{m/s}^2$, time taken $= 6 \, \text{s}$. Find the final velocity.

2   Initially at rest, acceleration $= 3 \, \text{m/s}^2$, time taken $= 4 \, \text{s}$. Find the distance travelled.

3   Initially at rest, distance $= 100 \, \text{m}$, acceleration $= 2 \, \text{m/s}^2$. Find the final velocity.

4   Initial velocity $= 10 \, \text{m/s}$, time taken $= 10 \, \text{s}$, final velocity $= 30 \, \text{m/s}$. Find the distance travelled.

5   Acceleration $= 2 \, \text{m/s}^2$, time taken $= 15 \, \text{s}$, final velocity $= 40 \, \text{m/s}$. Find the initial velocity.

6   Initial velocity $= 10 \, \text{m/s}$, final velocity $= 70 \, \text{m/s}$, distance $= 120 \, \text{m}$. Find the time taken.

7   Initial velocity $= 20 \, \text{m/s}$, final velocity $= 50 \, \text{m/s}$, time taken $= 3 \, \text{s}$. Find the acceleration.

8   Initial velocity $= 20 \, \text{m/s}$, final velocity $= 10 \, \text{m/s}$, distance $= 100 \, \text{m}$. Find the acceleration.

9   Initial velocity $= 50 \, \text{m/s}$, acceleration $= 6 \, \text{m/s}^2$, distance $= 200 \, \text{m}$. Find the time taken.

10   Acceleration $= -4 \, \text{m/s}^2$, distance $= 2 \, \text{m}$, final velocity $= 3 \, \text{m/s}$. Find the initial velocity.

11   A skier increases his speed from $8 \, \text{m/s}$ to $20 \, \text{m/s}$ in $10 \, \text{s}$. How far does he travel in this time and what is his acceleration?

12   A car travelling at $54 \, \text{km/h}$ is brought to rest with uniform retardation

in 10 s. Find the retardation and the distance that the car has travelled in this time.

13 A car accelerates uniformly from 18 km/h to 72 km/h over a distance of 50 m. Find the acceleration, and the speed when the car has covered half the distance.

14 A raindrop running down a window increases its speed from 8 cm/s to 15 cm/s in 6 s. How far does it go and what is its acceleration?

15 A train is travelling at 162 km/h when the driver sees a tree across the track 50 m ahead. The brakes produce a retardation of 8 m/s². With what speed will the train hit the tree?

**16 A car, starting from rest and accelerating uniformly, travels 9.5 m in the tenth second after starting. Find the acceleration of the car and the distance it covers in the ninth second.

Exercise 1.4

Note: any quadratic equations in this exercise can be factorised. Unless the question specifies otherwise, use $g = 9.8\,\text{m/s}^2$

1 A stone falls from the top of a cliff. Find:
   (a) how far it will fall in 10 s
   (b) how long it takes to fall 100 m
   (c) its velocity after falling 100 m.

2 A stone is thrown from the top of a cliff downwards with a velocity of 40 m/s and it takes 3 seconds to reach the ground. How high is the cliff?

3 A ball is thrown vertically upwards with a velocity of 24.5 m/s. How high will it go? After what time will it first be at a height of 19.6 m?

4 Water is projected vertically from a fountain. It rises to a height of 10 m. With what speed is the water emitted?

5 A stone is dropped down a well. It takes 4 seconds to reach the bottom. How deep is the well?

6 A coin is thrown vertically upwards at 6 m/s and caught at the same height. How long was it in the air, how high did it go and what was its speed when it was caught?

7 Two brickies toss a coin to decide which of them will go down the ladder for the next load of bricks. The coin is thrown vertically upwards at 5 m/s. Unfortunately the thrower fails to catch the coin and it falls to the ground 10 m below. After what time will it reach the ground? (For this question take $g = 10\,\text{m/s}$.)

8 A ball is held 2 m above the ground. It is projected downwards with a

velocity of 3 m/s. The ball hits the ground and rebounds with half the speed it had just before the impact. How high does the ball reach after the first bounce?

9 A stone is thrown vertically upwards with a velocity of 35 m/s. Find the distance it ascends during the third second of its motion.

*10 A bullet is fired vertically upwards at a speed of 49 m/s. Find the length of time for which the bullet is at least 102.9 m above the point of firing.

*11 A stone is dropped from the top of a tower. One second later another stone is thrown downwards with an initial speed of 12 m/s. Find the height of the tower if the two stones land at the same time.

**12 A brick falls from rest from the top of an old chimney. During the last second it falls 9/25 of the whole distance. Find the height of the chimney.

Exercise 1.5

1 Three posts A, B and C are 2 km apart. A train, under uniform retardation, takes 100 s to travel from A to B and 150 s to travel from B to C. Show that the retardation is $\frac{4}{75}$ m/s². How far beyond C does the train travel before it stops?

2 A can manufacturer is carrying out speed trials on its new model. The car starts from rest and moves in a straight line with constant acceleration. In a certain 4 seconds of its motion it travels 80 m and in the next 5 seconds it travels 190 m. Find its acceleration and its speed at the start of the timing.

3 A ball is thrown vertically upwards with a speed of 30 m/s. Two seconds later another ball is thrown vertically upwards with a speed of 25 m/s. Find when the two balls are at the same height and find the velocity of each ball at this moment, giving all answers to 2 decimal places.

4 A cyclist X starts from rest with an acceleration of 3 m/s². At the same time as X sets off, another cyclist Y is 18 m behind him, travelling at 10 m/s. Y immediately starts in pursuit with an acceleration of 2 m/s². Prove that Y will overtake and pass X after 2 s and that X will overtake Y after a further 16 s.

*5 A lift makes the first part of its descent with uniform acceleration a and the remainder with uniform retardation $3a$. Prove that, if h is the distance travelled and t is the time taken, then

$$h = \frac{3}{8}at^2$$

Exercise 1.6

1 A train starts from A and accelerates uniformly at 0.2 m/s² for 2 minutes. It then travels at a constant speed for 10 minutes, after which it is brought to rest at B with a constant retardation of 1.5 m/s². Find the distance AB.

2 A pit-cage goes down a mineshaft 800 m deep in 80 s. For the first quarter of the distance the cage is uniformly accelerated and for the last quarter of the distance the cage is uniformly retarded. The acceleration and the retardation are equal. Find the speed of the cage during the middle part of the journey.

3 A train starts from rest and accelerates uniformly for 500 m. It then travels for 1500 m at a uniform speed, after which it is retarded uniformly, bringing it to rest in 250 m. If the time for the whole journey is 5 minutes, find the acceleration and the retardation.

4 A car travelling in a straight line accelerates uniformly from rest for x m. The car then travels the same distance at a constant speed taking 50 s, after which it decelerates uniformly until it stops. If the total distance travelled is $\frac{5}{2}x$, calculate the time taken for the journey.

*5 A lift ascends with a constant acceleration a, then travels with a constant speed and is finally brought to rest under a constant retardation a. If the total distance travelled is h and the total time taken is t, show that the time spent travelling at constant speed is

$$\left[t^2 - \frac{4h}{a} \right]^{1/2}$$

Miscellaneous Exercise 1A

1 If $p = 2i$, $q = 3i - 4j$, $r = i + 2j$ find:

(a) $|p|$
(b) the modulus of q
(c) $2p + r$
(d) the magnitude and direction of r if i is due east and j is due north
(e) the vector parallel to r and three times its magnitude.

2 If P has position vector $\begin{pmatrix} 2 \\ 3 \end{pmatrix}$ relative to the origin and Q has position vector $\begin{pmatrix} 1 \\ 0 \end{pmatrix}$ relative to the origin, find the position vector of P relative to Q.

3 If A has velocity $(6i + 8j)$ m/s and B has velocity $(5i - 12j)$ m/s which is moving faster?

4 A body moving with velocity $3i + pj$ has a speed of 6.7 m/s. Find the value of p correct to 1 decimal place.

, 5 A particle has initial position vector $(3i - 2j)$ m. If it moves with a constant velocity of $(-i + 3j)$ m/s find its position vector after (a) 1 s (b) 3 s.

. 6 A body, with initial position vector $(2i + j)$ m is moving with velocity $(pi + qj)$ m/s. After 2 seconds its position vector is $(8i + j)$ m/s. Find the values of p and q.

7   A ship leaves port P with velocity $(3\mathbf{i} + 2\mathbf{j})$ km/h. At the same instant a motor boat leaves port Q with a velocity of $(\mathbf{i} + a\mathbf{j})$ km/h. Port Q has a position vector of $10\mathbf{i}$ km relative to P. If the motor boat intercepts the ship, find the value of a and the time of interception.

8   Two trains are running on parallel tracks. When they are initially level with each other, one is moving with a uniform speed of 90 km/h and the second has a speed of 30 km/h but is accelerating at a uniform rate of 0.4 m/s². How long will it be before they are level again and how far will they have travelled?

Miscellaneous Exercise 1B

1   A ball A is thrown vertically upwards at 25 m/s from a point P. Three seconds later a second ball is also thrown vertically upwards from the point P at 25 m/s. Taking the acceleration due to gravity as 10 m/s², calculate:

(a)   how long A has been in motion when the balls meet
(b)   the height above P at which A and B meet.   [A: 1.4, 1.5]

2   A particle moving in a straight line with constant acceleration passes through points O, A and B at times $t = 0, 1$ and 2 seconds respectively, where A and B are on the same side of O and OA = 10 m, OB = 60 m. Find the magnitudes and senses (from O to B or B to O) of both the acceleration and the velocity of the particle at time $t = 0$.   [J: 1.5]

3   A train has a maximum allowed speed of 20 m/s. With its brakes fully applied the train has a deceleration of 0.5 m/s². If it can accelerate at a constant rate of 0.25 m/s² find the shortest time in which it can travel from rest in one station to rest in the next station 8 km away.   [A: 1.6]

4   The brakes of a train, which is travelling at 108 km/h, are applied as the train passes point A. The brakes produce a constant retardation of magnitude $3f$ m/s² until the speed of the train is reduced to 36 km/h. The train travels at this speed for a distance and is then uniformly accelerated at f m/s² until it again reaches a speed of 108 km/h as it passes point B. The time taken by the train in travelling from A to B, a distance of 4 km, is 4 minutes. Sketch the speed–time graph for this motion and hence calculate:

(a)   the value of f
(b)   the distance travelled at 36 km/h.   [L: 1.6]

5   A horizontal platform is descending vertically with constant speed 5.6 m/s. When the platform is at a distance two metres below a point P, a small ball is released from rest at P. Find the time that elapses before the ball hits the platform.   [J: 1.2, 1.4]

6   A vehicle travelling on a straight track joining two points A and B can accelerate at a constant rate 0.2 m/s² and can decelerate at a constant rate of 1 m/s². It covers a distance of 1.8 km from rest at A to rest at B by

accelerating uniformly to a fixed speed and then travelling at that speed until it starts to decelerate. Given that the journey from rest to rest takes $2\frac{1}{2}$ minutes find the steady speed, in m/s, at which the vehicle travels. Explain clearly the reason for your choice of the value for the speed.

[A: 1.6]

7   A particle moves along a straight line ABC. The particle starts from rest at A and moves from A to B with constant acceleration $2f$. It then moves from B to C with acceleration f and reaches C with speed V. The times taken in the motion from A to B and from B to C are each equal to T. Find T in terms of V and f. Show that

$$AB = \frac{2}{5}BC \qquad\qquad [\text{J}: 1.3, 1.5]$$

8   When a train accelerates, its acceleration is always constant and of magnitude $f\,\text{km/h}^2$. When the train decelerates its deceleration is always constant and a magnitude $r\,\text{km/h}^2$. On a journey of 43 km, the train accelerates from rest for a distance of 2 km to its maximum speed of 60 km/h. It then continues at this maximum speed, until 2 minutes before the end of the journey. During the last 2 minutes of the journey it decelerates to rest. Determine:

(a) f and r
(b)   the time taken to attain the maximum speed from rest
(c)   the distance moved whilst decelerating from 60 km/h to rest
(d)   the total time taken for the complete journey.

On a particular day, a section of the track is being repaired. Consequently, during its journey, the train is required to slow down from 60 km/h to rest and to wait for 2 minutes before accelerating back to 60 km/h. Show that the train arrives at its destination 5 minutes late.   (L: 1.6]

9   A vehicle P moves along a straight track which passes through two points A and B which are at a distance of 80 m part. A vehicle Q moves along a straight parallel track whose perpendicular distance from the first track is to be neglected. The vehicle P has an acceleration of $3\,\text{m/s}^2$ in the sense from A to B whilst Q has an acceleration of $1\,\text{m/s}^2$ in the sense from B to A. At time $t = 0$ s P passes through A moving towards B with a speed of 12 m/s. Find the distance of P from A six seconds later and its speed at this time.

Also at time $t = 0$ s Q passes B moving towards A with a speed of 8 m/s. Find an expression for the distance PQ at time t and determine when P and Q are at a distance of 32 m apart.   [A: 1.5]

*10   A particle A starts from the origin O with velocity u m/s and moves along the positive x-axis with constant acceleration f m/s^2, where $u > 0, f > 0$. Ten seconds later, another particle B starts from O with velocity u m/s and moves along the positive x-axis with acceleration $2f$ m/s^2. Find the time that elapses between the start of A's motion and the instant when B has the same velocity as A, and show that A will then have travelled twice as far as B.   (J. 1.5]

UNIT

FORCES AND ACCELERATION

2.1 Forces

Types of Force

1 **Attractions:** when a force is exerted on the body without any visible contact with the body
The **weight** of a body is the force with which the earth attracts it.

2 **Tensions or Thrusts:** the force acts along a string, spring or rod.

3 **Normal Reactions:** equal and opposite sources between two bodies in contact with each other. These act perpendicular to the common tangent between the bodies (i.e. along the common normal).

4 **Friction:** this force acts along the common tangent between the two surfaces. Friction opposes the tendency to motion.

Force Diagrams

Example 1   Draw diagrams showing the forces acting on the following bodies. In each case label the forces either W (weight), F (friction), T (tension), P (thrust) or R (normal reaction).

(a)   An apple falling to the ground.

(b)   An apple hanging from the tree.

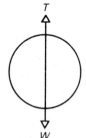

(c)   An apple lying on the ground.

(d)   An apple being pushed horizontal on rough ground.

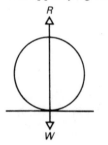

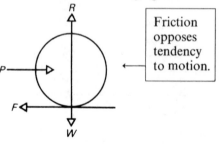

Friction opposes tendency to motion.

Example 2   Draw diagrams illustrating the forces acting on a toboggan in each of the following cases. The toboggan is lying on a slope which is inclined at $\alpha°$ to the horizontal. Label the forces as in example 1.

(a) the toboggan lies at rest on the slope

> The friction prevents the toboggan from sliding down the slope.

(b) the toboggan is being pulled up the slope by a rope inclined at $\beta°$ to the slope.

> Friction opposes the tendency to motion.

(c) the toboggan is prevented from sliding down the slope by a rope inclined at $\beta°$ to the slope.

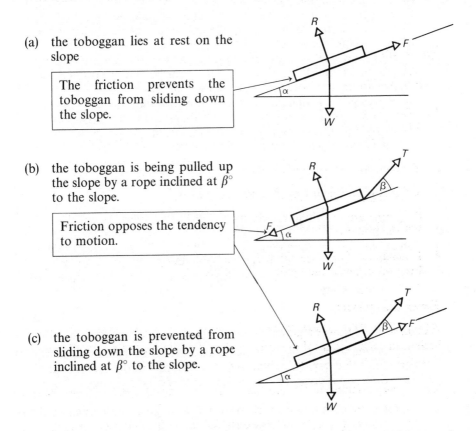

Example 3   A matchstick leans against a marble with the head of the match on the ground. All contacts are rough.
Draw diagrams showing:
(a) the forces acting on the marble   (b) the forces acting on the match.

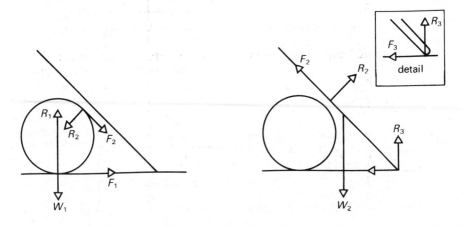

Note that:
(1) The pressure on the marble from the stick is the same as the pressure on the stick from the marble (R_2).
(2) If the contact between the ground and the marble was smooth, (i.e. no friction), the marble would move to the left. Hence the friction acts to the right.
(3) Friction prevents the stick sliding down and to the right. Hence F_3 acts to the left and F_2 on the stick acts upwards. Thus we deduce that F_2 acts in the opposite direction on the marble.
(4) Wherever possible diagrams are drawn in two dimensions only.
(5) These forces could be combined in one diagram but, in this case, it is suggested that the forces acting on the marble are drawn in a different colour to those acting on the stick.

Most Common Errors in Force Diagrams

(1) Diagrams are too small for clarity.
(2) Forces omitted.
(3) Forces put in more than once.
(4) Friction put in the wrong direction (or omitted).
(5) When more than one body is involved, it is not made clear which forces act on which body.

Conventions and Assumptions

(1) A **light string** (rope, cable, etc) means that the string has a weight which is negligible in comparison with the other weights in the problem. In this case strings are represented by straight lines (i.e. no sagging).

(2) The force which a string exerts on a body is called the **tension**. A light string attached to two bodies exerts the same tension at either end.

(3) A light rod attached to two bodies can exert either

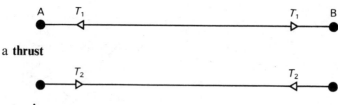

a **thrust**

or a **tension**.

(4) If a string passes over a smooth peg, or a light pulley (i.e. negligible mass) which rotates smoothly on its bearings, then the tension is considered to be the same on either side of the peg or pulley.

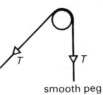

smooth peg

(5) If a string passes over a rough peg, then the friction alters the tension in the string as it passes over the peg. In this case this friction is not put onto the diagram, but manifests itself in the altered tension.

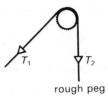

rough peg

(6) A string is frequently described as **inextensible**. In practice there is no such thing as an inextensible string, but it means that the amount of stretching is negligible.

● Exercise 2.1   See page 46

2.2 Newton's Laws of Motion

The First Law: A body will remain at rest, or continue to move in a straight line with constant velocity, unless it is made to change that state by external forces.

The Second Law: The acceleration of a body is proportional to, and in the same direction as, the resultant of the forces acting on the body.

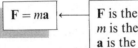

$$F = ma$$

F is the force in newtons
m is the mass in kg
a is the acceleration in m/s²

The Third Law: Action and reaction are equal and opposite.

A **newton** is the force which gives a mass of 1 kg an acceleration of 1 m/s². **Acceleration** is the rate of change of velocity with respect to time.

Mass and Weight

Mass is a measure of the inertia of a body; that is, the resistance that the body offers to having its velocity altered by the application of a force.
The **weight** of a body is the force with which the Earth (or another heavenly body) attracts it.

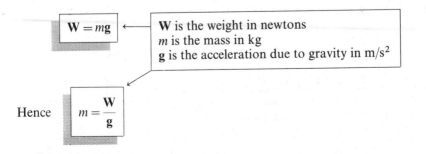

$$W = mg$$

W is the weight in newtons
m is the mass in kg
g is the acceleration due to gravity in m/s²

Hence $$m = \frac{W}{g}$$

The mass of a body is constant. The weight of a body varies accordingly as **g** varies.

g varies on the Earth's surface from about 9.78 m/s² at the Equator to about 9.83 m/s² at the Earth's poles, whereas its value on the Moon's surface is approximately 1.63 m/s². The mass of the Earth is 6×10^{24} kg. It has no weight.

The International System of Units (SI)

The basic unit of force is the newton	N
The basic unit of distance is the metre	m
The basic unit of time is the second	s
The basic unit of mass is the kilogramme	kg
The basic unit of energy is the joule	J
The basic unit of power is the watt	W
The basic unit of capacity is the litre	l

If all the data used in a calculation are in SI units then the data that are produced will be in SI units.

E.g. If *m* is in kg and *a* is in m/s² then $F = ma$ will give a force in newtons.

The Arrow Conventions

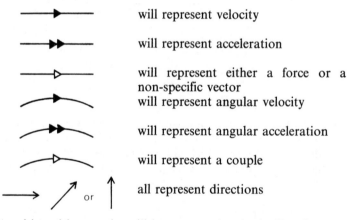

	will represent velocity
	will represent acceleration
	will represent either a force or a non-specific vector
	will represent angular velocity
	will represent angular acceleration
	will represent a couple
	all represent directions

In diagrams, quantities without units will be presumed to be in SI units.

e.g. ——▷—— is a force of 3 N and ——▶—— is a velocity of 3 m/s
         3   3

Example 4   What acceleration is produced when a force of 3 N acts on a mass of 500 g?

$$F = ma \Rightarrow 3 = 0.5a \Rightarrow \underline{a = 6 \, \text{m/s}^2}$$

> First put amounts
> into SI units.

Example 5 What force will give a mass of 8 kg a velocity of 30 m/s in 2 minutes, from rest?

We know the mass; we want to find the force; so first we must calculate the acceleration.

$u = 0$
$v = 30$
$a = ?$
$b = 2\,\text{min} = 120\,\text{s}$
$s = —$

Using $v = u + at$

$$30 = 0 + a \cdot 120 \Rightarrow a = \frac{30}{120} = \frac{1}{4}.$$

$F = ma \Rightarrow \qquad\qquad F = 8 \times \frac{1}{4}$

$\Rightarrow \qquad\qquad \underline{\text{Force} = 2\text{N}}$

Example 6 What acceleration is produced on a body of mass 2 kg by the resultant of two forces $(3\mathbf{i} + 2\mathbf{j})$ N and $(\mathbf{i} - \mathbf{j})$ N?

Resultant force $= (3\mathbf{i} + 2\mathbf{j})\,\text{N} + (\mathbf{i} - \mathbf{j})\,\text{N} = (4\mathbf{i} + \mathbf{j})\,\text{N}$

> resultant = sum of vectors – see 1.1

$\mathbf{F} = m\mathbf{a} \Rightarrow \qquad\qquad 4\mathbf{i} + \mathbf{j} = 2\mathbf{a}$

$\Rightarrow \qquad\qquad \underline{\mathbf{a} = 2\mathbf{i} + 0.5\mathbf{j}}$

Example 7 A lorry of mass 4 tonnes travels at a constant speed and its engine produces a pulling force of 2000 N. What is the resistance to its motion? If the pulling force is increased to 6000 N and the resistance remains the same, what is the acceleration of the lorry? What is the normal reaction between the lorry and the ground?

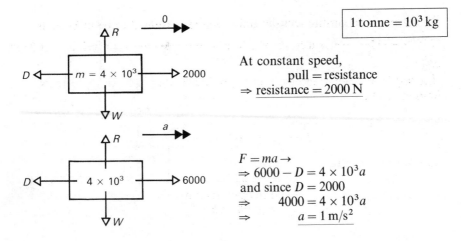

> 1 tonne $= 10^3$ kg

At constant speed,
$$\text{pull} = \text{resistance}$$
$$\Rightarrow \underline{\text{resistance} = 2000\,\text{N}}$$

$F = ma \rightarrow$
$\Rightarrow 6000 - D = 4 \times 10^3 a$
and since $D = 2000$
$\Rightarrow \qquad 4000 = 4 \times 10^3 a$
$\Rightarrow \qquad\qquad \underline{a = 1\,\text{m/s}^2}$

$F = ma\uparrow$

$$\Rightarrow R - W = 0$$
$$\Rightarrow \qquad R = W$$

Acceleration in $\uparrow$ direction is zero.

but $W = 4 \times 10^3 g$

$W = mg$

$\therefore R = 4 \times 10^3 \times 9.8 = \underline{39200 \text{ N}}$

Example 8   A box of mass 20 kg is lowered vertically by a rope. If the acceleration is $2\,\text{m/s}^2$, find the torsion in the rope.

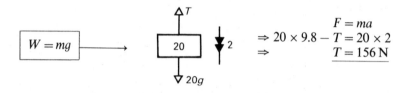

$F = ma$
$$\Rightarrow 20 \times 9.8 - T = 20 \times 2$$
$$\Rightarrow \qquad\qquad T = 156\,\text{N}$$

● **Exercise 2.2**   See page 47

2.3 Motion of Connected Particles

Example 9   Two weights, of masses 2 kg and 3 kg, are connected by a light inextensible string passing over a smooth fixed pulley. Find the acceleration of the weights and the tension in the string.

$F = ma$ for $3\,\text{kg}\downarrow$

$$3g - T = 3a \qquad (1)$$

$F = ma$ for $2\,\text{kg}\uparrow$

$$T - 2g = 2a \qquad (2)$$

Since the pulley is smooth the tension is the same throughout the string – see 2.1.

To solve these simultaneously it is easier to write (2) as

$$-2g + T = 2a$$

and (1) is $\qquad 3g - T = 3a$

Adding $\qquad\quad g \qquad = 5a$

$$\Rightarrow \qquad a = \tfrac{2}{5}\,\text{m/s}^2$$

Substituting for a in (2) $\Rightarrow T = 2g + \dfrac{2g}{5}$

$$\Rightarrow T = \frac{12g}{5}\,\text{N}$$

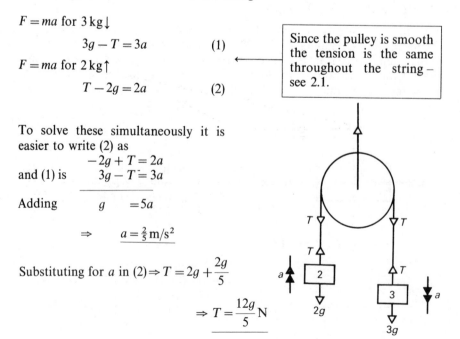

Note: Unless a specific value of g is given, it is usual to give answers in terms of g. Answers may be left in fraction form.

Example 10   A mass of 6 kg, resting on a smooth table, is connected by a light horizontal string passing over a smooth pulley at the edge of the table, to a mass of 4 kg hanging freely. Find the common acceleration, the tension in the string and the reaction between the 6 kg mass and the table.

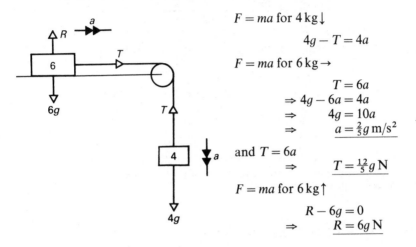

$F = ma$ for 4 kg ↓

$$4g - T = 4a$$

$F = ma$ for 6 kg →

$$T = 6a$$
$$\Rightarrow 4g - 6a = 4a$$
$$\Rightarrow \quad 4g = 10a$$
$$\Rightarrow \quad a = \tfrac{2}{5}g \,\text{m/s}^2$$

and $T = 6a$

$$\Rightarrow \quad T = \tfrac{12}{5}g\,\text{N}$$

$F = ma$ for 6 kg ↑

$$R - 6g = 0$$
$$\Rightarrow \quad R = 6g\,\text{N}$$

● Exercise 2.3   See page 48

2.4 Components of Forces

The **resultant** of two (or more) forces is the vector sum of the forces.

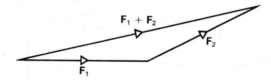

The **component** of a force in a given direction is a measure of the effect of the force in that direction. It is the product of the magnitude of the force and the cosine of the angle between the force and the given direction.

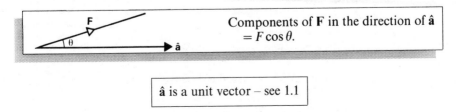

Components of **F** in the direction of **â**
$= F\cos\theta$.

â is a unit vector – see 1.1

Resolving a Force into Components in Two Mutually Perpendicular Directions

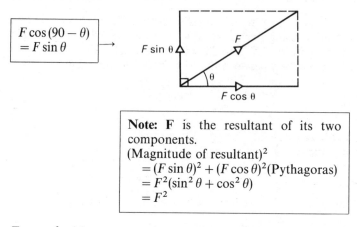

$$F\cos(90-\theta) = F\sin\theta$$

Note: F is the resultant of its two components.
(Magnitude of resultant)^2
$$= (F\sin\theta)^2 + (F\cos\theta)^2 \text{(Pythagoras)}$$
$$= F^2(\sin^2\theta + \cos^2\theta)$$
$$= F^2$$

Example 11   A force of 8 N acts at an angle of 30° below the positive x-axis.
(a)   Find its components in the directions of the x and y-axes.
(b)   Express the force in the form $a\mathbf{i} + b\mathbf{j}$

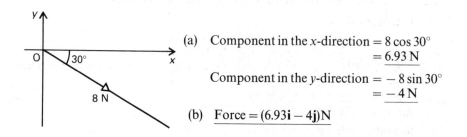

(a)   Component in the x-direction $= 8\cos 30°$
$$= 6.93\,\text{N}$$

Component in the y-direction $= -8\sin 30°$
$$= -4\,\text{N}$$

(b)   Force $= (6.93\mathbf{i} - 4\mathbf{j})\text{N}$

Example 12   A body of mass m kg rests on an incline of angle $\alpha°$. Find the component of the weight:
(a)   down the plane   (b)   perpendicular to the plane

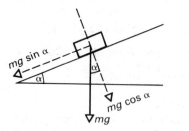

(a)   component down the plane
$$= mg\sin\alpha$$

(b)   component perpendicular to the plane
$$= mg\cos\alpha$$

Example 13   The following forces act in the given directions:

10 N at 30° above the positive x-axis
5 N along the negative y-axis.
and 8 N at 60° above the negative x-axis.

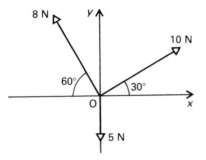

Find the sum of the components in the direction of:
(a)   the positive x-axis   (b)   the negative y-axis.

(a)   sum of components $= 10 \cos 30° - 8 \cos 60° = \underline{4.66 \text{ N}}$

(b)   sum of components $= 5 - 10 \sin 30° - 8 \sin 60° = \underline{-6.93 \text{ N}}$

●   Exercise 2.4   See page 49

2.5 Finding the Resultant from the Sums of the Components

Example 14   Find the resultant of the following forces:
$(\mathbf{i} - 2\mathbf{j})\text{N}, (3\mathbf{i} + \mathbf{j})\text{N}, (-\mathbf{i} - 2\mathbf{j})\text{N}$, giving the answer   (a)   in the form $a\mathbf{i} + b\mathbf{j}$ and
(b)   in magnitude and direction form.

(a)   Resultant $= (\mathbf{i} - 2\mathbf{j}) + (3\mathbf{i} - \mathbf{j}) + (-\mathbf{i} - 2\mathbf{j})$
      Resultant $= \underline{3\mathbf{i} - 3\mathbf{j}}$

(b)

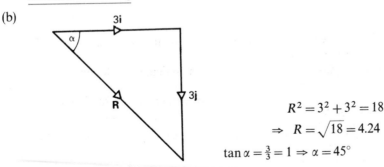

$$R^2 = 3^2 + 3^2 = 18$$
$$\Rightarrow R = \sqrt{18} = 4.24$$
$$\tan \alpha = \tfrac{3}{3} = 1 \Rightarrow \alpha = 45°$$

∴ resultant is 4.24 N at $-45°$ with the $\mathbf{i}$ direction.

Example 15   The following forces act at a point: 4 N on a bearing of 030°,
6 N on a bearing of 135°, 5 N on a bearing of 260°.

(a)   Find the sum of the components in an easterly direction.
(b)   Find the sum of the components in a northerly direction.
(c)   Find the resultant of all the forces.

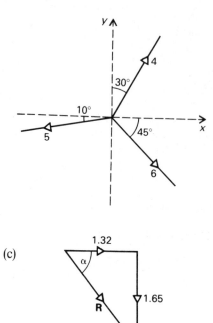

(a)   Sum of components →

$$X = 4\sin 30° + 6\cos 45° - 5\cos 10°$$
$$\Rightarrow X = 1.32\,\text{N}$$

(b)   Sum of components ↑

$$Y = 4\cos 30° - 6\sin 45° - 5\sin 10°$$
$$\Rightarrow Y = -1.65\,\text{N}$$

(c)

> Since X is positive it is put in as → but the negative sign of Y changes the ↑ to ↓. Taking care with the directions ensures that the resultant is in the correct direction. A diagram like this is essential.

Resultant = **R**

where
$$R^2 = X^2 + Y^2 = 1.32^2 + 1.65^2$$
$$\Rightarrow R = 2.11\,\text{N}$$

and
$$\tan \alpha = \frac{1.65}{1.32}$$

$$\Rightarrow \quad \alpha = 51.3° \Rightarrow \text{bearing} = \alpha + 90° = 141.3°$$
resultant is 2.11 N on a bearing of 141.3°

Example 16   ABCD is a rectangle with AB = 3 units and BC = 2 units. Forces of 8 N, 14 N, $6\sqrt{13}$ N, 15 N act along the lines AB, CB, AC, CD respectively, in the direction indicated by the order of the letters. Find the magnitude of the resultant and the direction it makes with AB.

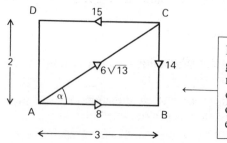

> It does not matter where the letters go provided that they go round the rectangle in the order ABCD (cyclic order). But, since we want the direction of the resultant to AB, it is easier to have AB across the page.

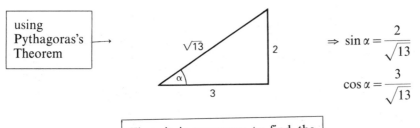

using Pythagoras's Theorem

$\Rightarrow \sin \alpha = \dfrac{2}{\sqrt{13}}$

$\cos \alpha = \dfrac{3}{\sqrt{13}}$

First, it is necessary to find the direction AC makes with either AB or BC.

Resolving parallel to AB

$$X = 8 - 15 + 6\sqrt{13}\cos \alpha$$

$$= 8 - 15 + 6\sqrt{13} \cdot \frac{3}{\sqrt{13}} = 11$$

Resolving parallel to AD

$$Y = 6\sqrt{13}\sin \alpha - 14$$

$$= 6\sqrt{13} \cdot \frac{2}{\sqrt{13}} - 14 = -2.$$

$$R^2 = 11^2 + 2^2$$
$$\Rightarrow R = 11 \cdot 18$$

$$\tan \beta = \frac{2}{11} \Rightarrow \beta = 10.3°$$

∴ resultant is 11.18 N at an angle of 10.3° with AB

- Exercise 2.5   See page 51

2.6 Equilibrium

Terminology

A **particle** is a body whose dimensions may be neglected.

A body is in a state of **equilibrium** when two or more forces act on it and motion does not take place.

Important Property of Bodies in Equilibrium

When a body is in equilibrium under a set of forces, then the sum of the components of the forces, in any direction, is zero.

Example 17 The particle O is in equilibrium under the set of forces in the diagram. By resolving in two directions, find the values of P and Q.

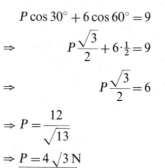

Resolving parallel to Ox

$$P\cos 30° + 6\cos 60° = 9$$

$$\Rightarrow \qquad P\frac{\sqrt{3}}{2} + 6·\tfrac{1}{2} = 9$$

$$\Rightarrow \qquad P\frac{\sqrt{3}}{2} = 6$$

$$\Rightarrow P = \frac{12}{\sqrt{13}}$$

$$\Rightarrow P = 4\sqrt{3}\,\text{N}$$

Resolving parallel to Oy

$$Q + P\sin 30° = 6\sin 60°$$

$$\Rightarrow \qquad Q + P·\tfrac{1}{2} = 6·\frac{\sqrt{3}}{2}$$

$$\Rightarrow \qquad Q + 2\sqrt{3} = 3\sqrt{3}$$

$$\Rightarrow \qquad Q = \sqrt{3}\,\text{N}$$

Example 18 The forces $(2\mathbf{i} + 3\mathbf{j})\,\text{N}$, $(-\mathbf{i} + 2\mathbf{j})\,\text{N}$, $(p\mathbf{i} + q\mathbf{j})\,\text{N}$ are in equilibrium, find the values of p and q.

If the forces are in equilibrium, their resultant is zero. Hence

$$\begin{pmatrix} 2 \\ 3 \end{pmatrix} + \begin{pmatrix} -1 \\ 2 \end{pmatrix} + \begin{pmatrix} p \\ q \end{pmatrix} = \begin{pmatrix} 0 \\ 0 \end{pmatrix}$$

$$\Rightarrow p = -1 \text{ and } q = -5$$

Example 19 A particle of 10 kg is suspended from two strings. In equilibrium, the particle hangs with the strings at 60° and 45° to the vertical. Find the tensions in the strings.

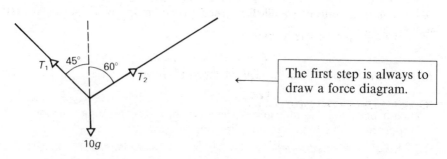

The first step is always to draw a force diagram.

Resolving horizontally

$$T_2 \sin 60° = T_1 \sin 45°$$

$$\Rightarrow \qquad T_2 \frac{\sqrt{3}}{2} = \frac{T_1}{\sqrt{2}} \qquad (1)$$

Resolving vertically

$$T_2 \cos 60° + T_1 \cos 45° = 10g$$

$$\Rightarrow \qquad T_2 \cdot \tfrac{1}{2} + T_1 \frac{1}{\sqrt{2}} = 10g \qquad (2)$$

(1) and (2)

$$\Rightarrow \qquad \frac{T_2}{2} + \frac{T_2\sqrt{3}}{2} = 10g$$

$$\Rightarrow \qquad \underline{T_2 = 7.32g \text{ N}}$$

(1)

$$\Rightarrow \qquad T_1 = \frac{\sqrt{6}}{2} T_2$$

$$\Rightarrow \qquad \underline{T_1 = 8.97g \text{ N}}$$

Example 20   A particle is in equilibrium on a 30° plane under the action
of the forces shown in the diagram.
Find the value of P and θ.

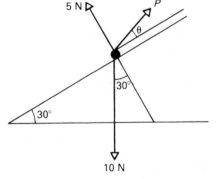

The following technique is a very
important one and will be used
often.

Resolving parallel to plane

$$P \cos \theta = 10 \sin 30°$$
$$\Rightarrow P \cos \theta = 5 \qquad (1)$$

Resolving perpendicular to plane

$$P \sin \theta + 5 = 10 \cos 30°$$
$$\Rightarrow \quad P \sin \theta = 3.66 \qquad (2)$$

Dividing (2) by (1) $\Rightarrow \tan \theta = \dfrac{3.66}{5}$

$$\Rightarrow \underline{\theta = 36.2°}$$

Substituting in (1) $\Rightarrow P = \dfrac{5}{\cos 36.2°}$

$$= 6.2\,\text{N}$$

● Exercise 2.6   See page 52

2.7 More Particles in Equilibrium

Example 21   A particle of mass m rests on a smooth plane inclined at α to the horizontal, where $\tan \alpha = \frac{3}{4}$. It is held at rest by a force P, pulling up the plane and parallel to the line of greatest slope of the plane. Find the value of P and the normal reaction between the particle and the plane.

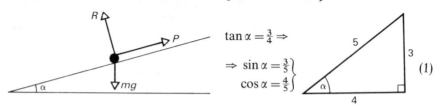

$\tan \alpha = \frac{3}{4} \Rightarrow$

$\left.\begin{array}{l} \Rightarrow \sin \alpha = \frac{3}{5} \\ \cos \alpha = \frac{4}{5} \end{array}\right\}$ (1)

Resolving parallel to the plane:
$$P = mg \sin \alpha$$
$$(1) \Rightarrow P = \tfrac{3}{5} mg$$

Resolving perpendicular to the plane:
$$R = mg \cos \alpha$$
$$(1) \Rightarrow R = \tfrac{4}{5} mg$$

Example 22   ABCD is a light inextensible string fastened to two fixed points A and D. It has two equal weights attached to B and C and is in equilibrium, with AB and CD inclined at 30° and 60° respectively to the vertical. Prove that the tension in BC is equal to either weight and that BC is inclined at 60° to the vertical.

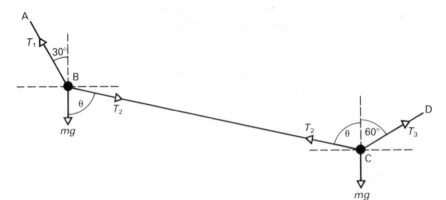

Resolving horizontally at B:
$$T_1 \sin 30° = T_2 \sin \theta$$
$$\Rightarrow \tfrac{1}{2} T_1 = T_2 \sin \theta \tag{1}$$

Resolving vertically at B:

$$T_i \cos 30° = T_2 \cos \theta + mg$$

$$\Rightarrow \frac{\sqrt{3}}{2} T_1 = T_2 \cos \theta + mg \qquad (2)$$

Resolving horizontally at C:

$$T_2 \sin \theta = T_3 \sin 60°$$

$$\Rightarrow \frac{\sqrt{3}}{2} T_3 = T_2 \sin \theta \qquad (3)$$

Resolving vertically at C:

$$T_3 \cos 60° + T_2 \cos \theta = mg$$

$$\Rightarrow \tfrac{1}{2} T_3 + T_2 \cos \theta = mg \qquad (4)$$

Since we do not want T_1, we eliminate T_1 from (1) and (2).

---

(1) and (2) $\Rightarrow \sqrt{3}\, T_2 \sin \theta = T_2 \cos \theta + mg \qquad (5)$

Similarly, we eliminate T_3 from (3) and (4).

$$\frac{T_2}{\sqrt{3}} \sin \theta + T_2 \cos \theta = mg \qquad (6)$$

(5) $\Rightarrow \qquad\qquad T_2 = \dfrac{mg}{\sqrt{3} \sin \theta - \cos \theta} \qquad (7)$

and (6) $\Rightarrow \qquad\qquad T_2 = \dfrac{\sqrt{3}mg}{\sin \theta + \sqrt{3} \cos \theta}$

Hence $\qquad\qquad \dfrac{mg}{\sqrt{3} \sin \theta - \cos \theta} = \dfrac{\sqrt{3}mg}{\sin \theta + \sqrt{3} \cos \theta}$

$$\Rightarrow \sin \theta + \sqrt{3} \cos \theta = 3 \sin \theta - \sqrt{3} \cos \theta$$

$$\Rightarrow \qquad\qquad 2 \sin \theta = 2\sqrt{3} \cos \theta$$

$$\Rightarrow \qquad\qquad \tan \theta = \sqrt{3}$$

$$\Rightarrow \qquad\qquad \theta = 60°$$

$\therefore$ BC is inclined at 60° to the vertical.

(7) $\Rightarrow \quad T_2 = \dfrac{mg}{\sqrt{3} \cdot \frac{\sqrt{3}}{2} - \frac{1}{2}} = mg$

$\therefore$ tension in BC equals one of the hanging weights.

- Exercise 2.7   See page 53

2.8 On the Move Again

The Lift Problem

Imagine a mass of m kg hung from a spring balance in a lift. The tension in the spring is given by the measurement on the scale of the balance.

Case 1:   Lift at rest

$$F = ma: T = mg$$

tension = weight

Case 2:   Lift accelerating upwards

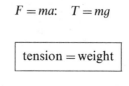

$$F = ma: T - mg = ma$$
$$\Rightarrow T \qquad = m(g + a)$$

apparent weight = tension = $mg + ma$ $\Rightarrow$ mass apparently weighs more

Case 3:   Lift accelerating downwards

$$F = ma: mg - T = ma$$
$$\Rightarrow \quad T \qquad = mg - ma$$

apparent weight = tension = $mg - ma$ $\Rightarrow$ mass apparently weighs less

So, a person in an aeroplane which is diving downwards with an acceleration equal to g undergoes a feeling of weightlessness.

Case 4:   Lift moving (up or down) with constant speed.

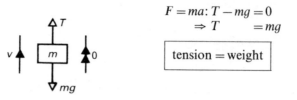

$$F = ma: T - mg = 0$$
$$\Rightarrow T \qquad = mg$$

tension = weight

Motion on an Inclined Plane

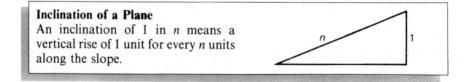

Inclination of a Plane
An inclination of 1 in n means a vertical rise of 1 unit for every n units along the slope.

Example 23   A train, travelling at a constant speed of 90 km/h on the flat, begins an ascent of 1 in 75. The pulling force that the engine exerts during the climb is constant and equal to $\frac{1}{100}$ of the weight of the train. The resistance is also constant and equal to $\frac{1}{150}$ of the weight of the train. Find how far the train will go before it comes to a standstill.

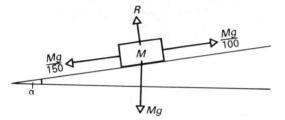

$\sin \alpha = \frac{1}{75}$ (1)

$F = ma$ parallel to the slope

$$\frac{Mg}{100} - \frac{Mg}{150} - Mg \sin \alpha = Ma$$

and (1) $\Rightarrow$ $a = \left[\dfrac{1}{100} - \dfrac{1}{150} - \dfrac{1}{75} \right] g$

$\Rightarrow$ $a = -\dfrac{g}{100}$ 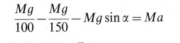 | Since the acceleration is constant we can use the **constant accelera-tion formulae.**

$u = 90\,\text{km/h} = 25\,\text{m/s}$
$v = 0$
$a = -g/100$
$t = -$
$s = ?$

$$v^2 = u^2 + 2as$$

$$\Rightarrow 0 = 25^2 - \frac{2g}{100}s$$

$$\Rightarrow s = \frac{625 \times 100}{2 \times 9.8} = 3189\,\text{m}$$

$\therefore$ train will come to a standstill after 3.19 km.

• Exercise 2.8   See page 54

2.9 Connected Particles and Inclined Planes

Example 24   A body of mass 4 kg is placed on a smooth plane of height 5 m and length 20 m. The body is connected by a light inextensible string, which passes over a smooth pulley at the top of the plane, to a mass of 3 kg hanging freely. If, initially, the 4 kg mass is at rest at the bottom of the plane and the 3 kg mass hangs just over the pulley, find:
(a)   the common acceleration and the tension in the string
(b)   how long it takes the hanging mass to reach the ground
(c)   how far up the slope the 4 kg mass goes.

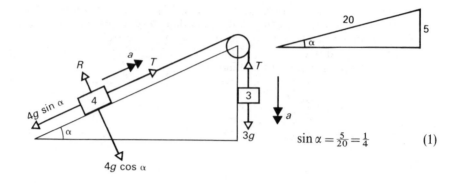

$$\sin \alpha = \tfrac{5}{20} = \tfrac{1}{4} \qquad (1)$$

Note that, in a force diagram, you can *either* put in the components of the weight *or* the weight itself but *not both*.

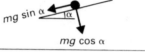   or

(a) $F = ma$ for 3 kg mass: $\qquad 3g - T = 3a \qquad\qquad (2)$
     $F = ma$ for 4 kg mass: $\qquad T - 4g \sin \alpha = 4a \qquad (3)$

   (1) and (3) $\Rightarrow \qquad\qquad T - g = 4a$
     (2) $\Rightarrow \qquad\qquad\qquad -T + 3g = 3a$
     _____

   Adding $\qquad\qquad\qquad\qquad 2g = 7a$
     $\Rightarrow \qquad\qquad$ acceleration $= a = 2g/7$
     _____

   (2) $\Rightarrow T = 3g - 3a = 3g - \dfrac{6g}{7}$

     $\Rightarrow$ Tension $= T = \dfrac{15g}{7}\,\text{N}$

> Since the acceleration is constant we can use the **constant acceleration formulae** to solve (b) and (c).

(b) $u = 0$
     $v = -$
     $a = \dfrac{2g}{7}$
     $t = ?$
     $s = 5$

$$s = ut + \tfrac{1}{2}at^2 \Rightarrow 5 = \frac{g}{7}t^2$$

$$\Rightarrow t^2 = \frac{35}{g} = \frac{35}{9.8}$$

$$\Rightarrow t = 1.89\,\text{s}$$

$\therefore$ it takes 1.89 s to reach the ground

(c)   During the first part of the motion, the 4 kg mass travels 5 m up the slope. When the string goes slack the mass decelerates. We need to find the speed of the masses at the point when the string goes slack.

$$v = u + at \Rightarrow v = 0 + \frac{2g}{7} \times 1.89 = 5.29 \text{ m/s}$$

After the string goes slack,

> Negative because the force is in the opposite direction to the acceleration.

$F = ma$ for 4 kg mass: $-4g \sin \alpha = 4a'$

$$\Rightarrow a' = -g \sin \alpha = -\frac{g}{4}$$

$u = 5.29$
$v = 0$
$a = -\dfrac{g}{4}$
$t = -$
$s = ?$

$$v^2 = u^2 + 2as$$

$$\Rightarrow 0 = (5.29)^2 - 2\frac{g}{4}s$$

$$\Rightarrow s = \frac{2 \times 5.29^2}{g} = 5.71 \text{ m}$$

∴ total distance travelled up the plane = $(5 + 5.71)$m
$$= 10.71 \text{ m}$$

• Exercise 2.9   See page 55

2.10 Related Accelerations

1. Systems of Pulleys

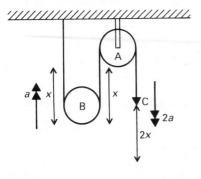

A is fixed. B can move up or down. For B to move up a distance x, the end of the string, C, must move down a distance $2x$. Hence, if B has an upward acceleration a then C must have a downward acceleration $2a$.

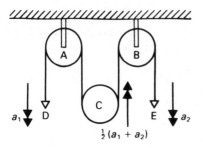

A and B are fixed. C can move up or down. If D moves down a distance x and E moves down a distance y then the portion of the string between A and B will shorten by an amount $x+y$. Thus C will move up a distance $\frac{1}{2}(x+y)$. Thus if the acceleration of D is $a_1\downarrow$ and if the acceleration of E is $a_2\downarrow$ then the acceleration of C is $\frac{1}{2}(a_1+a_2)$ upwards.

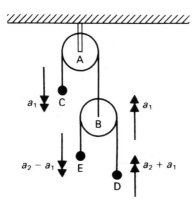

A is fixed. B is fixed to the end of a string suspended over A. Let B and C both move with acceleration a_1. Given that the mass at E is greater than the mass at D and, if B were fixed, D and E would move with acceleration a_2, then the accelerations will be as shown.

Note: The tensions in the two strings will be different.

Example 25

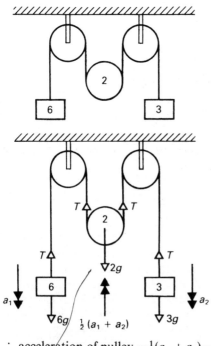

A movable pulley of 2 kg is suspended on a light inextensible string between two fixed pulleys as shown. The two ends of the string carry loads of 6 kg and 3 kg. The pulleys are smooth and the hanging points of the string are vertical. Find the acceleration of the movable pulley.

$F = ma$

for 3 kg: $3g - T = 3a_2$ (1)
for 6 kg: $6g - T = 6a_1$ (2)
for pulley: $2T - 2g = 2 \times \frac{1}{2}(a_1 + a_2)$ (3)

(1) and (2) $\Rightarrow 3g = 6a_1 - 3a_2$
           $\Rightarrow g = 2a_1 - a_2$ (4)
(1) and (2) $\Rightarrow 2(3g - 3a_2) - 2g = a_1 + a_2$
           $\Rightarrow 4g = 7a_2 + a_1$ (5)

Solving (4) and (5) simultaneously

      $\Rightarrow a_1 = \frac{11}{15}g \quad a_2 = \frac{7}{15}g$

$\therefore$ acceleration of pulley $= \frac{1}{2}(a_1 + a_2) = \frac{3}{5}g$ m/s^2

Point to note:
If the wrong direction has been chosen for any acceleration then the answer obtained will be negative.

2 Particles on Wedges that are Free to Move

Example 26   A particle of mass m lies on the smooth sloping face of a wedge which is itself standing on a smooth horizontal surface. If the sloping face of the wedge is inclined at $60°$ to the horizontal and the mass of the wedge is M, find the acceleration of the wedge.

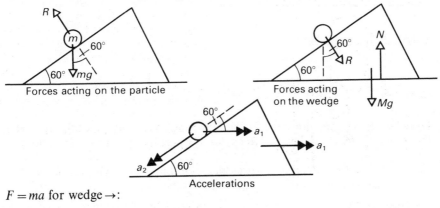

Forces acting on the particle   Forces acting on the wedge

Accelerations

$F = ma$ for wedge $\rightarrow$:

$$R \sin 60° = Ma_1$$

$$\Rightarrow R = \frac{2Ma_1}{\sqrt{3}} \tag{1}$$

$F = ma$ perpendicular to wedge for particle:

$$mg \cos 60° - R = ma_1 \sin 60°$$

$$\Rightarrow \frac{mg}{2} - R = ma_1 \frac{\sqrt{3}}{2} \tag{2}$$

For the particle, $F = ma$ was taken perpendicular to the wedge in order that a_2 need not appear in the equation.

(1) and (2) $\Rightarrow \dfrac{mg}{2} - \dfrac{2Ma_1}{\sqrt{3}} = ma_1 \dfrac{\sqrt{3}}{2}$

$$\Rightarrow \frac{mg}{2} = a_1\left[\frac{2M}{\sqrt{3}} + \frac{m\sqrt{3}}{2}\right] \longrightarrow \boxed{\text{multiplying through by } 2\sqrt{3}}$$

$$\Rightarrow \sqrt{3}mg = a_1[4M + 3m]$$

$$\Rightarrow a_1 = \frac{\sqrt{3}m}{4M + 3m}g$$

- Exercise 2.10   See page 56

2.11 Revision of Unit

● Miscellaneous Exercise 2A   See page 58

2.12+ A Level Questions

● Miscellaneous Exercise 2B   See page 59

Unit 2 Exercises

Exercise 2.1

Draw force diagrams for each of the following situations. In each case, label the forces either W (weight), T (tension), P (thrust) or R (normal reaction).

1   Show the forces acting on a ladder, standing on rough ground, leaning against a rough wall at an angle of θ to the wall.

2   Show the forces acting on a ladder leaning against a rough wall, standing on smooth ground with a rope tied between the bottom of the ladder and the bottom of the wall. The ladder is inclined at an angle α to the horizontal.

3   Show the forces acting on a plank, 4 m long, resting on two supports, each support being 1 m from an end of the plank.

4   Show the forces acting on a tea-chest, lying on horizontal rough ground, being pushed along by a force applied to the centre of one upper edge, the force being at an angle of 30° above the horizontal.

5   Show the forces acting on a plumb-bob which is pushed aside from its natural hanging position by a horizontal force.

6   Two cylinders, one of radius 1 m and the other of radius 1.5 m are lying with their axes parallel and 4.5 m apart on rough ground. A plank lies across the cylinders, perpendicular to the axes, jutting out 1 m beyond each point of contact. All contacts are rough. Draw diagrams to show:
    (a)   the forces acting on the plank
    (b)   the forces acting on the two cylinders.

7   A box is lying on rough horizontal ground. A rough plank is leaning against the middle of one upper edge of the box. Show:
    (a)   the forces acting on the box
    (b)   the forces acting on the plank.

8 A block rests on a rough horizontal table. Attached to the block is a light string which runs parallel to the table, over a smooth peg at the edge of the table and down to a weight hanging freely. On the diagram draw the forces acting on the block and the hanging weight.

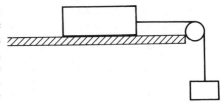

9

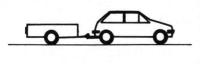

A light smooth pulley hangs from the ceiling. Over it hangs a light inextensible string with equal weights on either end. On one diagram show the forces acting on the pulley and the weights.

10 A heavy rough pulley hangs from the ceiling. Over it hangs a light inextensible string with unequal weights on either end. On one diagram show the forces acting on the pulley and the weights.

11

A car tows a camping trailer. The engine produces a thrust P and the car experiences an air resistance P. The trailer runs freely without any air resistance. Show the forces acting on:
(a) the trailer
(b) the car.

Exercise 2.2

1 Find the accelerations produced when a force of 10 N acts on:
(a) a mass of 20 kg   (b) a mass of 20 g.

2 A force of 2 N produces an acceleration of 400 m/s² on a body. What is the mass of the body?

3 An ink pellet of mass 0.8 gram is given an acceleration of 200 m/s². What is the force used?

4 Find the acceleration produced in a body of mass 4 kg which is acted upon by forces $(2\mathbf{i} + 4\mathbf{j})$N and $(-\mathbf{i} + 4\mathbf{j})$N.

5 Find the resultant force required to produce an acceleration of $(3\mathbf{i} - \mathbf{j})$ m/s² on a mass of 10 kg.

6 What force will give a mass of 42 kg a velocity of 36 km/h in 3.5 minutes?

7   An ice-yacht of mass 400 kg is blown along with a force of 250 N. How long will it take to reach a velocity of 18 km/h from rest?

8   A box of mass 40 kg is dragged across the floor by a horizontal force of 80 N. What is the frictional resistance if:
    (a)   the box moves with constant velocity
    (b)   the box moves with constant acceleration 1 m/s²?

9   The total mass of a man and his sailboard is 60 kg. If the wind produces a force of 250 N and the water a resistance of 100 N, find the acceleration of the sailboard and man.

10 A mass of 10 kg falls with an acceleration of 7.5 m/s². What resistance is acting on the mass?

11 A load of mass 2 kg is hung on the end of a string. In each of the following situations calculate the tension in the string:
    (a)   the body is raised with an acceleration of 3 m/s²
    (b)   the body is lowered with an acceleration of 3 m/s²
    (c)   the body is raised with a velocity of 3 m/s.

12 A hot air balloon rises from the ground with uniform acceleration. It reaches a height of 100 m in 20 s. If the mass of the balloon and basket is 200 kg, find the lifting force.

13 A 20 000 tonne ship slows, with its engines stopped, from 18 km/h to 15 km/h in a distance of 500 m. If the resistance of the water can be assumed to be uniform, find its value.

**14   An engine and train weigh 200 tonnes. The engine exerts a pull of 4×10^4 N. The resistance to motion is $\frac{1}{100}$ of the weight of the train and its braking force is $\frac{1}{4}$ of the weight. The train starts from rest and accelerates uniformly until it reaches a speed of 36 km/h. At this point the brakes are applied until the train stops. Find the time taken to the nearest second.

Exercise 2.3

A clear force diagram **must** be drawn for each of the following questions.

1   Two weights, of masses 3 kg and 7 kg are connected by a light inextensible string passing over a smooth fixed pulley. Find the common acceleration and the tension in the string.

2   Two weights, of masses 10 grams and 25 grams are connected by a light inextensible string passing over a smooth fixed pulley. Find the common acceleration and the tension in the string.

3   A truck, of mass 50 kg, runs on smooth horizontal rails. A light rope is attached to the truck and runs horizontally until it passes over a light smooth pulley and drops down a vertical shaft. At the other end of the

rope is a 10 kg bucket which is falling down the shaft. Find the common acceleration of the bucket and the truck and the tension in the rope. If the bucket falls 300 m from rest to the bottom of the shaft, with what velocity will it hit the bottom?

4 The truck in the previous question is at rest with the bucket at the bottom of the shaft. 20 kg of rock is put into the bucket. What horizontal force must be applied to the truck to bring the bucket up with an acceleration of 1 m/s². (Take $g = 10$ m/s².)

5 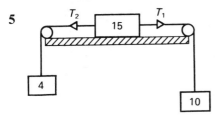 A body of mass 15 kg lies on a smooth table. Two bodies of mass 10 kg and 4 kg are attached to the 15 kg body by light inextensible strings which pass over smooth pulleys at the edges of the table. Find the acceleration of the system and the tensions in the strings.

*6 Two bodies A and B, of masses m_1 and m_2 respectively, are connected by a light inextensible string passing over a smooth pulley. If $m_1 > m_2$, find the acceleration of the bodies and show that the tension in the string is

$$\frac{2m_1 m_2 g}{m_1 + m_2}$$

*7 A 1 tonne car tows a caravan of mass 800 kg along a level road, using a rigid tow bar. The resistance to the car's motion is 150 N and the caravan experiences a resistance of 250 N. If the engine exerts a pulling force of 2.5 kN, find the tension in the tow bar and the acceleration.
The engine is disengaged and the brakes are applied. The braking force is 350 N. Find the retardation. What is the force in the tow bar?
[Note: 1 kN = 1 kilonewton = 1000 N]

Exercise 2.4

1 For each of the following forces, find the components in the directions of the x- and y-axes:

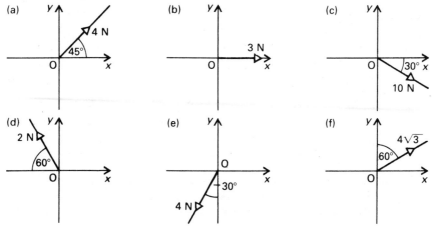

2 Express the following forces in the form $a\mathbf{i} + b\mathbf{j}$:

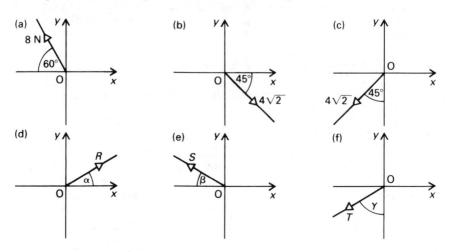

3 A body of mass 10 kg lies on an inclined plane of angle 60°. Find the components of the weight:

(a) down the plane
(b) normal to the plane.

4

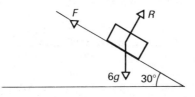

A body of mass 6 kg lies on an inclined plane of angle 30°. If the body is at rest on the plane then the friction between the two surfaces must balance the pull of the weight down the plane. Find the value of the frictional force.

For each of the following sets of forces, find the sum of the components in the direction of:

(a) the positive *x*-axis   (b) the positive *y*-axis.

5

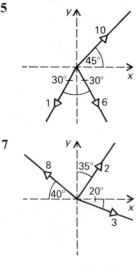

6

7

8

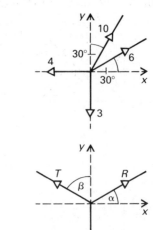

For each of the following sets of forces, find the sum of the components in the direction of:

(a) $\overrightarrow{AB}$ (b) $\overrightarrow{CD}$.

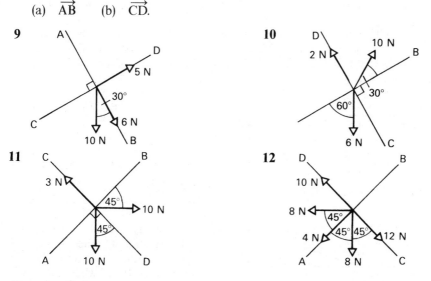

9

10

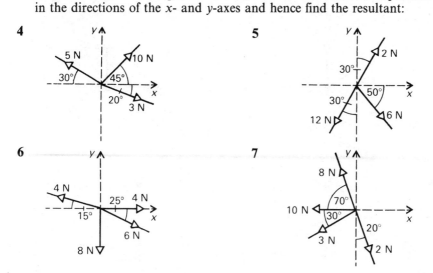

11

12

Exercise 2.5

1   The resultant of the forces $(2\mathbf{i} + 3\mathbf{j})$N, $(4\mathbf{i} + \mathbf{j})$N and $(a\mathbf{i} + b\mathbf{j})$N is $(3\mathbf{i} + 2\mathbf{j})$N. Find the values of a and b.

2   Find the magnitude and direction of each of the following forces:

   (a) $3\mathbf{i} + 4\mathbf{j}$ (b) $3\mathbf{i} - 4\mathbf{j}$ (c) $-3\mathbf{i} - 4\mathbf{j}$.

3   ABCD is a rectangle. Forces of 10 N, 8 N, 6 N, 4 N act along the lines DC, BA, DA, BC respectively in the direction indicated by the order of the letters. Find the magnitude of the resultant and the angle it makes with AB.

For each of the following sets of forces, find the sums of the components in the directions of the *x*- and *y*-axes and hence find the resultant:

4

5

6

7

8 PQRS is a rectangle. Forces of $5\sqrt{2}\,\text{N}$, $3\sqrt{2}\,\text{N}$, $1\,\text{N}$ act along the lines PQ, RS, QR respectively, in the direction indicated by the order of the letters. Find the magnitude of the resultant and the angle it makes with PS.

*9 ABCD is a square. E is the midpoint of BC. F is the midpoint of CD. Forces of $4\,\text{N}$, $6\,\text{N}$, $10\,\text{N}$, $8\,\text{N}$, $3\,\text{N}$ act along AB, AE, CA, AF, AD respectively in the directions indicated by the order of the letters. Find the resultant.

*10

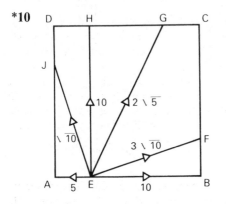

In the diagram, ABCD is a square of 4 cm. The points E, F, G, H and J lie on the sides AB, BC, CD, DA respectively in such a way that $AE = BF = CG = HD = DJ = 1\,\text{cm}$. Forces of magnitude $10\,\text{N}$, $3\sqrt{10}\,\text{N}$, $2\sqrt{5}\,\text{N}$, $10\,\text{N}$, $\sqrt{10}\,\text{N}$ and $5\,\text{N}$ act at the point E in the directions EB, EF, EH, EJ and EA respectively. Calculate the magnitude of the resultant of these forces and show that its line of action makes an acute angle θ with EB where $\tan\theta = \frac{4}{3}$.

[J]

Exercise 2.6

In each of the following questions, the particle is in equilibrium under the set of forces in the diagram. By resolving in two directions, find the unknown forces and angles:

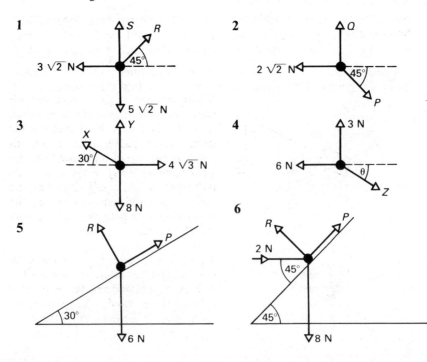

7   Each of the following sets of forces is in equilibrium. Find the value of p and q in each case:

(a) $(3\mathbf{i} + 5\mathbf{j})\text{N}, (3\mathbf{i} - 2\mathbf{j})\text{N}, (p\mathbf{i} + q\mathbf{j})\text{N}$
(b) $(p\mathbf{i} - 2\mathbf{j})\text{N}, (-3\mathbf{i} + 4\mathbf{j})\text{N}, (2\mathbf{i} - q\mathbf{j})\text{N}$
(c) $(2p\mathbf{i} - p\mathbf{j})\text{N}, (-4\mathbf{i} + 3p\mathbf{j})\text{N}, (p\mathbf{i} - q\mathbf{j})\text{N}, (2q\mathbf{i} - 5\mathbf{j})\text{N}$.

8   A particle of 2 kg hangs in equilibrium from two strings which make angles of 40° and 20° to the vertical. Find the tension in the strings.

9   A particle of mass 8 kg is placed on a smooth plane of inclination 60°. What force applied (a) parallel to the plane (b) horizontally, will maintain the particle in equilibrium?

10   A particle of mass 4 kg is attached to one end of a string. The other end of the string is fixed. An upward vertical force of 3.5 N and a horizontal force of 14 N act upon the particle, so that it rests in equilibrium with the string at θ to the vertical. Calculate the tension in the string and the angle θ.

Exercise 2.7

1

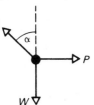

A particle is in equilibrium under the three forces P, Q, W as shown in the diagram.
Show that $P = W \tan \alpha$.

2   A particle is in equilibrium under the forces shown in the diagram.
Show that $P = W(2 - \sqrt{3})$.

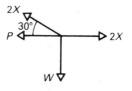

3

A body of mass 3 kg rests on a smooth plane inclined at θ to the horizontal. It is attached to a light inextensible string which passes over a smooth pulley at the top of the plane to a 2 kg mass hanging freely at the other end. If the system is in equilibrium, find the angle θ, the tension in the rope and the normal reaction between the 3 kg mass and the plane.

4

A particle of mass m rests on a smooth plane inclined at 60° to the horizontal. A force P, inclined up the plane at θ to the plane, holds the particle in equilibrium. If $R = mg/4$ find the value of θ and show that $P = \sqrt{13}mg/4$.

5 A string hangs freely from a fixed point and has a mass of 5 kg attached to the free end. The breaking strain of the string is 80 N. A horizontal force *P* acts on the mass, dragging it sideways. In the position of equilibrium when *P* = 20 N, find the tension in the string and the angle that the string makes with the vertical.

In the equilibrium position when the string is on the point of breaking, find the magnitude of the force *P* and the angle that the string makes with the vertical.

6

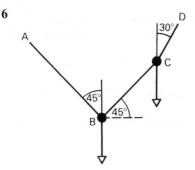

A light string ABCD is suspended from A and D which are on the same level. A weight of mass 10 kg hangs at B and a weight of mass *M* kg hangs at C. AB is at 45° to the vertical, DC is at 30° to the vertical, and BC slopes upwards at 45° to the horizontal. Find the tensions in AB, BC and CD and show that the value of *M* is $5(\sqrt{3}-1)$.

*7

A particle of mass *m* rests on a smooth plane inclined at an angle *α* to the horizontal. It is held in equilibrium by a force *P* acting along the line of greatest slope at an angle *θ* to the plane. If the normal reaction is *kmg* show that

$$\tan\theta = \frac{\cos\alpha - k}{\sin\alpha}$$

and find the value of *P* in terms of *k* and *α*.

Exercise 2.8

1 A boy, of mass 50 kg, stands on the floor of a lift. Find the reaction between the boy and the floor when the lift is (a) ascending (b) descending, with a uniform acceleration of 4 m/s².

2 A mass of 2 kg rests on a horizontal plane which is made to ascend (a) with a constant velocity of 3 m/s (b) with a constant acceleration of 3 m/s². Find in each case the reaction of the mass with the floor.

3 A mass of 500 grams rests on the scale pan of a set of scales which is drawn upwards with a constant acceleration. The reaction between the mass and the pan is found to be 5 N. Find the acceleration of the scale pan.

4 A woman whose true weight is 650 N appears to weigh 600 N in a moving lift. What is the acceleration of the lift at the moment of weighing? (Take *g* = 10 m/s²)

5   The resistance to the motion of a train is equal to $\frac{1}{150}$ of the weight of the train. The train is travelling on level track at 72 km/h and it comes to the foot of an incline of 1 in 120. If the power is turned off, how far will the train go up the slope before it stops?

6   A railway goods truck accidentally becomes uncoupled from the rest of the train whilst at rest on an incline of 1 in 140. It acquires a speed of 27 km/h in 10 minutes as it moves down the slope. If the resistance to the motion of the truck is k N/kg mass of the truck, find the value of k.

• *7   In a lift moving upwards with a constant acceleration a, a mass appears to have a weight of 100 N. When the lift moves downwards with a constant acceleration of $2a$, the mass appears to weigh 70 N. Find the actual mass and the upward acceleration of the lift. (Take $g = 10$ m/s^2).

*8   A toboggan is found to travel with uniform speed down a slope of 1 in 70. If the toboggan starts from the bottom of the same slope with a speed of 18 km/h how far will it travel *up* the slope before coming to rest?

Exercise 2.9

1.   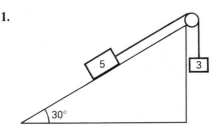   A body of mass 5 kg is placed on a smooth plane inclined at 30° to the horizontal. It is connected by a light inextensible string, which passes over a smooth pulley at the top of the plane, to a mass of 3 kg hanging freely. Find the common acceleration and the tension in the string.

2   If, in question **1**, the hanging weight is changed from 3 kg to 2 kg, find the common acceleration.

3   A mass of 3 kg lies at the bottom of a smooth inclined plane 12 m long and 4 m high. It is attached by a light string to a mass of 2 kg which hangs just over the top of the plane. The system is allowed to move. How long does the 2 kg mass take to reach the ground? If the 2 kg mass then stays on the ground, how much longer does the 3 kg mass continue to move up the plane?

4   Two masses of 500 grams and 7.5 kg are connected by a light inextensible string 1.8 m long and lie on a smooth table 80 cm high. The string is taut and perpendicular to the edge of the table when the lighter mass is pushed gently off the edge of the table. Find:

(a)   the time that elapses before the 500 gram mass hits the floor
(b)   the further time that elapses before the other mass reaches the edge of the table.

5   A body of mass M kg rests on the surface of a smooth plane inclined at an angle α to the horizontal. It is connected by a light inextensible string passing over a smooth pulley at the top of the plane to a mass m kg, hanging

freely. Given that the hanging mass moves downwards, find the acceleration and show that the tension is

$$\frac{Mmg}{M+m}(1+\sin\alpha)$$

For what values of $\sin\alpha$ does the hanging mass move downwards?

6

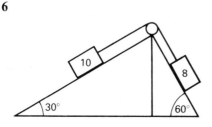

Two smooth planes, inclined at 30° and 60° to the horizontal, are placed back to back. Masses of 10 kg and 8 kg are placed on the planes and connected by a light inextensible string passing over a smooth pulley at the top of the plane. Without using a calculator, show that the resulting acceleration is

$$(4\sqrt{3}-5)\frac{g}{18}$$

Exercise 2.10

In this exercise, all pulleys are frictionless, all surfaces are smooth, all strings are light and inextensible and their hanging parts are vertical.

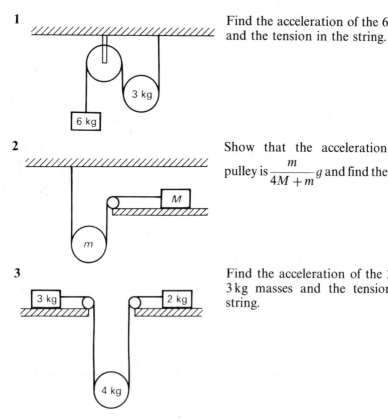

1   Find the acceleration of the 6 kg mass and the tension in the string.

2   Show that the acceleration of the pulley is $\dfrac{m}{4M+m}g$ and find the tension.

3   Find the acceleration of the 2 kg and 3 kg masses and the tension in the string.

4 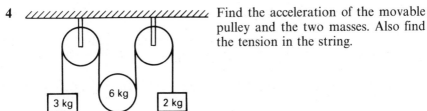 Find the acceleration of the movable pulley and the two masses. Also find the tension in the string.

5 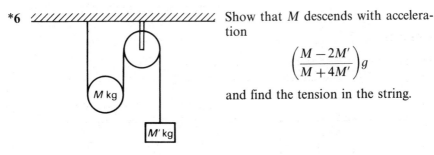 Find the acceleration of the wedge, the reaction between the particle and the wedge and the reaction between the wedge and the ground.

*6 Show that M descends with acceleration

$$\left(\frac{M-2M'}{M+4M'}\right)g$$

and find the tension in the string.

*7 A wedge of mass M is placed on a smooth horizontal table and a particle of mass m is allowed to slide down a face of the wedge which is at an angle α to the horizontal. Find the horizontal force that must be applied to the wedge to keep it from moving and show that the reaction between the wedge and the table is

$$(M + m\cos^2\alpha)g$$

*8 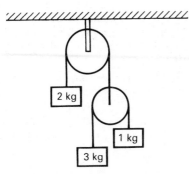 The weight of the lower pulley is negligible. By applying Newton's second law to the 2 kg mass and then to that part of the system which consists of the lower pulley and its two hanging weights, show that the acceleration of the lower pulley is $g/3$ and find the tension in the upper string. Hence show that the tension in the lower string is g N.

- Miscellaneous Exercise 2A

1 Find the acceleration produced on a body of mass 3 kg by forces $(4\mathbf{i} - \mathbf{j})$N and $(2\mathbf{i} + 4\mathbf{j})$N.

2 A mass of 5 kg falls with an acceleration of 6 m/s². What resistance is acting on the mass?

3 A 1 tonne car tows a caravan of mass 800 kg along a level road. The resistance to motion is negligible and the engine exerts a pulling force of 2.5 kN (1 kN = 1000 N). Find the acceleration produced and the tension in the towbar.

4

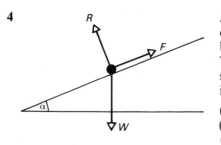

A body of weight W rests on an incline of angle α. It is prevented from slipping by a frictional force F up the plane. The normal reaction is R. Find the sum of the components of W, F and R in each of the following directions:

(a)   vertically upwards
(b)   horizontally and to the right
(c)   parallel to the plane and up the plane
(d)   upwards and normal to the plane.

5

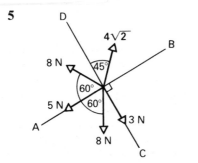

(a)   Find the sum of the components:
    (i)   in the direction of AB
    (ii) in the direction of CD.
(b)   Find the magnitude and the direction of the resultant of the system of forces.

6 Two masses of 2 kg and 3 kg are connected by a light inextensible string passing over a smooth pulley and the parts of the string not touching the pulley are vertical. Find the acceleration of the system. The 3 kg weight hits the ground and the string goes slack after the 3 kg mass has fallen 1.5 m from rest. How much higher will the 2 kg mass go after the string goes slack?

7

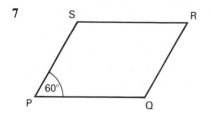

PQRS is a rhombus with angle P equal to 60°. Forces of 3 N, 4 N, 1 N, 2 N, 6 N act along PQ, QR, RS, SP, PR in the directions indicated by the letters. Find the resultant of the system of forces.

8 A cable car of mass 2 tonnes is drawn up a slope of 1 in 5 with an acceleration of $g/100$ against a constant frictional force of $\frac{1}{50}$ of the weight of the car. Find the tension in the cable and find the speed of the car 2 minutes from the start of the motion.

9 A particle of mass 4 kg hangs in equilibrium from two strings which make angles of 30° and 60° to the vertical. Find the tensions in the strings.

10 A mass of 2 kg is placed on a horizontal surface moving downwards with a constant acceleration. The reaction between the mass and the surface is found to be 5.6 N. Find the acceleration.

● **Miscellaneous Exercise 2B**

1 A smooth peg, of negligible diameter, is fixed at a height $3l$ above a horizontal table and a light inextensible string of length $4l$ hangs over the peg. A particle of mass m is attached to one end of the string and a particle of mass $2m$ is attached to the other end. The system is held at rest with the particles hanging at the same level and at a distance l from the table. The parts of the string not in contact with the peg are vertical. The system is then released from rest. Find, in terms of g and l, the speed u with which the particle of mass $2m$ strikes the table.   [J:2.3]

2 A man stands on a weighing machine which rests on the floor of a stationary lift and the dial of the machine shows 80 kg. The lift then ascends, during which time the dial shows a constant value of 84 kg. Taking g as $10\,\text{m/s}^2$, find the upward acceleration of the lift, and show that the speed of the lift at the end of six seconds is 3 m/s.

At this time the lift begins to slow down and the dial reading changes to a constant value of 74 kg. Find the retardation of the lift and the distance travelled, during the retardation, before it comes to rest.   [J:2.8]

3 A particle P of mass $8m$ rests on a smooth horizontal rectangular table and is attached by light inelastic strings to particles Q and R of mass $2m$ and $6m$ respectively. The strings pass over light smooth pulleys on opposite edges of the table so that Q and R can hang freely with the strings perpendicular to the table edges. The system is released from rest. Obtain the equations of motion of each of the particles and hence, or otherwise, determine the magnitude of their common acceleration and the tensions in the strings.

After falling a distance x from rest the particle R strikes an inelastic floor and it is brought to rest. Determine the further distance y that Q ascends before momentarily coming to rest. (It is to be assumed that the lengths of the strings are such that P remains on the table and Q does not reach it.)
        [A:2.3, 1.3]

4 A particle of mass m hangs freely from a light inextensible string, the other end of which is attached to a lift which moves upwards with a constant acceleration f, and the tension in the string is T. Express T in terms of f, g and m.

In a lift at rest at the bottom of a vertical shaft a particle of mass m is suspended from a light inextensible string. The lift then rises with constant acceleration for a time t_1 and then slows down, with a constant retardation (less than g), to rest after a further time t_2. During the period of acceleration the tension in the string is T_1 and during the period of retardation it is T_2. Show that

$$mg(t_1 + t_2) = T_1 t_1 + T_2 t_2 \qquad \text{[J:2.8]}$$

5   A car of mass 900 kg pulls a trailer of mass 350 kg along a straight level road against a total resistance of 1250 N. Given that the car is using its full power of 45 kW, show that its acceleration is 1.4 m/s² when its speed is 54 km/h.

Find also the tension in the coupling between the car and the trailer at this speed, assuming that the total resistance is divided between the car and the trailer in the ratio of their masses. [J:2.2]

6

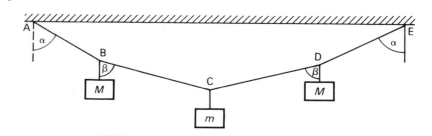

The diagram represents a light inextensible string ABCDE in which $AB = BC = CD = DE$ and to which are attached masses M, m and M at the points B, C and D respectively. The system hangs freely in equilibrium with the ends A and E of the string fixed in the same horizontal line. AB and DE each make an acute angle α with the vertical such that $\tan \alpha = \frac{3}{4}$. BC and CD make an acute angle β with the vertical such that $\tan \beta = \frac{12}{5}$.

(a)   By considering the forces acting at C, calculate the tension in BC in terms of m and g.

(b)   By considering the forces acting at B calculate the tension in AB in terms of M and g.

Show also that $10M = 11m$. [J:2.6, 2.7]

*7   Two points A and B are at the same level, and they are connected by a road consisting of two straight sections AC and CB, each of uniform gradient. The point C is a height h vertically above the line AB, and the inclinations to the horizontal of AC and CB are α and β respectively. A car of mass m starts from rest at A, accelerates uniformly to reach a speed $2U$ at C and then moves with constant retardation, attaining a speed U at B. Show that the time taken to move from A to B is

$$\frac{h}{u}(\operatorname{cosec} \alpha + \tfrac{2}{3}\operatorname{cosec} \beta)$$

The resistance to the motion of the car is a constant force R. Find the force exerted by the engine during the climb from A to C. [J:2.8, 2.9]

*8

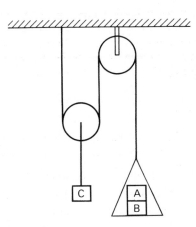

One end of a light inextensible string is attached to a ceiling. The string passes under a smooth light pulley carrying a weight C and then over a fixed smooth light pulley. To the free end of the string is attached a light scale pan in which two weights A and B are placed with A on top of B as shown. The portions of the string not in contact with the pulleys are vertical. Each of the weights A and B has a mass m and the weight C has a mass km. If the system is released from rest find the acceleration of the moveable pulley and show that the scale pan will ascend if $k > 4$. When the system is moving freely find:

(a)   the tension in the string
(b)   the reaction between the weights
       A and B.   [L:2.10]

UNIT FRICTION

3.1 Introducing Friction

Laws of Friction

1 When sliding does not take place, the frictional force is just sufficient to prevent the relative motion of the two surfaces. Thus it acts parallel to the surfaces in contact and in a direction so as to oppose the motion.

> Friction opposes the tendency to motion.

2 The frictional force has a limiting value and this is proportional to the normal reaction between the two surfaces.

> $$F \leqslant \mu R$$
> μ is the **coefficient of friction**
> μR is the **limiting friction**

3 The value of μ depends only on the roughness of the two surfaces, not on the area or shape of the surfaces in contact.
4 When sliding does take place, the frictional force is at its limiting value and in the opposite direction to the motion.

The Coefficient of Friction

$$\mu \geqslant 0$$

For a perfectly smooth surface $\mu = 0$.

Problems Involving Friction

Case 1: If a body is in limiting equilibrium, that is, it is on the point of sliding, then

$$F = \text{limiting friction} = \mu R.$$

Case 2: If a body is sliding, then

$$F = \text{sliding friction} = \mu R$$

Case 3: If a body is not sliding and not on the point of sliding, then

$$F < \mu R$$

Example 1   A box of mass 10 kg rests on a rough horizontal plane, the coefficient of friction between the box and the plane being 0.8. A horizontal force P acts on the box. When (a) $P = 50$ N (b) $P = 80$ N, find the frictional force induced and state whether the box moves. If the box moves find its acceleration.

There is no motion perpendicular to the plane.

$\Rightarrow$ Resolving vertically $R = 10g = 98$ N

$\Rightarrow$ limiting friction $= \mu R = 0.8 \times 98$ N $= 78.4$ N

(a)   If $P = 50$ N then P is less then μR
$\therefore$ the box does not move.

     The frictional force is equal and opposite to P
$\Rightarrow$ frictional force $= 50$ N.

(b)   If $P = 80$ N, then P is greater than μR
$\therefore$ the box moves.

     The frictional force $=$ sliding friction $= \mu R = 78.4$ N
$F = ma \rightarrow$:

$$\begin{aligned} P - F \quad &= ma \\ \Rightarrow 80 - 78.4 \quad &= 10a \\ \Rightarrow \text{acceleration} = a &= 0.16 \, \text{m/s}^2 \end{aligned}$$

Applied Force not Horizontal

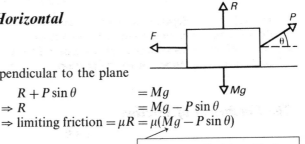

There is no motion perpendicular to the plane

$\Rightarrow$ Resolving vertically: $R + P \sin \theta \qquad = Mg$
$\Rightarrow R \qquad\qquad\quad = Mg - P \sin \theta$
$\Rightarrow$ limiting friction $= \mu R = \mu(Mg - P \sin \theta)$

The force tending to cause motion is $P \cos \theta$.

> **Note** that limiting friction is reduced when the force is not horizontal.

Example 2   A 12 kg toboggan rests on rough horizontal ground. The rope pulling it is inclined at $30°$ above the horizontal and the coefficient of friction is $\sqrt{3}/4$. Find the magnitude of the force which must be applied along the rope to move the toboggan.

Resolving vertically:

$$R + P \sin 30° = 12g$$

$$\Rightarrow R \qquad = 12g - \frac{P}{2}$$

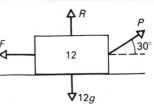

At the point of limiting equilibrium,

$$P \cos 30° = F = \mu R$$

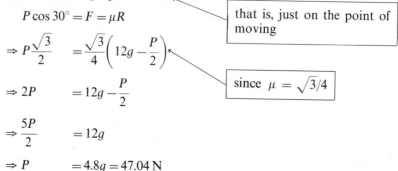

that is, just on the point of moving

$$\Rightarrow P\frac{\sqrt{3}}{2} \quad = \frac{\sqrt{3}}{4}\left(12g - \frac{P}{2}\right)$$

since $\mu = \sqrt{3}/4$

$$\Rightarrow 2P \quad = 12g - \frac{P}{2}$$

$$\Rightarrow \frac{5P}{2} \quad = 12g$$

$$\Rightarrow P \quad = 4.8g = 47.04\,\text{N}$$

Hence, the toboggan is on the point of moving when $P = 47.04\,\text{N}$.
$\Rightarrow$ the toboggan will move if P is greater than 47.04 N.

Example 3   If the toboggan in the previous example is pulled with a force of 64 N, what speed will it have gained from rest in $\frac{1}{2}$ minute?

$F = ma \rightarrow$

$$64 \cos 30° - F = 12a \tag{1}$$

Resolving vertically:

$$R + 64 \sin 30° = 12g$$

$$\therefore F = \mu R = \frac{\sqrt{3}}{4}(12g - 32) \tag{2}$$

(1) and (2) $\Rightarrow 32\sqrt{3} - \sqrt{3}(3g - 8) = 12a$
$\quad\Rightarrow 40\sqrt{3} - 3\sqrt{3g} \quad = 12a$
$\quad\Rightarrow \qquad\qquad a = 1.53\,\text{m/s}^2$
$v = u + at \Rightarrow \qquad v = 0 + 1.53 \times 30$
$\qquad\qquad \Rightarrow \text{speed} = 45.9\,\text{m/s}$

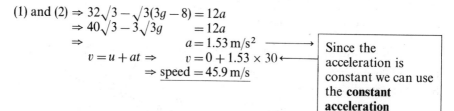

Since the acceleration is constant we can use the **constant acceleration formulae**.

• Exercise 3.1   See page 72

3.2 Rough Inclined Planes

Case 1: No other forces present.
For equilibrium:
Resolving perpendicular to plane:

$$R = Mg \cos \alpha$$

Resolving parallel to plane:

$$F = Mg \sin \alpha$$

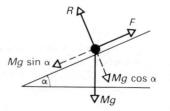

For no slipping:

$$F \leqslant \mu R$$
$$\Rightarrow Mg \sin \alpha \leqslant \mu Mg \cos \alpha$$
$$\Rightarrow \quad \tan \alpha \leqslant \mu$$

Thus, for equilibrium $\tan \alpha \leqslant \mu$
and if $\tan \alpha > \mu$, the body slides down the slope.

Case 2: Preventing motion down the plane when $\tan \alpha > \mu$.
For equilibrium:
Resolving perpendicular to plane:

$$R + P \sin \beta = Mg \cos \alpha$$

Resolving parallel to plane:

$$P \cos \beta + F = Mg \sin \alpha$$

and, for no slipping $\quad F \leqslant \mu R$

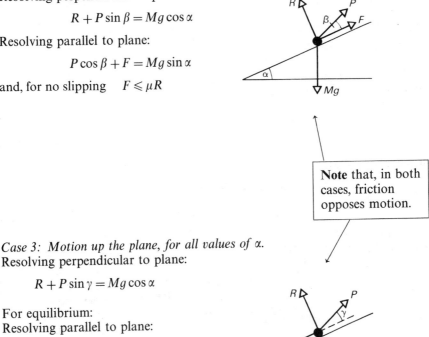

Note that, in both cases, friction opposes motion.

Case 3: Motion up the plane, for all values of α.
Resolving perpendicular to plane:

$$R + P \sin \gamma = Mg \cos \alpha$$

For equilibrium:
Resolving parallel to plane:

$$P \cos \gamma = F + Mg \sin \alpha$$

In Motion
$F = ma$ parallel to plane:

$$P \cos \gamma - F - Mg \sin \alpha = Ma$$

Note: You are not allowed to quote any of these results, but must derive them yourself, using the techniques shown here.

Example 4   A body of mass 20 kg is on the point of sliding down a rough inclined plane when supported by a force of 98 N, acting parallel to the plane. When this force is increased to 147 N, it is on the point of moving up the plane. Find the coefficient of friction.

On the point of slipping down
Resolving perpendicular to plane:

$$R_1 = 20g \cos \alpha$$

Resolving parallel to plane:

$$F_1 + 10g = 20g \sin \alpha$$

Friction is limiting $\Rightarrow F_1 = \mu R_1$
$\Rightarrow 20g \sin \alpha - 10g = \mu 20g \cos \alpha$ (1)

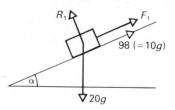

On the point of moving up
Resolving perpendicular to plane:

$$R_2 = 20g \cos \alpha$$

Resolving parallel to plane:

$$F_2 + 20g \sin \alpha = 15g$$

Friction is limiting $\Rightarrow F_2 = \mu R_2$
$\Rightarrow 15g - 20g \sin \alpha = \mu 20g \cos \alpha$ (2)

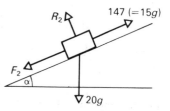

Subtracting (1) from (2) $\Rightarrow 25g - 40g \sin \alpha = 0$
$$\Rightarrow \qquad \sin \alpha = \tfrac{5}{8}$$
$$\Rightarrow \qquad \alpha = 38.7°$$

$$(1) \Rightarrow \mu = \tan \alpha - \frac{1}{2 \cos \alpha}$$

$$\Rightarrow \mu = 0.16$$

● Exercise 3.2   See page 73

3.3 Motion under Frictional Forces

Example 5   A and B are two particles of masses 2 kg and 3 kg respectively. They lie on a rough horizontal surface and are connected by a light inextensible string which is taut. The coefficient of friction is 0.1. A force of 7 N is applied to A in the direction of BA produced. Find the acceleration of the particles and the tension in the string.

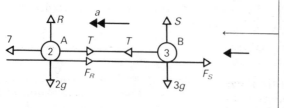

When there are more than one frictional forces, it is a good idea to label them in such a way that it is easy to see which frictional force is related to which normal reaction. i.e. F_R is related to R.

Resolving vertically for A:
$$R = 2g$$
Resolving vertically for B:
$$S = 3g$$

Sliding friction $\Rightarrow F_R = \mu R \Rightarrow F_R = 0.2g$
and $F_S = \mu S \Rightarrow F_S = 0.39$

Method 1

$F = ma \leftarrow$ for B:
$$T - F_S = 3a$$
$$\Rightarrow T \quad\quad = 0.3g + 3a \tag{1}$$

$F = ma \leftarrow$ for A:
$$7 - T - F_R = 2a$$
$$\Rightarrow 7 - T \quad\quad = 0.2g + 2a \tag{2}$$

Substituting for T from (1) in (2)
$$\Rightarrow 7 - 0.3g - 3a = 0.2g + 2a$$
$$\Rightarrow \quad 5a \quad\quad = 7 - 0.5g$$
$$\Rightarrow \quad 5a \quad\quad = 2.1$$
$$\Rightarrow \text{acceleration} = a = 0.42 \text{ m/s}^2$$

and (1)
$$\Rightarrow \underline{T = 4.2 \text{ N}}$$

Method 1 is a variation on the technique you first met in 2.3. However there is a second method which can be very useful (particularly in questions where you do not need to find the tension). Instead of applying $F = ma$ to the two particles separately, $F = ma$ is applied to the whole system. In this case the tensions cancel each other out. They are known as **internal forces.**

Method 2

$F = ma \leftarrow$ for the system
$$\Rightarrow 7 - F_R - F_S = 5a$$
$$\Rightarrow 7 - 0.5g \quad\quad = 5a$$
$$\Rightarrow a = 0.42 \text{ m/s}^2$$

$F = ma$ for B:
$$T - F_S = 3a$$
$$\Rightarrow T = 4.2 \text{ N}$$

Example 6 A body of mass 2 kg rests on a rough horizontal surface whose coefficient of friction is 0.2. If the horizontal surface is given an acceleration of 1.5 m/s², will the 2 kg mass slip?

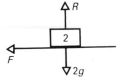

If the mass does not slip, then the frictional force induced is sufficient to give the 2 kg mass an acceleration of 1.5 m/s²

Resolving vertically:

$$R = 2g \quad \Rightarrow \quad F_{max} = \mu R = 0.4g = 3.92\,\text{N}$$

Let force needed to give $2\,\text{kg}$ mass an acceleration of $1.5\,\text{m/s}^2$ be P.
Then $F = ma \Rightarrow P = 2 \times 1.5 = 3\,\text{N}$

Since the force required is less than F_{max}, the mass will move at $1.5\,\text{m/s}^2$.
That is, it will not slip.

- Exercise 3.3   See page 73

3.4 Problems Involving Algebraic Manipulation

Example 7

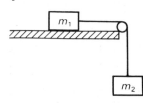

The mass m_1 is lying on a rough horizontal surface and m_2 is hanging freely over a smooth pulley. The coefficient of friction is μ. Find the acceleration of the system, when moving, and show that the condition for motion to occur is $m_2 > \mu m_1$. Find the tension in the string.

Resolving vertically for m_1:

$$R = m_1 g \qquad (1)$$

Friction is limiting

$$\therefore F = \mu R = \mu m_1 g \qquad (2)$$
$$(\text{using } (1))$$

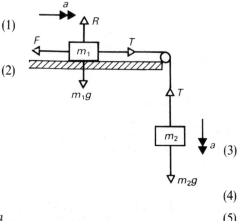

Method 1
$F = ma \downarrow$ for m_2:

$$m_2 g - T = m_2 a$$

$F = ma \rightarrow$ for m_1:

$$T - F = m_1 a \qquad (4)$$

(3) and (4) $\Rightarrow m_2 g - F = (m_2 + m_1)a \qquad (5)$

Method 2
$F = ma$ for system:

$$m_2 g - F = (m_1 + m_2)a \qquad (5)$$

For explanation of method see 3.3.

Both Methods:
(3) and (5) $\Rightarrow m_2 g - \mu m_1 g = (m_1 + m_2)a$

$$\Rightarrow \text{acc}^n = a \quad = \frac{(m_2 - \mu m_1)g}{m_1 + m_2} \qquad (6)$$

Condition for motion to occur is that $a > 0 \Rightarrow \underline{m_2 > \mu m_1}$ QED

To find the tension, we find $F = ma$ for m_2 as in method 1, above

$$\Rightarrow T = m_2 g - m_2 a = m_2 g - m_2 g \left(\frac{m_2 - \mu m_1}{m_1 + m_2} \right) \qquad \text{using (6)}$$

$$\Rightarrow T = \frac{m_2 g}{m_1 + m_2} [m_1 + m_2 - (m_2 - \mu m_1)]$$

$$\Rightarrow \underline{T = \frac{m_1 m_2 g (1 + \mu)}{m_1 + m_2}}$$

- Exercise 3.4   See page 75

3.5 Angle of Friction

At the point of slipping

$$\tan \lambda = \frac{F_{max}}{R}$$

$$\Rightarrow \tan \lambda = \frac{\mu R}{R}$$

∴ For limiting friction $\boxed{\tan \lambda = \mu}$

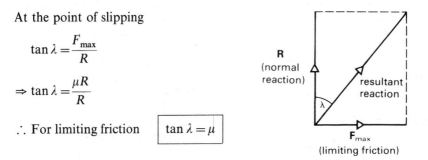

R
(normal reaction)

resultant reaction

λ

$\mathbf{F}_{max}$
(limiting friction)

Note that λ is physically of little significance but it is useful mathematically for solving certain problems.

Example 8   A block of mass 5 kg is just on the point of sliding, when pulled by a horizontal force of 10 N across a rough horizontal surface. Find the angle of friction.

Resolving vertically:

$$R = 5g$$

Resolving horizontally:

$$F = 10$$

F is limiting ∴ $\tan \lambda = \mu = \dfrac{F}{R}$

$$\Rightarrow \tan \lambda = \frac{10}{5g} \Rightarrow \underline{\lambda = 11.5°}$$

Example 9   A block of mass 5 kg rests on a rough horizontal surface. If the coefficient of friction is 0.4 find the least force required to move it and the angle at which it acts.

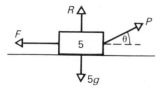

Let the pulling force be P at an angle θ above the horizontal.
Let the block be on the point of sliding.
Let $\mu = \tan \lambda = 0.4$ $\qquad\qquad$ (1)
Resolving vertically:

$$R + P \sin \theta = 5g \qquad\qquad (2)$$

Resolving horizontally:

$$P \cos \theta = F \qquad\qquad (3)$$

Limiting friction $\Rightarrow F = \mu R = \tan \lambda(5g - P \sin \theta)$ $\qquad$ (using (1) and (2)) (4)

(3) and (4)

$$
\begin{aligned}
\Rightarrow P \cos \theta && &= \tan \lambda(5g - P \sin \theta) \\
\Rightarrow P \cos \theta \cos \lambda && &= 5g \sin \lambda - P \sin \theta \sin \lambda \\
\Rightarrow P[\cos \theta \cos \lambda + \sin \theta \sin \lambda] && &= 5g \sin \lambda \\
\Rightarrow P \cos(\theta - \lambda) && &= 5g \sin \lambda \\
\Rightarrow P && &= \dfrac{5g \sin \lambda}{\cos(\theta - \lambda)}
\end{aligned}
$$

multiplying both sides by $\cos \lambda$ and using $\tan \lambda = \dfrac{\sin \lambda}{\cos \lambda}$

P is a minimum when $\cos(\theta - \lambda)$ takes its maximum value. The maximum value of $\cos(\theta - \lambda)$ is 1 and it occurs when $\theta = \lambda$
$\therefore$ Minimum value of $P = 5g \sin \lambda$
and (1) $\Rightarrow \tan \lambda = 0.4 \Rightarrow \lambda = 21.8° \Rightarrow \sin \lambda = 0.3714$
$\therefore$ Minimum value of P is 18.2 N at an angle $21.8°$ above the horizontal.

since $P_{\min}$ occurs when $\theta = \lambda$

Example 10   A mass of 10 kg is to be dragged up a plane inclined at $20°$ to the horizontal, whose coefficient of friction is 0.2. Find the direction and magnitude of the least force required.

Let the force be P at an angle θ to the plane.
Let the mass be on the point of sliding.

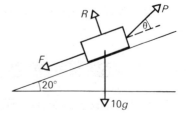

Let $\mu = \tan \lambda = 0.2$ (1)

Resolving perpendicular to the plane:

$$R + P \sin \theta = 10g \cos 20° \qquad (2)$$

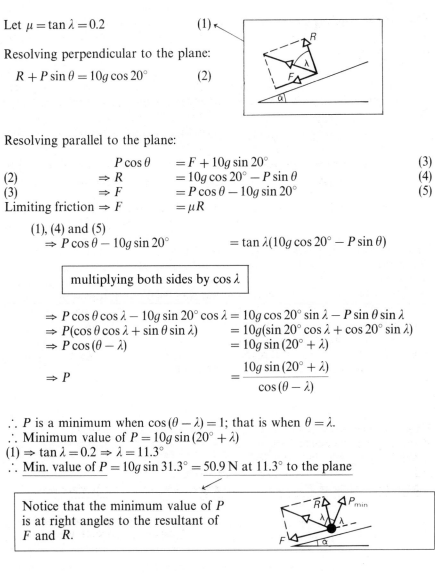

Resolving parallel to the plane:

	$P \cos \theta$	$= F + 10g \sin 20°$	(3)
(2)	$\Rightarrow R$	$= 10g \cos 20° - P \sin \theta$	(4)
(3)	$\Rightarrow F$	$= P \cos \theta - 10g \sin 20°$	(5)
Limiting friction	$\Rightarrow F$	$= \mu R$	

(1), (4) and (5)

$$\Rightarrow P \cos \theta - 10g \sin 20° = \tan \lambda (10g \cos 20° - P \sin \theta)$$

> multiplying both sides by $\cos \lambda$

$$\Rightarrow P \cos \theta \cos \lambda - 10g \sin 20° \cos \lambda = 10g \cos 20° \sin \lambda - P \sin \theta \sin \lambda$$
$$\Rightarrow P(\cos \theta \cos \lambda + \sin \theta \sin \lambda) = 10g(\sin 20° \cos \lambda + \cos 20° \sin \lambda)$$
$$\Rightarrow P \cos(\theta - \lambda) = 10g \sin(20° + \lambda)$$
$$\Rightarrow P = \frac{10g \sin(20° + \lambda)}{\cos(\theta - \lambda)}$$

$\therefore$ P is a minimum when $\cos(\theta - \lambda) = 1$; that is when $\theta = \lambda$.

$\therefore$ Minimum value of $P = 10g \sin(20° + \lambda)$

(1) $\Rightarrow \tan \lambda = 0.2 \Rightarrow \lambda = 11.3°$

$\therefore$ Min. value of $P = 10g \sin 31.3° = \underline{50.9 \text{ N at } 11.3° \text{ to the plane}}$

> Notice that the minimum value of P
> is at right angles to the resultant of
> F and R.

- Exercise 3.5   See page 76

3.6 Revision of Unit

- Miscellaneous Exercise 3A   See page 76

3.7 + A Level Questions

- Miscellaneous Exercise 3B   See page 77

Unit 3 Exercises

Exercise 3.1

1 A body of mass 20 kg rests on a rough horizontal plane whose coefficient of friction is 0.5. Find the least force which would just move the body if applied:

   (a) horizontally   (b) at 45° to the horizontal.

2 A box of mass 40 kg is resting on a rough horizontal surface and can just be moved by a horizontal force of 90 N. Find the coefficient of friction.

3 A block of mass 4 kg rests on a rough horizontal plane, whose coefficient of friction is $\frac{3}{4}$. A horizontal force of 35 N acts on the block. Find the induced frictional force.

4 A trunk of mass 20 kg rests on a rough horizontal surface whose coefficient of friction is 0.8. If it is pushed with a force of 300 N from an angle of 30° above the horizontal, what is the induced frictional force? Will the trunk move?

5 In each of the following situations, the forces cause a body to accelerate along a rough horizontal plane. R is the normal reaction, F is the frictional force and, in each case, the coefficient of friction is 0.5. Find the acceleration of the body in each case.

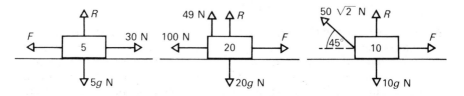

6 A body of mass m travels across a rough horizontal surface under a retardation of $2\,\text{m/s}^2$. Show that the coefficient of friction is $2/g$.

7 A pebble skids across a horizontal sheet of ice. If it travels 20 m in coming to rest from a speed of 5 m/s find the coefficient of friction between the pebble and the ice.

*8 A box of mass 2 kg rests on a rough horizontal surface whose coefficient of friction is 0.4. A light inextensible string is attached to the box in order to pull the box along.

   (i) If the tension in the string is T N, find the value that T must exceed if the box is to move when the string is: (a) horizontal   (b) 30° above the horizontal   (c) 30° below the horizontal.

   (ii) If the string is pulled at 40° above the horizontal, what value of T would cause an acceleration of $3\,\text{m/s}^2$?

Exercise 3.2

1

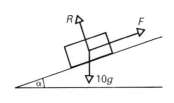

A body of mass 10 kg lies on a rough plane of inclination α. Find the magnitude of the frictional force F and show whether the body can rest in equilibrium if (a) $\alpha = 10°$ (b) $\alpha = 20°$ (c) $\alpha = 30°$. The coefficient of friction is 0.5.

2 A body of mass 8 kg rests in limiting equilibrium on a rough plane inclined at 30° to the horizontal. Find the coefficient of friction.

3 A force of 49 N acting parallel to the plane will just prevent a mass of 10 kg from sliding down the plane, whose inclination to the horizontal is $\tan^{-1}\frac{3}{4}$. Find the coefficient of friction.

4 A body of mass 6 kg is on the point of slipping down a plane inclined at 30° to the horizontal. If the same plane is raised until it is at 60° to the horizontal, find the magnitude of the least force parallel to the plane which is required to keep the mass in equilibrium.

5 A body of mass 20 kg lies on a rough inclined plane of inclination $\tan^{-1}\frac{3}{4}$ whose coefficient of friction is $\frac{1}{5}$. Find the least force acting parallel to the plane required:

(a) to prevent the body sliding down
(b) to pull the body up the plane.

*6 A box of mass 610 kg is placed on a rough inclined plane of slope $\sin^{-1}\frac{11}{61}$ and coefficient of friction $\frac{1}{6}$. A rope is attached to the box and the direction of the rope makes an angle $\sin^{-1}\frac{5}{13}$ with the upper surface of the plane. If the tension in the rope is T N, then for the box to remain in equilibrium, show that $a \leqslant T \leqslant b$, giving the values of a and b to the nearest integer.

Exercise 3.3

1 A rough plane is 15 m long and 9 m high. A particle slides down the plane from rest at the highest point. If the coefficient of friction is 0.5, find the time taken to reach the bottom of the plane and the velocity at that point.

2 Two particles P and Q are of masses 3 kg and 2 kg respectively. They are connected by a light inextensible string and lie at rest on a horizontal table with the string taut. A force of 40 N is applied to P in the direction QP produced and a force of 10 N is applied to Q in the direction PQ produced. Find the acceleration of the particles and the tension in the string:

(a) if the table is smooth
(b) if the coefficient of friction is 0.5.

3   A cream cake, of mass 500 grams, lies on a horizontal plate. The coefficient of friction between the cake and the plate is $\frac{1}{7}$. If the plate is moved horizontally with an acceleration of $1.6 \, \text{m/s}^2$, will the cake slide across the plate?

4   A mass of 4 kg rests on a rough horizontal table whose coefficient of friction is $\frac{1}{2}$. The mass is attached to one end of a light inextensible string which passes over a smooth pulley at the edge of the table and which carries a mass of 3 kg hanging freely. The system moves off from rest. Find the distance travelled by the masses and the velocity acquired in 7 seconds.

5

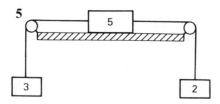

A mass of 5 kg lies on a rough horizontal table. It is connected by a light inextensible string to two hanging masses of 3 kg and 2 kg. When the system is released from rest it moves with an acceleration of $0.35 \, \text{m/s}^2$. Show that the coefficient of friction is $\frac{9}{70}$.

6

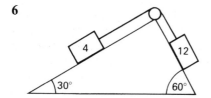

Masses of 4 kg and 12 kg are placed on a double inclined plane, as shown in the diagram, and are connected by a light inextensible string passing over a smooth pulley at the top of the plane. If the coefficient of friction is 0.5 find the resulting acceleration.

*7

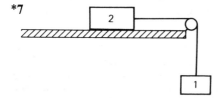

A 2 kg mass lies on a rough horizontal table, whose coefficient of friction is $\frac{1}{4}$. It is attached, by a light inextensible string, to a mass of 1 kg hanging freely. The 2 kg mass is 2 m from the pulley and the 1 kg mass is 1.5 m from the floor.

The system is released from rest. Find:

(a)   the acceleration of the system
(b)   the time taken for the 1 kg mass to reach the floor
(c)   the velocity with which the 2 kg mass hits the pulley.

*8   A particle slides down a rough plane inclined at 45° to the horizontal, whose coefficient of friction is $\frac{3}{4}$. Show that the time taken to descend any distance x is twice the time taken if the plane were smooth.

Exercise 3.4

1

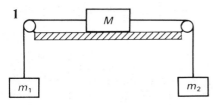

M is lying on a rough horizontal surface and m_1 and m_2 are hanging freely. The coefficient of friction is μ. Show that, if m_1 moves downwards, the acceleration of the system is

$$\frac{g(m_1 - m_2 - \mu M)}{m_1 + m_2 + M}$$

and state the condition that m_1 will move downwards. Find the tension in the string connecting m_1 and M.

2 A force P acting parallel to and up a rough plane of inclination α is just sufficient to prevent a body of mass m from sliding down the plane. A force $3P$ acting parallel to and up the same plane causes the same mass to be on the point of moving up the plane. If μ is the coefficient of friction show that $2\mu = \tan \alpha$.

3

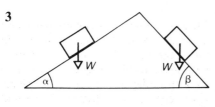

Two particles, each of weight W, are connected by a light inextensible string passing over a smooth pulley at the top of a double inclined plane, as shown in the diagram. The plane of inclination α is rough with a coefficient of friction μ. The plane of inclination β is smooth.

If the weight on the smooth plane is just on the point of moving down, show that

$$\sin \beta = \sin \alpha + \mu \cos \alpha$$

*4 A mass M is lying on a rough plane of inclination α to the horizontal. It is connected by a light inextensible string, which passes over a smooth pulley at the top of the plane, to a mass m which is hanging freely. The coefficient of friction is μ. When the system is free to move find the acceleration and the tension in the string if the hanging mass   (a)   descends   (b)   ascends.

*5 A horizontal force $4Q$ applied to a mass m on a rough plane of inclination α, causes the mass to be on the point of moving up the plane. A horizontal force Q applied to the same mass on the same plane is just sufficient to prevent the mass from sliding down the plane. μ is the coefficient of friction. Show that

$$5\mu = 3(1 + \mu^2) \sin \alpha \cos \alpha$$

Exercise 3.5

1   A particle of mass 3 kg is just on the point of sliding down a rough slope of 25°. Find the angle of friction.

2   A particle of mass 2 kg lies on a rough slope of 40°. A force of 20 N is applied up the slope and parallel to the slope. If the particle is just on the point of moving up the slope, find the angle of friction.

3   Find the least force required to move a mass of 20 kg along a rough horizontal plane when the coefficient of friction is $\frac{1}{4}$.

4   Find the magnitude and direction of the least force required to move a body of 80 kg up a rough plane of inclination 30° to the horizontal, whose coefficient of friction is $\tan^{-1}\frac{3}{4}$.

*5   The least force that will move a mass of M kg up a plane of inclination α is P. Show that $P = Mg\sin(\lambda + \alpha)$ where λ is the angle of friction. Show that the least force acting parallel to the plane which will move the mass up the slope is $P\sec\lambda$.

**6   A body of weight W can just be prevented from sliding down a rough plane of inclination α by a horizontal force P. It can also just be prevented from sliding by a force Q acting parallel to the plane. Show that the cosine of the angle of friction is

$$\frac{WP}{Q\sqrt{P^2 + W^2}}$$

Miscellaneous Exercise 3A

1   A tea-chest of mass 50 kg rests on a rough horizontal plane whose coefficient of friction is 0.7. A horizontal force of P N is applied to the box. Find the induced frictional force and deduce whether the tea-chest moves given that   (a) $P = 300$ N   (b) $P = 400$ N. If it moves, find its acceleration.

2   If the force applied in question 1 is replaced by a force Q N at 45° above the horizontal, find the least integral value of Q required to move the tea-chest.

3

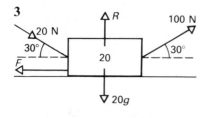

In this diagram, R is the normal reaction and F the frictional force between a body and horizontal ground. The coefficient of friction is 0.25. Find the magnitudes of R and F and the acceleration of the body.

4   A body of mass m slides across a rough horizontal surface under a retardation of a m/s². Show that the coefficient of friction is a/g.

5 A body of mass m rests in limiting equilibrium on a rough plane inclined at an angle α to the horizontal. Find the coefficient of friction.

6 A stone slides in a straight line across a frozen pond. If it starts to move with a speed of 6 m/s and slides 30 m before coming to rest, calculate the coefficient of friction to 2 decimal places.

7 A particle slides with an acceleration of 3.6 m/s² down a line of greatest slope of a rough plane of inclination $\tan^{-1} 0.75$. Calculate the coefficient of friction.

8 A body of mass 5.5 kg is placed on a rough plane of inclination $\cos^{-1}\frac{4}{5}$. The coefficient of friction is $\frac{2}{3}$. Find the magnitude of the horizontal force that will just prevent the body sliding down the plane. (Let $g = 10$ m/s²)

9 After an accident, a mountaineer is lying unconscious on a mountain slope inclined at 40° to the horizontal. He is prevented from falling further by his rope which has caught on a rocky projection. His rescuer must untie the rope and take the strain of his weight. If the rope is pulled parallel to a line of greatest slope, what is the minimum force required to (a) prevent the injured person sliding down the slope, and (b) pull him up the slope. The mass of the mountaineer and equipment is 90 kg and the coefficient of friction is 0.8. Give your answers to the nearest newton.

10 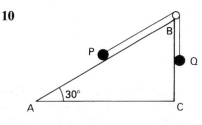 ABC is a fixed wedge whose sloping face, AB, is inclined at 30° to the horizontal and has coefficient of friction $1/(2\sqrt{3})$. BC is vertical and smooth. A particle P, of mass m, is placed on AB and a particle Q, of mass M, hangs alongside BC.

The two particles are connected by a light inextensible string which passes over a smooth pulley at B and is initially taut. The system is released from rest and particle Q descends. Find the tension in the string and the acceleration in terms of M, m and g. Show that the condition that Q descends is $M > 3m/4$.

*11 If the least force which will move a body up a plane of inclination α is twice the least force which will just prevent the body sliding down the plane, show that $\mu = \frac{1}{3}\tan\alpha$, where μ is the coefficient of friction.

Miscellaneous Exercise 3B

1 A stone is sliding in a straight line across a horizontal ice rink. Given that the initial speed of the stone is 6 m/s and that it slides 24 m before coming to rest, calculate the coefficient of friction between the stone and the ice.
[L: 3.3]

2 A rough plane is inclined at an angle α to the horizontal, where $\tan \alpha = \frac{3}{4}$. A particle slides with acceleration $3.5\,\text{m/s}^2$ down a line of greatest slope of this inclined plane. Calculate the coefficient of friction between the particle and the inclined plane.   [L: 3.3]

3 Two particles of masses $2m$ and m are attached to the ends A and B respectively of a light inextensible string which passes over a smooth pulley C at the top of a fixed rough plane inclined at an angle $60°$ to the horizontal. The particles are at rest with the mass $2m$ in contact with the plane and the mass m hanging freely so that AC is along a line of greatest slope of the plane and BC is vertical. Given that the mass $2m$ is in limiting eqilibrium and on the point of moving down the plane, find μ, the coefficient of friction.   [A: 3.2]

4 A small parcel lies on the horizontal rear seat of a car, the coefficient of friction between the parcel and the seat being $\frac{4}{5}$. When the car is travelling on a straight horizontal road with speed $20\,\text{m/s}$ the brakes are applied producing constant retardation. Find the minimum distance in which the car can be brought to rest without the parcel moving relative to the car. It may be assumed that the brakes can be applied so that any required constant retardation may be produced and g may be taken to be $10\,\text{m/s}^2$.   [A: 3.3, 1.2]

5 A particle of mass $2m$ is on a plane inclined at an angle $\tan^{-1}\frac{3}{4}$ to the horizontal. The particle is attached to one end of a light inextensible string. This string runs parallel to a line of greatest slope of the plane, passes over a small smooth pulley at the top of the plane and then hangs vertically carrying a particle of mass $3m$ at its other end. The system is released from rest with the string taut. Find the acceleration of each particle and the tension in the string when the particles are moving freely, given that:

(a) the plane is smooth,
(b) the plane is rough and the coefficient of friction between the particle and the plane is $\frac{1}{4}$.   [L: 3.3]

6 Three particles A, B, C are of masses $4, 4, 2\,\text{kg}$ respectively. They lie at rest on a horizontal table in a straight line, with particle B attached to the midpoint of a light inextensible string. The string has particle A attached to one end and particle C at the other end, and is taut.

   A force of $60\,\text{N}$ is applied to A in the direction CA produced and a force of $15\,\text{N}$ is applied to C in the opposite direction. Find the acceleration of the particles and the tension in each part of the string   (a)   if the table is smooth   (b)   if the coefficient of friction between each particle and the table is $\frac{1}{4}$. (Take g as $10\,\text{m/s}^2$.)   [L: 3.5]

7 Two beads A and B of masses $7m$ and $3m$ respectively are threaded on a rough horizontal wire, the coefficient of friction between each bead and the wire being μ.
The beads A and B are joined by a smooth light inextensible string along which a smooth bead C of mass $2m$ can slide freely. When A and B are at the greatest distance apart consistent with equilibrium, AC and BC each make an angle θ with the horizontal.

Show that:

(a)   B is about to slip but A is not
(b) $\mu = \frac{1}{4}\cot\theta$.

Find, in terms of m, g and θ, the greatest force that can be applied at A:

(c)   in the direction $\overrightarrow{AB}$
(d)   in the direction $\overrightarrow{BA}$
without disturbing equilibrium.   [A: 2.6, 2.7, 3.1]

8   A particle A of mass m is placed on a rough plane inclined at an angle $\tan^{-1}\frac{3}{4}$ to the horizontal. One end of a light inextensible string is attached to A and the string runs parallel to a line of greatest slope of the plane, passes over a small smooth pulley at the top of the plane and has a second identical particle B attached to its other end. The particles are released from rest when the string is taut and the part of the string between B and the pulley is vertical. Given that the coefficient of friction between A and the plane is $\frac{1}{4}$, show that B moves with acceleration $g/10$ and find the tension in the string.

After time t_0 the string breaks when A is at the point P on the plane and A comes momentarily to rest at a point Q on the plane. Show that $PQ = gt_0^2/160$.   [L:1.2, 3.3]

9   A particle of weight W is placed on a rough plane which is inclined at an angle α to the horizontal. The coefficient of friction between the particle and the plane is μ, where $\mu < \tan\alpha$. Show that the magnitude of the least horizontal force needed to maintain this particle in equilibrium is

$$\left(\frac{\tan\alpha - \mu}{1 + \mu\tan\alpha}\right)W$$

Indicate the direction of this force in a diagram.   [J: 3.2]

*10   A particle P of weight W rests on a rough horizontal table. The angle of friction between the particle and the table is λ. A light inextensible string passes over a smooth peg fixed above the table and has one end attached to P and the other to a load Q of weight L, as shown in the diagram. The part of the string between the particle and the peg is inclined at an angle $\alpha(\neq\pi/2)$ to the horizontal. The system is in equilibrium and L is then gradually increased. Prove that the system will be in limiting equilibrium when

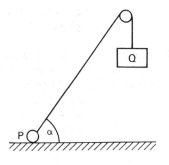

$$L = W\sin\lambda\sec(\alpha - \lambda)$$

In this case, find an expression, in terms of W, α and λ, for the magnitude of the total reaction exerted on P by the table.   [J: 3.5]

4 FORCES AND MOMENTS: RIGID BODY EQUILIBRIUM

4.1 Force Acting on a Particle

Parallelogram of Forces

Two forces acting on a particle at A are represented by the line segments AB and AD.

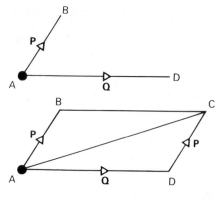

The parallelogram ABCD is completed by drawing BC and DC.

If $\overrightarrow{AB}$ represents the force **P**
then $\overrightarrow{DC}$ represents the force **P**
and $\overrightarrow{AC}$ represents the resultant of **P** and **Q**.

Example 1   Find graphically the magnitude of the resultant of the two forces shown below and find the angle that the resultant makes with the larger forces:

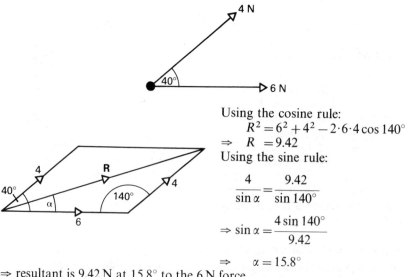

Using the cosine rule:
$$R^2 = 6^2 + 4^2 - 2 \cdot 6 \cdot 4 \cos 140°$$
$$\Rightarrow \quad R = 9.42$$
Using the sine rule:

$$\frac{4}{\sin \alpha} = \frac{9.42}{\sin 140°}$$

$$\Rightarrow \sin \alpha = \frac{4 \sin 140°}{9.42}$$

$$\Rightarrow \quad \alpha = 15.8°$$

$\Rightarrow$ resultant is 9.42 N at 15.8° to the 6 N force.

Note that the resultant could also be found using a scale drawing.

Angle Between Two Forces

If the angle between two forces is given as θ then this means that the forces are as shown in the diagram below.

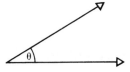

Example 2 Find the angle between a force of 3.2 N and a force of 8 N if their resultant has a magnitude of 10 N.

Using the cosine rule:

$$\cos\alpha = \frac{8^2 + 3.2^2 - 10^2}{2\cdot 8\cdot 3.2}$$

$\Rightarrow \qquad \alpha = 120.2°$

$\Rightarrow \qquad \theta = 180° - 120.2° = 59.8°$

$\Rightarrow$ angle between force $= 59.8°$

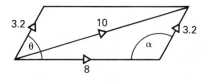

Triangle of Forces

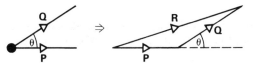

R is the resultant of **P** and **Q**

Polygon of Forces

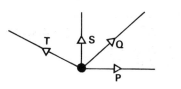

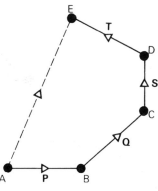

$\overrightarrow{AE}$ represents the resultant of **P**, **Q**, **S** and **T**

Example 3 Forces of 6 N, 4 N and 5 N act on a particle in the directions shown in the diagram. Find graphically the magnitude and direction of the resultant of these forces.

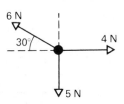

Let 1 cm ≡ 1 N
R is the resultant and, by measurement, it is 2.5 N and makes an angle of
240° with the 4 N force.

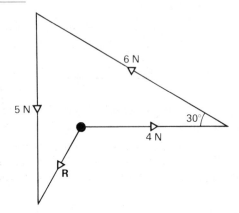

Three Forces in Equilibrium

If three forces are in equilibrium then their resultant is zero. Hence the forces
can be represented by the sides of a triangle.

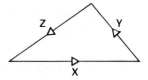

Conversely, if three forces acting at a point can be represented by the sides of
a triangle then the forces are in equilibrium.

Example 4

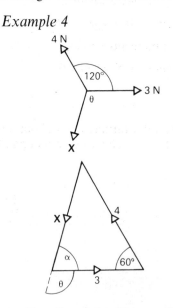

These three forces are in equilibrium.
Sketch the triangle of forces and calculate the values of X and θ.

Using the cosine rule:
$$X^2 = 3^2 + 4^2 - 2 \cdot 3 \cdot 4 \cos 60°$$
$$\Rightarrow X = \sqrt{13} = 3.6 \text{ N.}$$

Using the sine rule:
$$\frac{\sin \alpha}{4} = \frac{\sin 60°}{X}$$

$$\Rightarrow \sin \alpha = \frac{4 \sin 60°}{3.6}$$
$$\alpha = 74.2° \text{ and } \theta = 105.8°$$

• **Exercise 4.1**   See page 102

4.2 Moments

Definition

The **moment** of a force about a point is the product of the magnitude of the force and the perpendicular distance of the line of action of the force from the point.

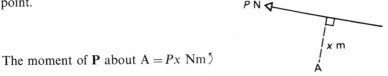

The moment of **P** about $A = Px$ Nm $\circlearrowleft$

In three-dimensional problems, that is in the real world, moments are always taken about an axis. However in this section, all the forces are coplanar and so, in the interest of simplicity, the phrase 'moment about a point' will be used.

Sense of Rotation

Moments can be considered in a clockwise or anticlockwise direction, but by convention, positive moments are anticlockwise. Otherwise, the sense of the rotation must be stated. $\circlearrowright$ can be used for clockwise, $\circlearrowleft$ for anti-clockwise.

Example 5   A 10 N force acts perpendicular to a rod AB, at A. Find the moment of the force about B, C and A if B is 2 m from A and C is $\frac{1}{2}$ m from A, on the rod.

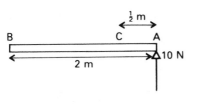

Taking moments about B:

Moment $= 10\,\text{N} \times 2\,\text{m} = \underline{20\,\text{Nm}\,\circlearrowleft}$

Taking moments about C:

Moment $= 10\,\text{N} \times \frac{1}{2}\text{m} = \underline{5\,\text{Nm}\,\circlearrowleft}$

Taking moments about A:

Moment $= 10\,\text{N} \times 0\,\text{m} = \underline{0}$

Algebraic Sum of Moments

If a set of coplanar forces act on a body, their total turning effect about any point may be calculated, by adding the moments of each force about that point.

Example 6   A rod AB in pivoted about an axis through X, as shown. Parallel forces of 6 N and 3 N act at A and B, respectively, perpendicular to AB. Find the combined turning effect about X of the two forces.

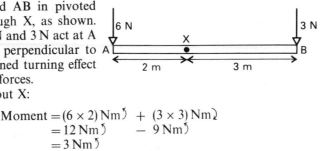

Taking moment about X:

$$\begin{aligned} \text{Moment} &= (6 \times 2)\,\text{Nm}\,\circlearrowleft \;+\; (3 \times 3)\,\text{Nm}\,\circlearrowright \\ &= 12\,\text{Nm}\,\circlearrowleft \qquad - \;9\,\text{Nm}\,\circlearrowleft \\ &= \underline{3\,\text{Nm}\,\circlearrowleft} \end{aligned}$$

Example 7   Find the moment, about the point O, of the forces shown in each of the following diagrams:

(a)

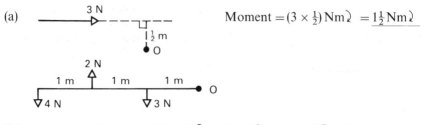

Moment $= (3 \times \tfrac{1}{2})\,\text{Nm}\,\circlearrowright\ = 1\tfrac{1}{2}\,\text{Nm}\,\circlearrowright$

(b)

$$\text{Moment} = [(3 \times 1)\,\circlearrowleft + (2 \times 2)\,\circlearrowright + (4 \times 3)\,\circlearrowleft\ ]\,\text{Nm}$$
$$\Rightarrow \text{Moment} = (3 - 4 + 12)\,\text{Nm}\,\circlearrowleft\ = 11\,\text{Nm}\,\circlearrowleft$$

Example 8   A force of $(2\mathbf{i} + 3\mathbf{j})$ N acts at the point with position vector $(\mathbf{i} + 2\mathbf{j})$ m. Find the moment of the force about the point X with position vector $(-2\mathbf{i} + 4\mathbf{j})$ m.

Moment $= [(2 \times 2) + (3 \times 3)]\,\text{Nm}\,\circlearrowleft$
$\Rightarrow$ Moment $= 13\,\text{Nm}\,\circlearrowleft$

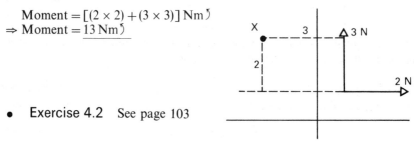

• Exercise 4.2   See page 103

4.3 Parallel Forces and Couples

Parallel Forces

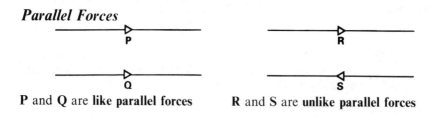

P and Q are **like parallel forces**   R and S are **unlike parallel forces**

Principle of Moments
The algebraic sum of the moments of a number of coplanar forces about any point in their plane is equal to the moment of their resultant about that point.

Resultant of Parallel Forces

Example 9   Two parallel forces of 1 N and 4 N are 2 m apart. Find the magnitude, direction and line of action of the resultant of these forces if they are (a) like   (b) unlike.

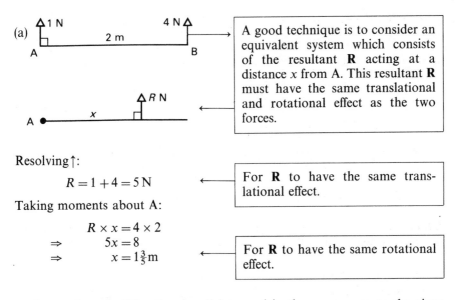

(a)

A good technique is to consider an equivalent system which consists of the resultant **R** acting at a distance x from A. This resultant **R** must have the same translational and rotational effect as the two forces.

Resolving ↑:

$$R = 1 + 4 = 5\,\text{N}$$

For **R** to have the same translational effect.

Taking moments about A:

$$R \times x = 4 \times 2$$
$$\Rightarrow \quad 5x = 8$$
$$\Rightarrow \quad x = 1\tfrac{3}{5}\,\text{m}$$

For **R** to have the same rotational effect.

∴ the resultant is 5N acting parallel to, and in the same sense as the given forces, at a distance of 1.6 m from the 1 N force.

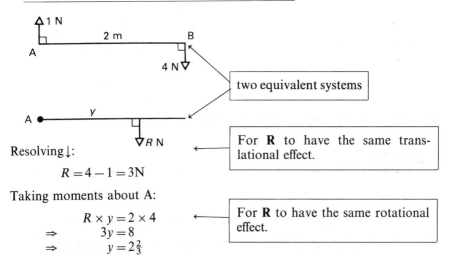

two equivalent systems

For **R** to have the same translational effect.

Resolving ↓:

$$R = 4 - 1 = 3\,\text{N}$$

Taking moments about A:

$$R \times y = 2 \times 4$$
$$\Rightarrow \quad 3y = 8$$
$$\Rightarrow \quad y = 2\tfrac{2}{3}$$

For **R** to have the same rotational effect.

∴ the resultant is 3 N acting parallel to, and in the same sense as, the 4 N force, at a distance of $2\tfrac{2}{3}$ m from the 1 N force and $\tfrac{2}{3}$ m from the 4 N force.

Position of the resultant of two parallel forces.
The resultant of two like parallel forces has its line of action between the forces.
The resultant of two unlike parallel forces has its line of action outside the forces.

Example 10   Find the resultant effect of two unlike forces of 5 N, a distance 2 m apart.

Resolving parallel to the forces:

$$R = 0$$

Taking moments about A:

Moment $= 10\,\text{Nm}$

Thus the two forces have no translational effect but they do have a rotational (turning) effect of moment $10\,\text{Nm}$.

Definition of a Couple (or a Torque)

Two unlike forces of equal magnitude but different lines of action are said to form a **couple** or a **torque**.

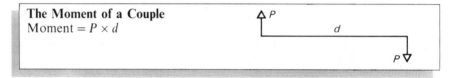

The Moment of a Couple
Moment $= P \times d$

The turning effect of a couple is independent of the point about which the turning is taking place.

Proof

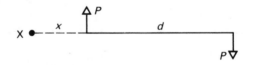

Taking moments about X:

$$\text{Moment} = P \times (d + x) - Px = Pd$$

This is true for all values of x.
Hence the moment of a couple is independent of the point about which the moment is taken.

Systems of Forces that are Equivalent to a Couple

If a system of forces is such that:

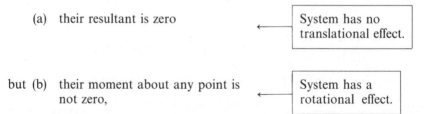

(a)   their resultant is zero

System has no translational effect.

but (b)   their moment about any point is not zero,

System has a rotational effect.

then the system is equivalent to a couple. It is often said that the system **reduces to a couple**.

Example 11 Find the resultant effect of the forces in the diagram.

Resolving↑:

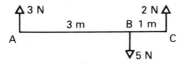

$$R = 3 + 2 - 5 = 0$$

Taking moments about A:

$$\text{Moment} = [(5 \times 3) - (4 \times 2)] \circlearrowleft = 7\,\text{Nm} \circlearrowright$$

The system has no translational effect but it has a rotational effect of moment 7 Nm.
Thus the system is equivalent to a couple of 7 Nm↺ .

Example 12 Show that the system of forces consisting of:

$(2\mathbf{i} + \mathbf{j})\text{N}$ acting at the point with position vector $(3\mathbf{i} + \mathbf{j})$m
$3\mathbf{i}\,\text{N}$ acting at the point with position vector $-2\mathbf{j}$m
$(-5\mathbf{i} - \mathbf{j})\,\text{N}$ acting at the point with position vector $(-\mathbf{i} + \mathbf{j})$m,

reduces to a couple and state its moment.

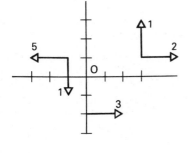

Resolving →:

$$X = 2 + 3 - 5 = 0$$

Resolving ↑:

$$Y = 1 - 1 = 0$$

$\Rightarrow$ resultant $= 0$

Taking moments about O:

$$G = (1 \times 3)\circlearrowright + (2 \times 1)\circlearrowright + (3 \times 2)\circlearrowright$$
$$+ (5 \times 1)\circlearrowright + (1 \times 1)\circlearrowright$$
$$F\,G = (3 - 2 + 6 + 5 + 1)\circlearrowright$$
$$\Rightarrow G = 13\,\text{Nm}\circlearrowright$$

∴ the system reduces to a couple of moment 13 Nm ↺

Note that, just as X and Y are often used to denote the sums of the components in two perpendicular directions, G is often used to denote the sum of the moments.

Example 13 Forces of 7 N, 2 N, 7 N and 2 N act along the sides PQ, QR, RS, SP respectively of a square of side 5 m, in the directions indicated by the order of the letters. Find the moment of the system of forces about   (a)   the point P   (b)   the centre of the square   (c)   the midpoint of PS. Show that the system forms a couple.

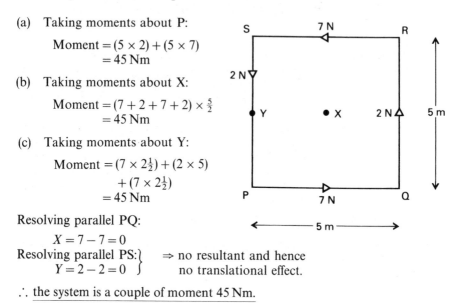

(a)   Taking moments about P:

$$\text{Moment} = (5 \times 2) + (5 \times 7)$$
$$= 45 \,\text{Nm}$$

(b)   Taking moments about X:

$$\text{Moment} = (7 + 2 + 7 + 2) \times \tfrac{5}{2}$$
$$= 45 \,\text{Nm}$$

(c)   Taking moments about Y:

$$\text{Moment} = (7 \times 2\tfrac{1}{2}) + (2 \times 5)$$
$$+ (7 \times 2\tfrac{1}{2})$$
$$= 45 \,\text{Nm}$$

Resolving parallel PQ:

$$X = 7 - 7 = 0$$

Resolving parallel PS: $\left. \begin{array}{l} \\ Y = 2 - 2 = 0 \end{array} \right\}$ ⇒ no resultant and hence no translational effect.

∴ the system is a couple of moment 45 Nm.

- Exercise 4.3   See page 105

4.4 Parallel Forces in Equilibrium

> **Conditions for Equilibrium**
> 1   The resultant force is zero.
> 2   The algebraic sum of the moments of the forces **about any point** is zero.

Example 14   A uniform rod AB, of mass 40 kg and length 10 m, rests on two supports, one at A and the other 2 m from B. Masses of 4 kg and 8 kg are attached at points which are 2 m and 6 m from A, respectively. Find the reaction at each of the supports.

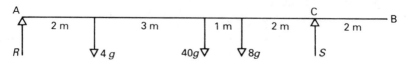

Taking moments about A:

$$8 \times S = (4g \times 2) + (40g \times 5) + (8g \times 6)$$
$$8S = (8 + 200 + 48)g$$
$$S = 32g$$

Taking moments about C:

$$8 \times R = (8g \times 2) + (40g \times 3) + (4g \times 6)$$
$$8R = (16 + 120 + 24)g$$
$$R = 20g$$

Points to note: 1 'Uniform rod' ⇒ weight acts in the middle.
        2 Taking moments about the point of action of an unknown force eliminates the force from the equation.
        3 Moments are taken about points, not forces. Thus we take moments about C, not S.
        4 The second reaction could have also been found by resolving vertically. In practice it is better to find the reactions separately and use the fact that the sum of the reactions must equal the sum of the weights as a check. Otherwise, a mistake in calculating the first will make both answers wrong.

Example 15   A uniform beam AB, of mass 20 kg and length 12 m, has masses 6 kg and 10 kg suspended from A and B respectively. At what point must the beam be supported if it is to rest horizontally?

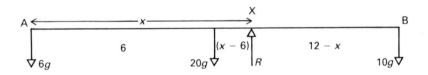

Method 1
Taking moments about X:

$$20g(x - 6) + 6gx = 10g(12 - x)$$
$$\Rightarrow \quad 20x - 120 + 6x = 120 - 10x$$
$$\Rightarrow \qquad\qquad 36x = 240$$
$$\Rightarrow \qquad\qquad x = 6\tfrac{2}{3}$$

Method 2
Resolving vertically:

$$R = 36g \qquad\qquad\qquad\qquad\qquad (1)$$

Taking moments about A:

$$Rx = 20g \times 6 + 10g \times 12$$
$$\Rightarrow 36gx = 120g + 120g \qquad\qquad \text{using (1)}$$
$$\Rightarrow \qquad x = 6\tfrac{2}{3}$$

The beam must be supported a distance $6\tfrac{2}{3}$ m from A

Example 16   A uniform beam, of mass 25 kg and 4 m long rests on two supports at equal distances from the ends. Find the maximum value of this distance so that a man of 80 kg may stand anywhere on the beam without causing it to tilt.

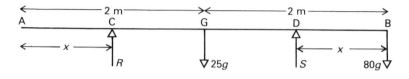

The man can only tilt the beam if he is between A and C or B and D. The maximum turning effect of his weight occurs when he is at one end of the beam. Let the man stand at one end of the beam and the beam be on the point of tilting.

At the point of tilting, $R = 0$
Taking moments about D:

$$80gx = 25g(2 - x)$$
$$\Rightarrow \quad 80x = 50 - 25x$$
$$\Rightarrow 105x = 50$$
$$\Rightarrow \quad x = \frac{50}{105} = \frac{10}{21}$$

∴ maximum distance of supports from each end is $\frac{10}{21}$ m.

Example 17   A uniform beam, of length 3.6 m and mass 10 kg is suspended in a horizontal position by two vertical strings, attached at each end. The breaking point of each string is 98 N. Within what distance from the centre can a weight of mass 6 kg be placed without breaking either string.

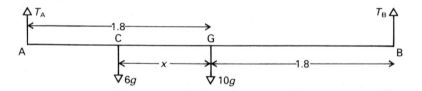

Let the weight be placed at the maximum distance, x, from G. In this position
$T_A = 98 = 10g$
Taking moments about B: ←———————— | Since we do not need to find T_B.

$$10g \times 1.8 + 6g \times (1.8 + x) = 10g \times 3.6$$
$\Rightarrow \qquad 18 + 10.8 + 6x = 36$ dividing by g
$\Rightarrow \qquad \qquad 6x = 7.2$
$\Rightarrow \qquad \qquad x = 1.2$

∴ 6 kg means can be placed within 1.2 m of the centre without breaking either string.

- Exercise 4.4   See page 107

4.5 Non-Parallel Forces in Equilibrium

Conditions for equilibrium are the same as those for parallel forces. Thus, equations can be found connecting the various forces both, by resolving and taking moments. However, in this section, we shall concentrate on finding unknown forces by taking moments.

Example 18 A bob of mass 2 kg is fastened to one end of a light inelastic string of length $2a$, whose other end is fixed. The string is held at 30° to the vertical by a horizontal force applied to the bob. Find the magnitude of this force.

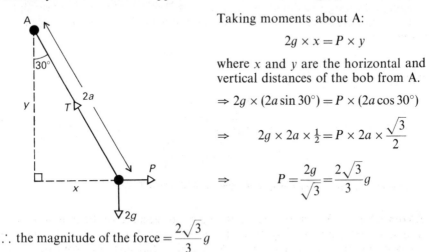

Taking moments about A:

$$2g \times x = P \times y$$

where x and y are the horizontal and vertical distances of the bob from A.

$$\Rightarrow 2g \times (2a \sin 30°) = P \times (2a \cos 30°)$$

$$\Rightarrow \quad 2g \times 2a \times \tfrac{1}{2} = P \times 2a \times \frac{\sqrt{3}}{2}$$

$$\Rightarrow \quad P = \frac{2g}{\sqrt{3}} = \frac{2\sqrt{3}}{3}g$$

∴ the magnitude of the force $= \dfrac{2\sqrt{3}}{3}g$

Note that taking moments about A gives an equation that does not involve T since the moment of T about A is zero.

Example 19 A uniform rod AB, of mass m and length $2a$ is hinged at A. The rod is held in equilibrium by a string attached to B. In the equilibrium position, the string makes an angle of 30° to BA and AB is at 45° to the horizontal with B above A. Find the tension in the string.

To avoid bringing in the forces at the hinge, moments are taken about A.

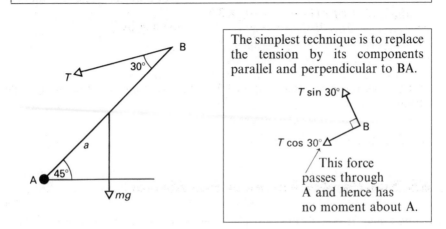

The simplest technique is to replace the tension by its components parallel and perpendicular to BA.

This force passes through A and hence has no moment about A.

Taking moments about A:

$$mga \cos 45° = 2aT \sin 30°$$

$$\Rightarrow \quad mga\frac{1}{\sqrt{2}} = 2aT \cdot \tfrac{1}{2}$$

$$\Rightarrow \quad T = mg \cdot \frac{1}{\sqrt{2}} = \frac{mg\sqrt{2}}{2}$$

$$\therefore \text{ tension} = \frac{mg\sqrt{2}}{2}$$

> It is usual to rationalise surds—that is, give the answer as $\dfrac{\sqrt{2}}{2}$ instead of $\dfrac{1}{\sqrt{2}}$.

• **Exercise 4.5** See page 108

This exercise is to be done without the use of calculators, to give the students practice in the use of $\cos 45° = \dfrac{1}{\sqrt{2}}$, $\sin 60° = \dfrac{\sqrt{3}}{2}$, etc.

4.6 Rigid Body Equilibrium: Rods and Ladders

The Most Common Method for Solving Rigid Body Equilibrium Problems

1 Draw a force diagram, showing on it all the information given in the question.
2 For each of two perpendicular directions, equate the components of the forces acting in one direction to those components acting in the opposite direction.
3 Take moments about one convenient point. Since the total moment is zero about any point, the sum of all the clockwise moments about the point must equal the sum of all the anticlockwise moments about the same point.

> **Points to consider:** The directions in which the forces are resolved and the points about which moments are taken may well simplify the solution of a particular problem. Usually, the best directions to resolve in are:
>   (a)   horizontally and vertically
> or (b)   parallel and perpendicular to an included plane,
> although there are exceptions to this.
> When choosing a point to take moments about, look for a point about which as many unknown forces as possible have zero moment.

Example 20   A uniform rod AB, of mass 12 kg and length 40 cm, is freely hinged to a vertical wall at A. A vertical force is applied upwards at B to keep AB in a horizontal position. Find the magnitude of P and the reaction at the hinge.

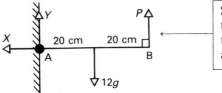

> Since the direction of the force at the hinge is unknown, it is put in as two perpendicular components X and Y.

Taking moments about A:

$$12g \times 20 = P \times 40 \quad \Rightarrow P = 6g \text{ N}$$

Resolving vertically:

$$Y + P = 12g \qquad \Rightarrow Y = 6g \text{ N}$$

Resolving horizontally:

$$X = 0$$
$$\Rightarrow \text{Reaction at the hinge} \quad = 6g \text{ N} \uparrow$$

Limiting Equilibrium

If there is a frictional force acting, it is necessary to decide whether the body is in limiting equilibrium. Only when the body is in limiting equilibrium, i.e. when the body is about to slip, does the frictional force have its maximum value μR.

Example 21   A uniform ladder of mass 8 kg and length 4 m rests in equilibrium at $60°$ to the horizontal with its upper end against a smooth vertical wall and its lower end on rough horizontal ground, with coefficient of friction μ. Find the magnitudes of the normal reactions at the wall and ground and the value of the frictional force. Deduce the minimum value of μ for the ladder to remain in equilibrium.

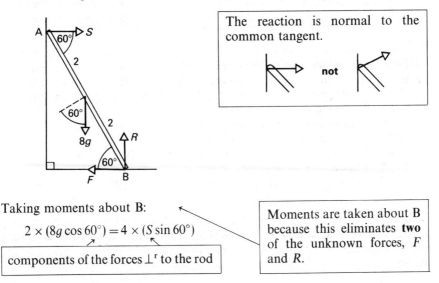

> The reaction is normal to the common tangent.

Taking moments about B:

$$2 \times (8g \cos 60°) = 4 \times (S \sin 60°)$$

components of the forces $\perp^r$ to the rod

> Moments are taken about B because this eliminates **two** of the unknown forces, F and R.

$$\Rightarrow S = \frac{8g\cos 60°}{2\sin 60°} = 4g\cot 60° = \frac{4g}{\sqrt{3}} = \frac{4\sqrt{3}}{3}g.$$

Resolving vertically:

$$R = 8g$$

Resolving horizontally:

$$F = S$$

$\Rightarrow$ Normal reaction at wall $= \dfrac{4\sqrt{3}}{3}g\,\text{N}$

Normal reaction at ground $= 8g\,\text{N}$

Frictional force $= \dfrac{4\sqrt{3}}{3}g\,\text{N}$

For equilibrium, $F \leqslant \mu R$

$$\Rightarrow \quad \mu \geqslant \frac{F}{R}$$

$$\Rightarrow \mu_{\min} = \frac{F}{R} = \frac{4\sqrt{3}}{3}g/8g = \frac{\sqrt{3}}{6}$$

- **Exercise 4.6**   See page 109

To be done without using calculators.

4.7 Rigid Body Equilibrium: More Complex Rod and Ladder Problems

Alternative Method of Finding The Reaction at a Hinge

In the last section, the force at a hinge was represented by two unknown horizontal and vertical components X and Y.

Alternatively the force on the hinge may be represented by a single force R acting at an angle θ to the vertical.

Example 22   A uniform rod AB of mass 4 kg and length 4 m is freely hinged at A to a vertical wall. The rod is kept in a position 30° above the horizontal by a force P applied at B at 60° to BA. Find the magnitude of the reaction at the hinge and the angle that the reaction makes to the vertical.

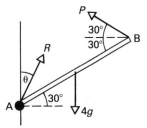

Taking moments above A:

$$4g \cos 30° \times 2 = P \sin 60° \times 4$$
$$\Rightarrow \qquad 4g\sqrt{3} = 2P\sqrt{3}$$
$$\Rightarrow \qquad P = 2g \qquad\qquad (1)$$

Resolving horizontally:

$$R \sin \theta = P \cos 30°$$
$$(1) \Rightarrow \qquad R \sin \theta = \sqrt{3}g \qquad\qquad (2)$$

Resolving vertically:

$$R \cos \theta + P \sin 30° = 4g$$
$$(1) \Rightarrow \qquad R \cos \theta = 4g - g \;\; = 3g \qquad\qquad (3)$$

Dividing (2) by (3):

$$\frac{R \sin \theta}{R \cos \theta} = \frac{\sqrt{3}g}{3g}$$
$$\Rightarrow \qquad \tan \theta = \frac{1}{\sqrt{3}}$$
$$\Rightarrow \qquad \underline{\theta = 30°}$$

Substituting for θ in (2)

$$\Rightarrow \qquad R \cdot \tfrac{1}{2} = \sqrt{3}g$$
$$\Rightarrow \qquad \underline{\underline{R = 2\sqrt{3}g \text{ N}}}$$

Non-Uniform Rods

Example 23   A non-uniform rod AB of mass 5 kg and length 2 m is in equilibrium in a horizontal position. It is supported by two strings, one at A making an angle of 30° with AB and the other at B, making an angle of 40° with BA. Find the tensions in the strings and the distance of the centre of gravity from A.

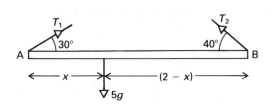

Resolving horizontally:

$$T_1 \cos 30° = T_2 \cos 40°$$

$$\Rightarrow T_1 \qquad = \frac{\cos 40°}{\cos 30°} T_2 \qquad\qquad (1)$$

Resolving vertically:

$$T_1 \sin 30° + T_2 \sin 40° = 5g \qquad\qquad (2)$$

(1) and (2) $\Rightarrow T_2 \dfrac{\cos 40°}{\cos 30°} \sin 30° + T_2 \sin 40° \qquad = 5g$

$\Rightarrow T_2[\cos 40° \sin 30° + \sin 40° \cos 30°] = 5g \cos 30°$
$\Rightarrow T_2 \sin 70° \qquad\qquad\qquad\qquad = 5g \cos 30°$

$\Rightarrow T_2 = 5g \dfrac{\cos 30°}{\sin 70°} = \underline{45.16\,\text{N}}$

(1) $\qquad \Rightarrow T_1 = 5g \dfrac{\cos 40°}{\sin 70°} = \underline{39.95\,\text{N}}$

Taking moments about A:

$$5gx = T_2 \sin 40° \times 2$$
$$\Rightarrow \quad x = \underline{1.18\,\text{m}}$$

More Complex Ladder Problems

Example 24   A ladder of weight W and length $3a$ rests in limiting equilibrium against a rough vertical wall and on rough horizontal ground. The centre of gravity of the ladder is a distance a from the bottom. If the coefficients of friction at the top and bottom of the ladder are $\frac{1}{2}$ and $\frac{1}{4}$ respectively, find the angle the ladder makes with the floor. Find also how far up the ladder a man of weight $2W$ can ascend without the ladder slipping.

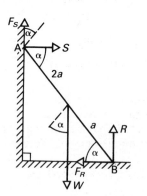

Since equilibrium is limiting

$$F_R = \tfrac{1}{4}R \qquad (1)$$
$$\text{and} \qquad F_S = \tfrac{1}{2}S \qquad (2)$$

Resolving horizontally:

$$F_R = S \qquad (3)$$

Resolving vertically:

$$R + F_S = W \qquad (4)$$

(1) and (3) $\qquad \Rightarrow \qquad \tfrac{1}{4}R = S \qquad\qquad\qquad (5)$
(2) and (4) $\qquad \Rightarrow \qquad R + \tfrac{1}{2}S = W \qquad\qquad (6)$

(5) and (6) $\qquad \Rightarrow R + \tfrac{1}{2}(\tfrac{1}{4}R) = W$

$$\Rightarrow \quad R = \tfrac{8}{9}W \quad \text{and} \quad S = \tfrac{2}{9}W$$
$$\text{Hence (1) and (2)} \Rightarrow F_R = \tfrac{2}{9}W \quad \text{and} \quad F_S = \tfrac{1}{9}W \left.\right\} \qquad (7)$$

Taking moments about **B**:

$$a(W \cos \alpha) = 3a(S \sin \alpha) + 3a(F_S \cos \alpha)$$

components of forces $\perp^r$ to ladder

$(7) \Rightarrow aW \cos \alpha = 3a \times \dfrac{2W}{9} \sin \alpha + 3a \times \tfrac{1}{9}W \cos \alpha$

$\Rightarrow \quad \cos \alpha = \tfrac{2}{3} \sin \alpha + \tfrac{1}{3} \cos \alpha \quad \longleftarrow \boxed{\div \text{ by } Wa}$

$\Rightarrow \quad \tfrac{2}{3}\cos \alpha = \tfrac{2}{3} \sin \alpha$

$\Rightarrow \quad \tan \alpha = 1 \quad \longleftarrow$

$\Rightarrow \quad \alpha = 45° \qquad\qquad \boxed{\tan \alpha = \dfrac{\sin \alpha}{\cos \alpha}}$

As soon as the man steps on the ladder, his weight increases the value of the normal reaction, R, at the bottom of the ladder and hence the ladder is no longer in limiting equilibrium with the ground.

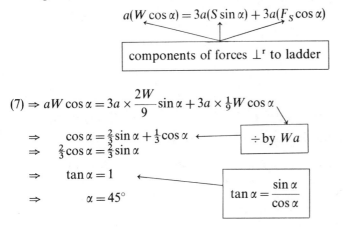

Assume that, when the man has climbed a distance x along the ladder, it is on the point of slipping.

Hence $\qquad\qquad F_{R'} = \tfrac{1}{4}R' \qquad$ (8)

and $\qquad\qquad F_{S'} = \tfrac{1}{2}S' \qquad$ (9)

Resolving horizontally

$$F_{R'} = S' \qquad (10)$$

Resolving vertically

$$R' + F_{S'} = 3W \qquad (11)$$

(8) and (10) $\Rightarrow S' = \tfrac{1}{4}R'$

(9) and (11) $\Rightarrow R' + \tfrac{1}{2}S' = 3W$

$\qquad \Rightarrow R' = \tfrac{8}{3}W$ and $S' = \tfrac{2}{3}W$

$\qquad\quad F_{R'} = \tfrac{2}{3}W$ and $F_{S'} = \tfrac{1}{3}W$ $\left.\right\}$ (12)

Taking moment about **B**:

$$a(W \cos \alpha) + x(2W \cos \alpha) = 3a(S' \sin \alpha) + 3a(F_{S'} \cos \alpha)$$

and (12) $\Rightarrow aW \cos \alpha + 2Wx \cos \alpha = 3a\tfrac{2}{3}W \sin \alpha + 3a\tfrac{1}{3}W \cos \alpha$

$\Rightarrow \cos \alpha(a + 2x) \qquad = 2a \sin \alpha + a \cos \alpha \quad \longleftarrow \boxed{\div W}$

$\Rightarrow \qquad a + 2x \qquad = 2a \tan \alpha + a$

$\Rightarrow \qquad\quad 2x \qquad = 2a \tan \alpha$

$\Rightarrow \qquad\quad x \qquad = a \tan \alpha$

but $\tan \alpha = 1$ $\therefore$ <u>man can ascend a distance a.</u>

● Exercise 4.7   See page 111

4.8   The Concurrency Principle: General Techniques for Solving Equilibrium Problems

Rigid Body Subject to Three Forces Only
If a rigid body is in equilibrium under the action of three forces only, the lines of action of these forces must all be parallel or concurrent.

concurrent ≡ meet at a point

Proof
Let F_1, F_2, F_3 be the forces. If they are not all parallel then two of them must meet at some point O. Assume that F_1 and F_2 are not parallel and meet at O. Then the resultant of F_1 and F_2 passes through O. Since the body is in equilibrium, this resultant must be equal and opposite to F_3. Hence F_3 must pass through O. Thus, if all three forces are not parallel they are concurrent.

Use of the Concurrency Principle

When a body is in equilibrium under the action of three forces, sometimes the direction of only two of the forces is known. The concurrency principle can then be used to determine the line of action of the third force.

Example 25   A smooth sphere has a light string attached to a point on its surface. The other end of the string is attached to the vertex of a smooth cone fixed with its axis vertical and its vertex upwards. Draw a diagram showing the forces acting on the sphere.

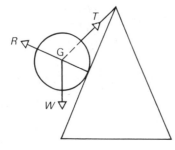

It is known that the normal reaction, R, passes through the centre of the sphere, G. Since there are only three forces acting on the sphere the line of action of the third force, T, must also pass through G.

General Techniques for Solving Equilibrium Problems.

In **4.5** problems were solved by taking moments only. In **4.6** and **4.7** problems were solved by resolving in two directions and taking moments. As equilibrium problems because more complex it will be useful to have a range of possible approaches.

Points to consider:
1 If moments are taken about three non-colinear points, this will result in three independent equations which can be solved to evaluate three unknowns. However, taking moments about a fourth point will not produce a fourth independent equation, but a combination of some of the first three equations.
2 Resolving components and equating their sum to zero in two non-parallel directions results in two independent equations. Resolving in a third direction will not produce a third independent equation.
3 Moreover, the equations obtained by resolving are not independent of the equations obtained by taking moments. Whether resolving and/or taking moments, only three independent equations can be found.

> Hence the techniques for deriving three independent equations are:
> 1 Resolve in two non-parallel (usually, but not always, perpendicular) directions and take moments about one point.
> 2 Resolve in one direction and take moments about two points.
> 3 Take moments about three non-colinear points.

The choice of technique, just as with the choice of points for taking moments, will depend on the individual question. To avoid bringing an unwanted unknown into equations, take moments about points on its line of action and/or resolve perpendicular to its direction. In many cases it is not necessary to derive **three** independent equations. One or two will be sufficient.

Example 26   A uniform rod of length 40 cm and mass 2 kg rests in equilibrium in the horizontal position, suspended by two strings of length 29 cm attached to each end of the rod and to a point O. Find the tension in each string.

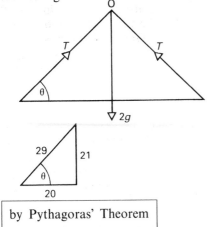

Since there are only three forces acting on the rod, the weight must act through O. By symmetry the tensions is the two strings are the same.
Resolving vertically:

$$2T \sin \theta = 2g$$

$$\Rightarrow T \times \frac{21}{29} = g$$

$$\Rightarrow T = \frac{29}{21} g \, \text{N.}$$

by Pythagoras' Theorem

Example 27   A uniform beam rests with its ends on two smooth planes inclined at 30° and 60° to the horizontal. A weight equal to twice the weight of the beam is placed on the beam. If the beam is in a horizontal position, find the position of the weight.

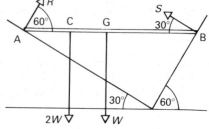

Let $AB = 2a$
Let $AC = x$
Resolving horizontally:

$$R \cos 60° = S \cos 30°$$

$$\Rightarrow R \cdot \frac{1}{2} = S \cdot \frac{\sqrt{3}}{2}$$

$$\Rightarrow R = \sqrt{3} S \qquad (1)$$

Resolving vertically:

$$R \sin 60° + S \sin 30° = 3W$$

$$\Rightarrow \frac{R\sqrt{3}}{2} + S\frac{1}{2} = 3W$$

$$\Rightarrow R\sqrt{3} + S = 6W \qquad (2)$$

(1) and (2) $\Rightarrow$ $S = 3\frac{W}{2}$ and $R = 3\sqrt{3}\frac{W}{2}$.

Taking moments about A:

$$2Wx + aW = S \sin 30° \times 2a$$

$$\Rightarrow 2Wx + aW = \frac{3W\,a}{2}$$

$$\Rightarrow 2Wx = \frac{aW}{2}$$

$$\Rightarrow x = \frac{a}{4}$$

∴ weight is 1/8 of the way along the rod from the end on the 30° plane.

• **Exercise 4.8**   See page 112

4.9 Rigid Bodies and Non-Limiting Friction

If a body is not slipping, or on the point of slipping,

$$F \leqslant \mu R$$

$$\Rightarrow \mu \geqslant \frac{F}{R}$$

Or, the minimum value of μ, for equilibrium to exist, $= \dfrac{F}{R}$

So, if a question asks for the minimum value of μ, or asks you to prove that $\mu \geqslant$ a constant value, then you use $F \leqslant \mu R$.

Example 28   A uniform ladder, of weight W, rests against a smooth vertical wall with its foot on rough horizontal ground. The ladder makes an angle α with the ground. Show that the minimum value of the coefficient of friction is $\frac{1}{2} \cot \alpha$.

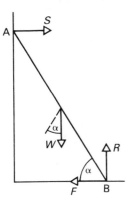

Resolving vertically:
$$R = W$$
Resolving horizontally:
$$S = F$$
Taking moments about B:
$$2 \sin \alpha = W \cos \alpha$$
$$\Rightarrow S = \frac{W}{2} \cot \alpha.$$

For equilibrium, $F \leqslant \mu R \Rightarrow \mu \geqslant \dfrac{F}{R}$

$$\Rightarrow \mu \geqslant \left(\frac{W}{2} \cot \alpha \right) / W$$

$$\Rightarrow \mu \geqslant \frac{1}{2} \cot \alpha.$$

$\therefore$ the minimum value of $\mu = \dfrac{1}{2} \cot \alpha$ 　　　　　　　　　　　　　　[QED]

- Exercise 4.9   See page 114

4.10 Revision of Unit

- Miscellaneous Exercise 4A   See page 115

4.11+ A Level Questions

- Miscellaneous Exercise 4B   See page 117

Unit 4 Exercises

Exercise 4.1

1 In each of the following diagrams two forces are shown. Calculate the magnitude of their resultant and the angle it makes with the larger of the two forces, giving your answers to 1 decimal place:

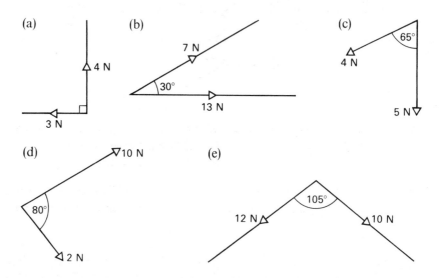

2 Forces of 3 N and 2 N act on a particle and the angle between the forces is 150°. Find the magnitude of the resultant and the angle it makes with the 3 N force.

3 Find the angle between two forces of magnitude 10 N and 4 N if their resultant has magnitude 8 N.

4 The angle between two forces **P** and **2P** is 60°. Find the magnitude of the resultant and the angle it makes with **P**.

5 Find the magnitude and direction of the resultant of two forces each of magnitude 5 N when the angle between the forces is 40°.

6 Find the magnitude and direction of two forces **P** and **P** making an angle of 120° with each other.

7 (a) Find, by drawing a polygon of forces, the resultant of the following forces:
       5 N due east, 3 N north-east, 6 N north-west, 2 N due south.
   (b) Taking the same forces in a different order and drawing another polygon of forces, show that the resultant is the same.

8 Each of the following systems of forces are in equilibrium. In each case, sketch the triangle of forces and calculate the magnitudes of X and Y and the size of the angle α.

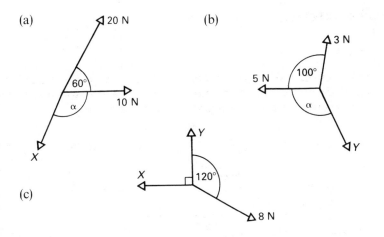

(a)   20 N

(b)   3 N

60°

10 N

α

X

(c)

100°

5 N

α

Y

Y

X

120°

8 N

*9 Two forces, of magnitude P N and Q N have a resultant of $\sqrt{21}$ N when they are perpendicular to each other. When the angle between the forces is 30° the resultant has magnitude $\sqrt{39}$ N. Calculate P and Q.

*10 The resultant of a force $4P$ N in a direction N60°E and a force 10 N in a direction due south is a force of $2P\sqrt{3}$ N. Find the value of P and the direction of the resultant.

These two forces and a third force of 25 N act at a point. The resultant of the three forces acts due south. Find the direction in which the third force is applied and find the magnitude of the resultant.

Exercise 4.2

In each of questions 1–10 find the moment about the point X of the forces shown in the diagram:

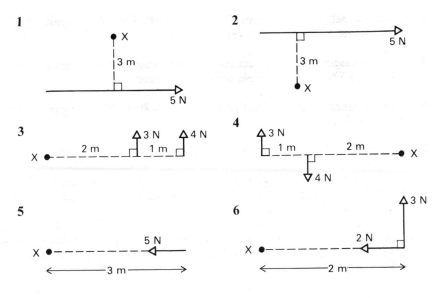

1

X

3 m

5 N

2

5 N

3 m

X

3

X

2 m

3 N

1 m

4 N

4

3 N

1 m

2 m

X

4 N

5

X

5 N

3 m

6

X

2 N

2 m

3 N

7

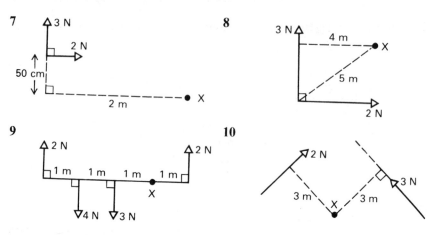

8

9

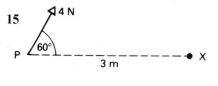

10

11 Find the moment about the origin of a force of $4\mathbf{i}$ N acting at the point with position vector $(3\mathbf{i} + 4\mathbf{j})$ m.

12 Find the moment about the origin of a force of $(4\mathbf{i} + \mathbf{j})$ N acting at the point with position vector $(2\mathbf{i} - \mathbf{j})$ m.

13 Find the moment about the point with position vector $(2\mathbf{i} - \mathbf{j})$ m of a force of $(3\mathbf{i} - \mathbf{j})$ N acting at the point with position vector $(3\mathbf{i} + 2\mathbf{j})$ m.

14

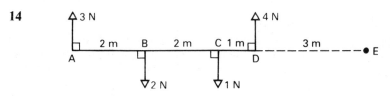

Find the moment of this system of forces about the point:
(a) A   (b) B   (c) C   (d) D   (e) E.

15

Replace the 4 N force by its components parallel and perpendicular to PX. Hence find the moment of the 4 N force about X.

16

Find the moment of the 10 N force about A.

17 A force of $(3\mathbf{i} + \mathbf{j})$ N acts at the point with position vector $(\mathbf{i} + \mathbf{j})$ m. A second force of $(3\mathbf{i} - \mathbf{j})$ N acts at the point with position vector $(-\mathbf{i} - \mathbf{j})$ m. Find the moment of the system of forces about the origin.

18 A force of $(3\mathbf{i} + 2\mathbf{j})$ N acts at the point with position vector $(2\mathbf{i} - \mathbf{j})$ m and a second force of $(-3\mathbf{i} + \mathbf{j})$ N acts at the point with position vector $(-3\mathbf{i} - \mathbf{j})$ m. Find the moment of the system about the point with position vector   (a) $\mathbf{i} + 2\mathbf{j}$ (b) $-5\mathbf{j}$.

In each of questions 19–22 find the moment of the system of forces about A:

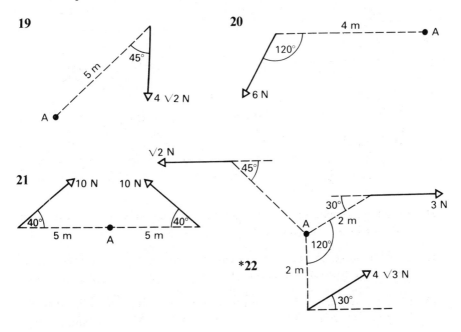

19

5 m 45° ∇4 √2 N A

20 4 m A 120° ▷6 N

21 √2 N 45° 10 N 10 N 40° 40° 5 m A 5 m 30° 3 N A 2 m 120° *22 2 m ∇4 √3 N 30°

Exercise 4.3

1 Parallel forces of 40 N and 70 N act along lines which are 22 cm apart. Find the magnitude, direction and distance of the line of action of their resultant from the 40 N force, if they are (a) like (b) unlike.

2 Find which of these systems reduce to a couple and, in these cases, find the moment of the couple:

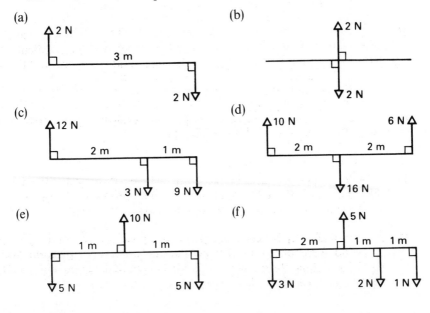

(a) △2 N 3 m 2 N▽

(b) △2 N ▽2 N

(c) △12 N 2 m 1 m 3 N▽ 9 N▽

(d) △10 N 6 N△ 2 m 2 m ▽16 N

(e) 10 N△ 1 m 1 m ▽5 N 5 N▽

(f) △5 N 2 m 1 m 1 m ▽3 N 2 N▽ 1 N▽

(g) 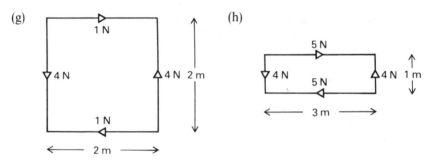 (h)

3 Two like forces 8 N and P N have a resultant of 20 N whose line of action is 6 m from the 8 N force. Find P and the distance of its line of action from the 8 N force.

4 Find the magnitude of two unequal like parallel forces acting at a distance of 4 m apart, given that they are equivalent to a force of 100 N acting at a distance of 1 m from one of these forces.

5 Two equal and opposite forces of 20 N are applied to a corkscrew along parallel lines 6 cm apart. Find the moment of the couple.

6 A force of $(-3\mathbf{i} + 5\mathbf{j})$ N is applied at the point with position vector $(2\mathbf{i} + 4\mathbf{j})$ m. A second force of $(3\mathbf{i} - 5\mathbf{j})$ N is applied at the point with position vector $2\mathbf{i}$ m. Show that the system is equivalent to a couple and find its moment.

7 A force of $(2\mathbf{i} - 3\mathbf{j})$ N is applied at the point with position vector $(\mathbf{i} - 2\mathbf{j})$ m. A second force of $(-2\mathbf{i} + 3\mathbf{j})$ N is applied at the point with position vector $(-2\mathbf{i} + 4\mathbf{j})$ m. Show that the system is equivalent to a couple and find its moment.

8 Two like horizontal forces of X N and 4 N act at points P and Q respectively where PQ is vertical and of length 6 m. Their resultant is a force of Y N and it acts at a point R between P and Q such that $QR = 4$ m. Find X and Y.

9 Two unlike parallel forces of 8 N and Y N act at points P and Q respectively where PQ is perpendicular to the forces and of length 4 m. The resultant of the two forces is a force of Z N which acts at a point S on PQ produced, such that $QS = 1$ m. Find Y and Z.

10 Find the magnitude, direction and line of action of the resultant of this system of forces.

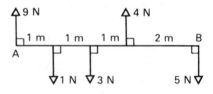

*11 Like parallel forces of magnitudes X, X and $2X$ act at the vertices P, Q, R of a triangle PQR. Show that their resultant passes through the midpoint of the line joining R to the midpoint of PQ.

**12 X and Y are like parallel forces. If Y is moved parallel to itself through a distance x, prove that the resultant of X and Y moves through a distance $Yx/(X + Y)$.

Exercise 4.4

1 A uniform rod, 1.8 m long and of mass 20 kg, rests horizontally on supports at its ends. A weight of mass 6 kg is hung from a point 1.2 m from one end. Find the reaction of the supports.

2 A uniform rod AB is 10 m long and has mass 40 kg. It rests horizontally on a support at A and another support 2 m from B. A weight of mass 20 kg is attached at a point 4 m from B. Find the reaction at each of the supports.

3 Two bearers carry a load of 50 kg hung from a light pole of length 2.5 m, with the ends of the pole resting on one shoulder of each of the men. The load is hung at a point which is 0.5 m nearer the rear bearer. What is the force on each man's shoulder?

4 A uniform beam, 4 m long, with a load of mass 30 kg hanging from one end, balances in the equilibrium position about a point 80 cm from this end. Find the mass of the beam.

5 A uniform rod of length 150 cm and mass 3 kg, has masses of 1 kg, 2 kg, 3 kg, 4 kg suspended from it at distances of 30 cm, 60 cm, 90 cm, 120 cm respectively from one end. Find the position of the point about which the rod will balance.

6 A uniform plank, of mass 10 kg and length 6 cm, rests on two supports 1 m from each end. What is the maximum load that can be placed on either end without the plank tilting?

7 A uniform rod AB, of length 3 m and mass 6 kg, rests on two supports. A load of 1 kg is applied at B, a load of 5 kg applied at 1 m from B and a load of 4 kg at 2 m from B. One support is at A. If the reaction at this support is 40 N, where is the other support?

8 A uniform rod, of length 2 m and mass 1 kg, rests horizontally on two supports at its ends. Each support will bear the weight of a mass of 8 kg and no more. On what part of the rod can a load of mass 10 kg be placed without breaking either support?

9 A uniform plank of mass 100 kg is 10 m long and 4 m of it project over the side of a quay. What is the minimum load that must be placed on the end of the plank so that a woman of mass 50 kg can walk to the other end of the plank without tipping into the water?

*10 A rigid rod AB, of mass 10 kg and length 3 m, whose centre of gravity G is 125 cm from A, is suspended horizontally by two vertical strings attached to points C and D, 50 cm from each end. Each string could just support the weight of the rod. Masses of m_1 kg and m_2 kg are placed at A and B respectively. Show that the tension in one of the strings is

$$(15 + 5m_2 - m_1)\frac{g}{4}$$

and find the tension in the other string. Also find the values of m_1 and m_2 when both the strings are on the point of breaking.

Exercise 4.5

(This exercise is to be done entirely without the use of calculators)

1   In each of the following diagrams, one end of a light inelastic string is fastened to a fixed point A. A 1 kg bob is fastened to the other end of the string. The string is pulled aside from the vertical and the bob held in equilibrium by the application of a force P. Find the magnitude of P in each case.

(a)   (b)   (c)   (d)

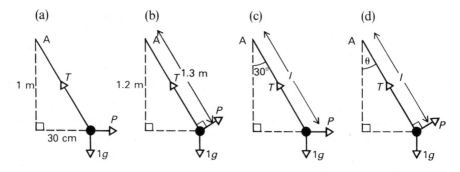

2   In each of the following diagrams, a uniform rod AB, of mass 3 kg and length 4 m, is hinged at A. The rod is held in equilibrium by a string fastened at B. Find the tension in the string in each case.

(a)   (b)

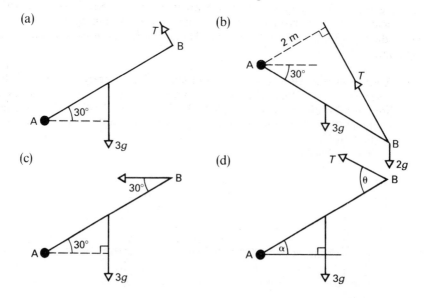

(c)   (d)

3   A uniform rod AB, of mass 5 kg and length 2 m, is hinged at A to a fixed point. A load of 3 kg is attached at B. A string is attached to the midpoint of AB and to a point C 2 m vertically above A. The rod rests in equilibrium with B above A and at 60° to the vertical. Find the tension in the string.

4   A uniform rod AB of length 2a and mass M is hinged at A to a fixed point. A load of mass 2M is attached at B. The rod is kept in equilibrium in the horizontal position by a string attached to the midpoint G of the rod and to a point C vertically above A. The length of the string is 2a. Find the tension in the string.

5   A uniform rod AB of length 2a and mass m hangs freely, with its end A hinged to a fixed point. The rod is pulled aside by a horizontal force P applied at a distance $\frac{3}{2}a$ from A, measured along the rod. In equilibrium it makes an angle of 45° with the vertical. Show that

$$P = 2mg/3$$

6   A uniform ladder, of length 4 m and mass 100 kg, rests on rough ground with its other end against a smooth wall. By taking moments about the bottom of the ladder, find the normal reaction between the wall and the ladder, if the ladder makes an angle of 30° with the wall.

7   A uniform rod AB, of mass M and length l has its end A resting on rough horizontal ground and is kept in equilibrium by a string attached to B. If the rod makes an angle of 45° with the horizontal and the string makes an angle of 60° with BA, show that the tension in the string is $Mg/\sqrt{6}$.

Exercise 4.6

(This exercise is to be done entirely without the use of calculators)

In this exercise, in any questions which involve a rigid body in contact with a vertical wall or horizontal ground, the plane containing the body is perpendicular to the wall or the ground. Hence the forces in the diagrams are coplanar.

✓ 1   A uniform rod AB of mass 6 kg and length 2 m is freely hinged at A to a vertical wall. A force P applied at B keeps the rod in equilibrium in a horizontal position. X and Y are the horizontal and vertical components of the reaction at the hinge. Find the magnitude of X, Y and P in each of the following cases:

(a)   (b)

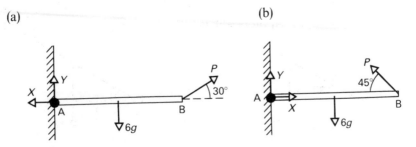

2 A uniform rod CD of mass 8 kg and length 4 m is freely hinged at C to a vertical wall. A force P applied at D keeps the rod in equilibrium. Find the force P and the magnitude of the reaction at the hinge in each of the following cases:

(a)   (b)

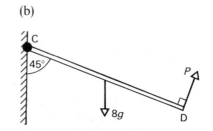

3

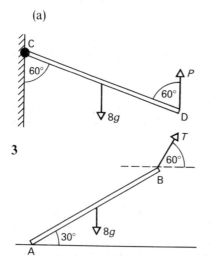

A uniform rod AB of mass 8 kg and length 2 m rests with its end A on a rough horizontal floor. A string attached to B keeps the rod in equilibrium. Find the magnitude of the tension in the string, the frictional force and the normal reaction between the rod and the floor.

4 A uniform ladder AB of mass 20 kg and length 6 m rests with A against a smooth vertical wall and B on a smooth horizontal floor. The ladder is kept in equilibrium by a light horizontal string attached at one end to B and at the other end to the base of the wall. The ladder makes an angle α with the horizontal. Find the tension in the string when (a) $\alpha = 45°$ (b) $\alpha = 60°$.

5 A uniform ladder AB of mass 10 kg and length $2a$ rests with A against a smooth vertical wall and B on a rough horizontal floor. The ladder makes an angle β with the vertical. Find:
(i)   the normal reaction between the ladder and the wall
(ii) the normal reaction between the ladder and the floor
(iii) the frictional force
(iv) the least possible value of the coefficient of friction when $\beta = 30°$.

6 A uniform ladder AB, of weight WN and length $4a$ rests with A against a smooth vertical wall and B on a smooth horizontal floor. A light horizontal string is attached to a point C on the ladder and to the wall. The ladder is in equilibrium at an angle $\tan^{-1}2$ to the horizontal. Find the tension in the string when BC is   (a) a (b) $2a$.

7 A uniform rod AB of weight W is freely hinged at A to a vertical wall. The rod is in equilibrium in each of the following cases:

(a)   Prove that $X = \frac{1}{2}W\cot\theta$

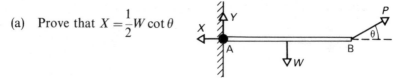

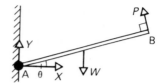

(b) Prove that $2Y = W(2 - \cos^2 \theta)$.

8 A uniform ladder AB of weight W and length $2a$ rests in equilibrium with A against a smooth vertical wall and B on a rough horizontal floor, at an angle θ to the horizontal. If F is the frictional force prove that

$$2F \tan \theta = W$$

*9 A uniform ladder is in limiting equilibrium with its top against a smooth vertical wall and its foot on rough horizontal ground. The ladder makes an angle θ with the vertical. Prove that

$$2\mu = \tan \theta$$

where μ is the coefficient of friction.

*10

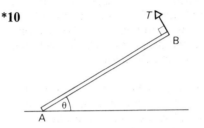

A uniform rod AB is supported at θ to the horizontal by a light string attached at B and perpendicular to B. A is resting on rough horizontal ground and λ is the angle of friction. If the rod is in limiting equilibrium prove that

$$\tan \lambda = \frac{\sin 2\theta}{3 - \cos 2\theta}$$

Exercise 4.7

1 A uniform rod AB of mass $10\,\text{kg}$ and length $4\,\text{m}$ is freely hinged at A to a vertical wall. A force P applied at B keeps the rod in equilibrium. Find the magnitude of P and the magnitude and the angle to the vertical of the reaction at the hinge in each of the following cases:

(a)

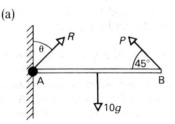

(b)

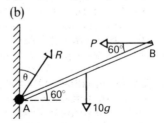

2

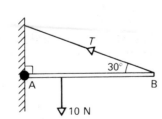

A non-uniform beam AB of weight $10\,\text{N}$ and length $2\,\text{m}$ has the end A hinged to a vertical wall. The rod is supported in the horizontal position by a string attached to B making an angle of $30°$ with BA. If the tension in the string is $6\,\text{N}$, find the distance from A to the centre of gravity of the beam and the magnitude and direction of the reaction at the hinge.

3

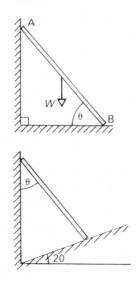

A uniform ladder of weight W and length $2l$ rests in limiting equilibrium with its upper end against a rough vertical wall and its lower end on a rough horizontal floor. The coefficients of friction are $\frac{1}{5}$ at the wall and $\frac{1}{3}$ at the floor. Find the angle that the ladder makes with the floor.

4

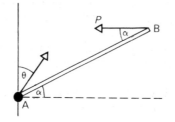

A uniform ladder rests in equilibrium with its top end against a smooth vertical wall and its foot resting on a smooth plane inclined at $20°$ to the horizontal. The ladder makes an angle θ with the vertical. Find the value of θ.

5   A non-uniform ladder of length $6\,m$ has its centre of gravity at a point $2\,m$ from the bottom of the ladder. The ladder rests in limiting equilibrium on rough horizontal ground with coefficient of friction $\frac{1}{3}$ and against a rough wall with coefficient of friction $\frac{1}{4}$. If the ladder makes an angle θ with the wall show that $\tan\theta = \frac{6}{5}$.

6

A uniform rod AB of weight W is hinged at A to a vertical wall. The rod is kept in equilibrium at an angle α above the horizontal by a horizontal force P applied at B. The reaction at the hinge makes an angle θ to the vertical.
Prove that $\tan\theta \tan\alpha = \frac{1}{2}$.

*7   A uniform ladder rests in limiting equilibrium on rough horizontal ground and against a rough wall. The contacts at the wall and ground have coefficients of friction $\frac{1}{4}$ and μ respectively. If the ladder is inclined at $60°$ to the horizontal find the value of μ.

*8   A uniform ladder of length $10\,m$ and weight $W\,N$ stands and on rough horizontal ground with coefficient of friction $\frac{1}{3}$. The top of the ladder rests against a smooth wall an angle of $\tan^{-1}\frac{10}{19}$ to the wall. How far up the ladder can a person of weight $2W$ climb before slipping occurs?

The least horizontal force that must be applied to the foot of the ladder so that this person can climb to the top of the ladder is kW. Find the valued of k.

Exercise 4.8

1   A non-uniform rod AB is suspended from a point O by two strings OA and OB attached to its ends. OA is $30\,cm$ long, OB is $20\,cm$ long and AB, which is 40 cm long, rests in equilibrium with B higher than A. Draw a diagram showing all the forces acting on the rod.

2 A picture of mass 4 kg has a cord 1.5 m long fastened to two rings 90 cm apart. The rings are equidistant from the sides and the top of the picture. The cord is hung over a smooth peg on the wall. Find the tension in the cord.

3 A sphere of mass 6 kg and radius 50 cm is hung by a string 80 cm long from a point on a smooth vertical wall. Find the tension in the string.

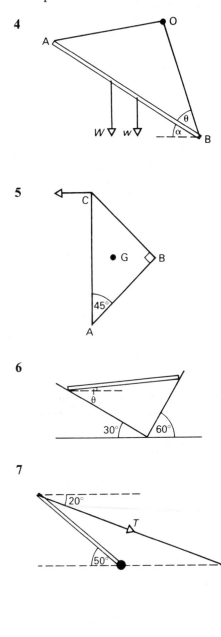

4 A uniform rod AB of length 4a and weight W is hung from a point O by two strings of equal length attached one to each end of the rod. A weight w is hung a distance a from B. The rod rests in equilibrium at an angle α to the horizontal. By taking moments about each end of the rod show that the ratio of the tensions in the strings is

$$2W + 3w : 2W + w$$

5 A uniform lamina ABC of weight W, in the form of a right angled isosceles triangle with hypotenuse AC = 2 m, is freely hinged at A and held with C vertically above A by a horizontal string attached to C. Find the tension in the string and the magnitude and angle to the horizontal of the reaction at A.
(The centre of gravity of the triangle, G, is $\frac{2}{3}$ m from B)

6 A uniform rod 4 m long and of mass 10 kg is placed on two smooth planes, inclined at 30° and 60° to the horizontal. Find the pressure on each plane and the inclination of the rod to the horizontal when it is in equilibrium.

7 A uniform flagstaff of mass 120 kg and 12 m long is being raised by a rope attached to the top. The bottom of the flagstaff is attached to a fixed swivel on the ground. At the point when the inclination of the flagstaff is 50° to the horizontal, the rope makes 20° with the horizontal. At this point, find the tension in the rope and the magnitude and direction of the reaction at the swivel.

8

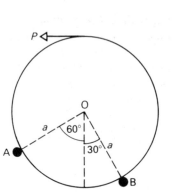

A circular disc, centre O, radius a and weight W, rests in a vertical plane on two rough pegs A and B. AO makes $60°$ with the vertical and BO makes $30°$ with the vertical. A tangential force P is applied at the highest point of the disc. The coefficient of friction is $\frac{1}{4}$. By taking moments about horizontal axes through O, A and B show that the greatest force P that can be applied without slipping taking place is

$$W(81 + 102\sqrt{3})/747$$

Exercise 4.9

1 A uniform rod AB, of weight W and length $2a$, rests with the end A on a rough horizontal table. The rod is maintained in equilibrium at an angle of $45°$ to the horizontal by a force acting at B in a direction perpendicular to the rod and in the same vertical plane as the rod. Find, in terms of W, the vertical and horizontal components of the force exerted by the table on the rod at A. Find the least possible value of the coefficient of friction. [J]

2 A uniform triangular lamina ABC has mass m and each side is of length $2a$. The lamina rests in a fixed vertical plane with A on a rough horizontal table and the side AC is vertical. Equilibrium is maintained by a force of magnitude T which acts along the side BC. Show that $T = mg/3$.
Find the magnitude of the reaction at A.
Show also that equilibrium is possible only when

$$\mu \geqslant \sqrt{3}/5,$$

where μ is the coefficient of friction between the lamina and the table. [L]

3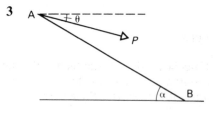

A uniform rod AB, of mass m, rests in equilibrium, at an angle α to the horizontal, with the end B standing on a rough horizontal plane. It is kept in equilibrium by a force of magnitude P, as shown in the diagram. The force and the rod are in the same vertical plane.

Show that

$$P = \frac{mg \cos \alpha}{2 \sin (\alpha - \theta)}$$

and find the normal reaction and the frictional force at B, in terms of m, g, and θ. Show that the least possible value of the coefficient of friction is

$$\frac{1}{2 \tan \alpha - \tan \theta}$$

4

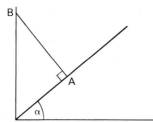

A uniform rod AB, of weight W, rests in equilibrium with A in contact with a plane inclined at α to the horizontal and B against a smooth vertical wall. The rod is perpendicular to a line of greatest slope of the plane. Find the normal and frictional forces at A and the normal reaction at B.

Hence show that the minimum value of the coefficient of friction, for which equilibrium is possible, is

$$\frac{\sin \alpha - \cos \alpha}{2 - \sin^2 \alpha}$$

5 A uniform rod AB of length $2a$ and weight $8W$ is smoothly hinged at the end A to a point on a fixed horizontal rough bar. A small ring C of weight W is threaded on the bar and is connected to the rod at B by a light inextensible string of length $2a$. The system is in equilibrium with the rod inclined at an angle θ to the horizontal. Find, in terms of W and θ, the tension in the string and the horizontal and vertical components of the force exerted by the bar on the ring.

In the case when the coefficient of friction between the bar and the ring is $\frac{1}{2}$, find the greatest possible value of the distance of the ring from A.   [J]

6 Two cylinders of radii a and b ($b > a$), are fixed with their axes parallel and in the same horizontal plane, and a plank rests across them with its long axis in a vertical plane perpendicular to the axes of the cylinders. The larger cylinder is smooth and the coefficient of friction between the other cylinder and the plank is μ. Show that the plank makes an angle θ with the horizontal, where $l \tan \theta = b - a$ and l is the distance between the points of contact of the plank and the cylinders.

Find, in terms of a, b and μ, the minimum value of the distance of the centre of mass of the plank from its point of contact with the larger cylinder.

[J]

Miscellaneous Exercise 4A

(Give all answers correct to one decimal place.)

1

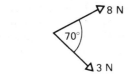

Calculate the magnitude of the resultant of these two forces and the angle it makes with the 8 N force.

2 Find the angle between two forces of magnitude 6 N and 9 N if their resultant has magnitude 5 N.

3

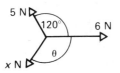

This system of forces is in equilibrium. Sketch the triangle of forces and calculate the magnitude of X and the angle θ.

4 Find the moment about the origin of a force of $(3\mathbf{i} - 2\mathbf{j})\,\text{N}$ acting at the point with position vector $(-\mathbf{i} + \mathbf{j})\,\text{m}$.

5 Parallel forces of 20 N and 40 N act along lines which are 60 cm apart. Find the magnitude, direction and distance of the line of action from the 20 N force, of their resultant, if they are (a) like   (b) unlike.

6 Find the magnitude of two unequal like parallel forces acting at a distance of 10 m apart, given that they are equivalent to a force of 50 N acting at a distance of 4 m from one of the forces.

7 A uniform beam AB of mass 20 kg and length 8 m rests on two supports, one at A and the other 2 m from B. Find the pressure on each of the supports.
    What is the maximum weight that could be hung from B?

8 A light inelastic string has one end fastened to a fixed point A. A bob of mass 2 kg is hung from the other end and the string is held at 30° to the vertical by means of a horizontal force P applied to the bob. Find the magnitude of P and the tension in the string.

9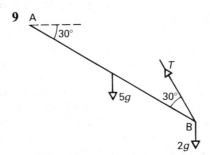
A uniform rod AB of mass 5 kg is hinged at A and a mass of 2 kg hangs from B. It is kept in a position of equilibrium at 30° below the horizontal by a light string attached to B at an angle of 30° to BA. Find the tension in the string and the horizontal and vertical components of the reaction at the hinge.

10 A uniform ladder AB of mass 10 kg and length 3 m rests with its end A against a smooth vertical wall and its end B on rough horizontal ground. When it is in limiting equilibrium it makes an angle of 40° with the wall. Show that $\mu = \frac{1}{2}\tan 40°$ where μ is the coefficient of friction.

11 A non-uniform beam AB is of length 10 m and mass 20 kg and its centre of gravity is 4 m from B. The beam is kept in a horizontal position by two strings at A and B. The string at A makes an angle of 30° with AB. Find the tensions in the strings and the angle that the string attached at B makes with BA.

12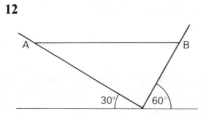
A uniform beam AB of length 1 m and weight W rests with its ends on two smooth inclined planes which make angles of 30° and 60° with the horizontal. A weight $2W$ can slide along the beam. Find the distance of the weight from A when the beam rests horizontally in equilibrium.

*13   A force of magnitude P acts along OA and a force of magnitude Q acts along OB, where the angle BOA is θ.

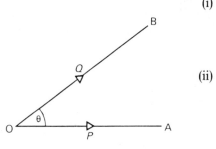

(i)   If the resultant of the two forces has magnitude P, show that:
    (a) $Q = -2P\cos\theta$
    (b)   the angle between OA and the resultant is $(2\theta - 180°)$.

(ii)   If the force along OA is increased to $5P/4$, but the resultant still has magnitude P, show that:
    (c) $Q = 1.5P$
    (d)   the angle between OA and the resultant is $83°$, to the nearest degree.

Miscellaneous Exercise 4B

1   Forces of magnitudes $1, 2, 3$ and 5 newtons act along the sides AB, BC, CD and DA, respectively, of a rectangle ABCD in which $AB = 4$ metres and $BC = 3$ metres. Two other forces of magnitudes P and Q newtons act along the diagonals AC and BD respectively. All six forces act in the senses indicated by the order of the letters. Given that the system reduces to a couple, determine the values of P and Q. Determine also the moment of the couple.   [J: 4.3]

2   The vertices of a triangular lamina, which is in the x-y plane, are at the origin O and the points A $(1, 2)$ and B$(-1, 3)$. Forces $\mathbf{i} + \mathbf{j}$ and $-3\mathbf{i} + \mathbf{j}$ are applied to the lamina at A and B, respectively, and a force $\mathbf{F}$, whose line of action is in the x-y plane, is applied at O. Given that the three forces form a couple, find the magnitude and the direction of $\mathbf{F}$.
Find also the magnitude and the sense of the additional couple that must be applied to the lamina in order to keep it in equilibrium.   [J: 4.3]

3   A uniform plank ABCD of length $6a$ and weight W rests on two supports B and C; $AB = BC = CD = 2a$.
A man of weight $4W$ wishes to stand on the plank at A, for which purpose he places a counterbalancing weight Y at a point P between C and D at a distance x from D. Show that the least value of Y required is

$$\frac{7Wa}{4a - x}$$

Show that, if the board is also to be in equilibrium when he is *not* standing on it, then the greatest value of Y that may be used is

$$\frac{Wa}{2a - x}$$
     [J: 4.4]

4   A uniform rod AB, of length $2l$ and weight W, is in equilibrium with the end A on a rough horizontal floor and the end B against a smooth vertical wall. The rod makes an angle $\tan^{-1}2$ with the horizontal and is in a vertical plane which is perpendicular to the wall. Find the least possible value of μ, the coefficient of friction between the floor and the rod. Given that $\mu = \frac{5}{16}$, find the distance from A of the highest point of the rod at which a particle of weight W can be attached without disturbing equilibrium.   [L: 4, 6, 4.7, 4.8, 4.9]

5

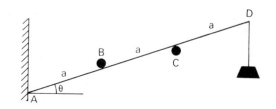

The diagram shows a thin light rod AD resting at an angle θ to the horizontal with its lower end A in contact with a smooth vertical wall which is perpendicular to the vertical plane containing the rod. The rod passes under a peg B and over a peg C, both pegs being fixed, smooth, horizontal and parallel to the wall. A weight W is suspended from D and the rod rests in equilibrium. Given that $AB = BC = CD = a$; find the reactions at B and C and show that

$$3\cos^2 \theta \geqslant 2 \qquad\qquad [J: 4.6, 4.7, 4.8]$$

6   A uniform rod AB, of length $4a$ and mass m, is smoothly hinged to a vertical wall at the end A. The rod is horizontal and is kept in equilibrium at right angles to the wall by a light inextensible string, one end of which is attached to the rod at the point C, where $AC = 3a$. The other end of the string is attached to the wall at the point D, which is vertically above A and at a distance $4a$ from it. Calculate, in terms of m and g:

(a)   the magnitude of the tension in the string
(b)   the magnitude of the force exerted by the hinge at A on the rod.

Calculate also the tangent of the angle made with the horizontal by the line of action of this force at the hinge.
Given that the string breaks when the magnitude of the tension in it exceeds $2mg$, calculate the magnitude of the greatest load which can be attached to the rod at the end B without the string breaking.
                                                        [L: 4.6, 4.7, 4.8]

7   A uniform beam ABC of weight W and length $3a$ rests horizontally on supports at A and B, where $AB = 2a$. A man of weight $2W$ stands on the beam at a distance x from A. The maximum load which the support at A can bear is $2W$. Show that equilibrium is possible only if $a \leqslant 4x \leqslant 9a$.
                                                        [J: 4.3]

8   A straight uniform rod CDEF, of length $6a$ and mass M, is suspended from a fixed point A by two light inextensible strings AD and AE, each of length $2a$, where D and E are the points of trisection of the rod. A particle of mass m is attached to the rod at a point distant x from the end C and the system hangs in equilibrium. Given that the tension in the string AD is twice the tension in the string AE, show that the rod is

inclined to the horizontal at an angle θ, where $\tan\theta = 1/(3\sqrt{3})$. Hence find, in terms of M, m, a, and g:

(a) the tensions in the strings
(b) the distance x.   [L: 4.6, 4.7, 4.8]

*9   The diagram shows a light rod AB inclined at an angle θ to the vertical and smoothly pivoted at its end A. The end B is pressed on a rough horizontal plane by a force of magnitude P applied at the mid-point C of the rod, and acting at an angle θ to the vertical as shown. The plane is moving horizontally in the direction shown. Find the force of friction in terms of θ, P and the coefficient of friction, μ, between the rod and the table.

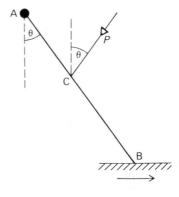

Find also the vertical component of the reaction exerted by the pivot on the rod.
[J: 4.6, 4.7, 4.8]

*10

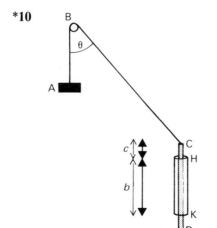

A body A of weight W is attached to one end of a light inextensible string ABC passing over a smooth peg B. A thin rod CD of weight w is attached to the other end of the string. The rod passes through a fixed cylindrical guide HK which keeps the rod vertical. The angle ABC is θ, HK $= b$, CH $= c$, $b + c <$ CD, and $w > W\cos\theta$. Find the vertical force applied at D which will maintain the system in equilibrium in the case when the guide is smooth.

Assuming that the only points of contact of the rod with the guide are at the top and bottom of the guide, and that the contacts are smooth, find the magnitude and direction of the forces exerted on the rod at these points.

In the case when the upper contact at H is rough and the lower contact at K is smooth, and no force is applied at D, show that the least value of the coefficient of friction between the rod and the guide at H for which equilibrium is possible is

$$\frac{b(w - W\cos\theta)}{W(b + c)\sin\theta}$$

[J: 4.8]

5 WORK, ENERGY, POWER, IMPULSE, MOMENTUM

5.1 Work and Energy

Work Done By a Constant Force

> **Work** is defined as the product of the constant force and the distance through which the point of application moves in the direction of the force.

If the point of application of a force of F newtons moves from A to B through a distance s metres in the direction of **F**, then the

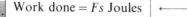

| Work done = Fs Joules | ← | J is the abbreviation for Joules. |

If the point of application of a force of **F** newtons moves from A to B through a distance s metres in a direction which is at an angle θ to the direction of **F**, then the

| Work done = $Fs\cos\theta$ Joules | ← | Note that this is equivalent to the work done by the component of **F** in the direction AB. |

Work Done Against Gravity

If a body of mass m is to be raised vertically at a constant speed, then a force of mg N must be applied to the body. If the body is raised a distance s metres, then the work done against gravity $= mgs$ J.

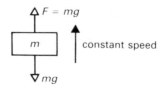

Work Done Against a Resisting Force

If a body is pulled at a constant speed against a resisting force, **R**, then the work done against the resistance $= R \times$ distance moved in the direction of **R**.

Motion at a Constant Speed

If a body is to be moved at constant speed a force **F**, which is equal in magnitude to the sum of the resistances, must be applied to the body.

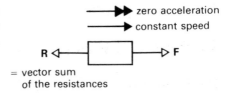

Example 1   Find the work done in raising a mass of 2 kg through a vertical distance of 5 m.

Vertical force required $= 2g$
Work done
$$\begin{aligned} &= F \times s \\ &= 2g \times 5 \\ &= 98 \, \text{J} \end{aligned}$$

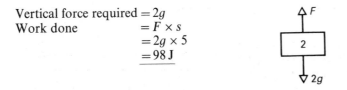

Example 2   A horizontal force pulls a body of mass 100 kg a distance of 10 m at a constant speed across a rough horizontal surface. Find the work-done against the friction if the coefficient of friction is $\frac{1}{2}$.

Resolving vertically: $R = 100g$
Sliding friction $\quad = \mu R = \frac{1}{2} \cdot 100g$
$\qquad\qquad\qquad\; = 50g$
Work done against friction $= 50g \times$ distance moved
$\qquad\qquad\qquad\qquad\; = 50g \times 10$
$\qquad\qquad\qquad\qquad\; = 4900 \, \text{J}$

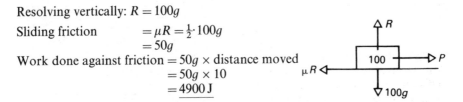

Example 3   A body of mass 10 kg lies on a rough plane inclined at $\tan^{-1} \frac{12}{5}$ to the horizontal. The coefficient of friction is $\frac{13}{20}$. If the body is pulled at a constant speed, by a force acting along the line of greatest slope, a distance of 2 m up the plane, find:

(a)   the work done against friction
(b)   the work done against gravity.

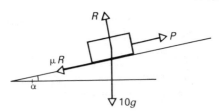

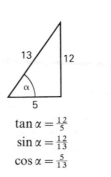

$\tan \alpha = \frac{12}{5}$
$\sin \alpha = \frac{12}{13}$ (1)
$\cos \alpha = \frac{5}{13}$

(a)   Resolving perpendicular to the plane: $R = 10g \cos \alpha = \dfrac{50g}{13}$ using (1)

$$\text{Sliding friction} = \mu R = \frac{13}{20} \times \frac{50g}{13} = \frac{5}{2}g$$

$$\text{Work done against friction} = \frac{5}{2}g \times \left\{ \begin{array}{l} \text{distance moved} \\ \text{along the plane} \end{array} \right.$$

$$= \frac{5}{2}g \times 2 = 49 \text{ J}$$

(b)   Work done against gravity $= mg \times$ vertical distance moved

$$= 10g \times 2 \sin \alpha$$

using (1) $= 10g \times 2 \times \dfrac{12}{13} = 180.9 \text{ J}$

Energy

The **energy** of a body is its capacity for doing work.

The **kinetic energy** (KE) of a body is the energy it possesses by virtue of its motion. It is measured by the amount of work that the body does in coming to rest.

Work done in coming to rest $= F \times s$

$$= (ma)s \quad \longleftarrow \quad \boxed{F = ma}$$
$$= m(as)$$

$$= m\left(\frac{v^2}{2}\right) \quad \longleftarrow \quad \boxed{\begin{array}{l} v^2 = 0 + 2as \\ \Rightarrow as = \dfrac{v^2}{2} \end{array}}$$

$$\boxed{\text{KE} = \tfrac{1}{2}mv^2}$$

The **potential energy** (PE) of a body is the energy it possesses by virtue of its position. It is measured by the amount of work that the body would do in moving from its actual position to some standard position.

$$\boxed{\text{PE} = mgh}$$ where $h =$ height above the standard position.

There is no fixed standard position of zero PE, although an arbitrary level may be used. It is the change in PE that is required in problems.

The SI unit of energy is the **Joule**. 1000 Joules $= 1$ kJ

> **Note**: Both work and energy are scalar quantities.

Example 4   Find the loss in kinetic energy of a snooker ball of mass 200 grams which slows down from 6 m/s to 1 m/s.

$$\text{Initial KE} = \frac{1}{2}mv^2 = \frac{1}{2}(0.2)6^2 = 3.6\,\text{J}$$

$$\text{Final KE} = \frac{1}{2}(0.2)1^2 = 0.1\,\text{J}.$$

$$\text{Loss in KE} = 3.6\,\text{J} - 0.1\,\text{J} = \underline{3.5\,\text{J}}$$

Example 5   Find the change in potential energy of a man of mass 80 kg who ascends a vertical distance of 2 m.

$$\text{Gain in PE} = mgh = 80g \times 2 = \underline{1568\,\text{J}}$$

An Alternative Way of Calculating the Work Done in Some Cases

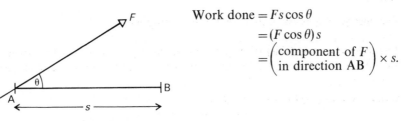

Work done $= Fs\cos\theta$

$$= (F\cos\theta)\,s$$

$$= \left(\begin{array}{c}\text{component of } F \\ \text{in direction AB}\end{array}\right) \times s.$$

Example 6   A bead is pulled along a horizontal wire by a string pulling at 45° above the horizontal. If the bead moves at a constant speed and the tension in the string is $8\sqrt{2}\,\text{N}$ find the work done in moving the bead 50 cm along the wire.

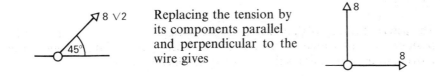

Replacing the tension by its components parallel and perpendicular to the wire gives

The component of the tension perpendicular to the wire does no work since the distance moved perpendicular to the wire is zero.

The work done by the tension = work done by the component parallel to the wire.

$$= 8 \times 0.5$$

$$= \underline{4\,\text{J}}$$

● Exercise 5.1   See page 144

5.2 The Work-Energy Principle: Conservation of Mechanical Energy

Energy

The energy of a body is its capacity for doing work.
When a body does some work, part of its energy is used up.
When work is done to a body, it gains energy.
Work and energy are mutually convertible and are measured in the same units.

When a body does some work,

work done = net loss in energy.

When work is done to a body,

work done = net gain in energy.

There are many forms of energy: heat, light, sound etc. In mechanics we are almost exclusively concerned with mechanical energy.

Mechanical Energy

The **mechanical energy** of a body is the sum of its kinetic and potential energies.

The total mechanical energy of a body, or system of bodies, will be changed if:
(i)   some mechanical energy is converted into another form of energy (heat, light, sound etc). This usually occurs when a sudden change in motion takes place, such as a jerk or a collision.
(ii) an external force (other than weight) causes work to be done.

> Work done by the weight is already included in the total mechanical energy as potential energy.

The Work–Energy Principle

Work done = change in mechanical energy

$\Rightarrow$ Work done = change in KE + change in PE

provide no sudden change in motion takes place.

Conservation of Mechanical Energy

If the external forces acting on a system do no work and there is no sudden change in motion, then the total mechanical energy of the system remains constant.

Hence   loss in KE = gain in PE
or   gain in KE = loss in PE

Example 7 A force pushes a body of mass 4 kg horizontally across a smooth surface increasing its speed from 2 m/s to 5 m/s.
(a)   Find the gain in KE
(b)   Find the work done by the force
(c)   If the force is constant and the distance travelled during the motion is 6 m, find the magnitude of the force.

(a)   Gain in KE = Final KE – Initial KE.

$$= \tfrac{1}{2} \cdot 4 \cdot 5^2 - \tfrac{1}{2} \cdot 4 \cdot 2^2$$

$$= 42\,\text{J}$$

(b)   There is no change in PE
∴ work done = gain in KE = 42 J

(c) $F \times 6 = 42$ ⟵ ⎹ Work done = force × distance.
⇒ Force = 7 N

Work Done Against Resistance

Work done against a resistance = loss in energy

Example 8 A body of mass 2 kg falls vertically past two points A and B which are 10 m apart. The speeds of the body as it passes A and B are 2 m/s and 12 m/s respectively. If the resistance to motion can be assumed to be constant, find its magnitude.

Loss in PE = $mgh = 2 \times 9.8 \times 10 = 196\,\text{J}$.
Gain in KE = $\tfrac{1}{2} \times 2 \times 12^2 - \tfrac{1}{2} \times 2 \times 2^2 = 140\,\text{J}$.

Total change in energy = loss of 56 J
= work done against the resistance.

If the resistance is R N

then $R = 10 = 56$
⇒ Resistance = 5.6 N.

Example 9 AB is a length of smooth wire. A bead of mass 100 grams is pushed gently from rest at A and it slides down the wire. If the vertical distance between A and B is 1 m find the speed of the bead when it leaves the wire at B.

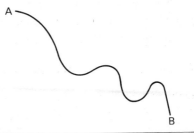

The only force acting on the body, other than the weight, is the normal reaction. Since it is always perpendicular to the direction of motion, it does no work. Therefore energy is conserved.

$$\text{loss in PE} = mgh = 0.1 \times 9.8 \times 1 = 0.98 \text{ J.}$$
$$\text{gain in KE} = \tfrac{1}{2} \times 0.1v^2 - \tfrac{1}{2} \times 0.1(0) = 0.05v^2 \text{ J.}$$

Using the Principle of Conservation of Energy:

$$0.05v^2 = 0.98$$
$$\Rightarrow \qquad v = 4.43 \text{ m/s}$$
$$\therefore \ \underline{\text{velocity at B} = 4.43 \text{ m/s.}}$$

Example 10   A rough inclined plane is 13 m long and 5 m high. A body of mass m is projected with a speed of 14 m/s from the bottom of the plane up a line of greatest slope, and just reaches the top of the plane. Use the work–energy principle to calculate the frictional force, and hence find the coefficient of friction.

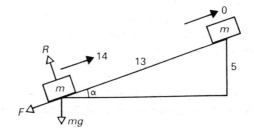

> If it *just* reaches the top of the plane then its speed at the top is zero.

$$\text{Loss in KE} = \tfrac{1}{2}m \times 14^2 = 98 \, m \, \text{J}$$
$$\text{Gain in PE} = m \times 9.8 \times 5 = 49 \, m \, \text{J}$$
$$\text{Overall loss in energy} = 49 \, m \, \text{J}$$
$$\text{Work done by friction} = \text{overall loss in energy}$$

$$\Rightarrow F \times 13 = 49 \, m$$

$$\Rightarrow \underline{\text{Frictional force} = F = \frac{49}{13} m \, \text{N}}$$

$$\sin \alpha = \frac{5}{15}$$
$$\Rightarrow \cos \alpha = \frac{12}{13} \qquad (1)$$

Resolving perpendicular to plane:

$$R = mg \cos \alpha = \frac{12}{13} mg \qquad \text{using (1)}$$

Sliding friction:

$$F = \mu R \Rightarrow \frac{49}{13} m = \mu \frac{12}{13} m \times 9.8$$

$$\Rightarrow \qquad \underline{\mu = \frac{5}{12}}$$

● **Exercise 5.2**   See page 146

5.3 Power

> **Power** = rate at which work is done = work done in one second.

Units of Power

1 Joule/second = 1 watt (W)
1000 Joules/second = 1 kilowatt (kW)

Example 11   A crane raises a load of 4000 kg at a constant speed of 0.6 m/s. At what rate is the crane's engine working?

The load rises 0.6 m in one second.

Rate of work = work done in one second
$\qquad$ = increase in PE in one second
$\qquad$ = $4000g(0.6)$ J/s
$\qquad$ = 23520 J/s = 23520 W = 23.52 kW

Example 12   A pump discharges 3000 litres of water per minute with a speed of 10 m/s. What is the power of the pump? (1 litre of water has a mass of 1 kg.)

$$\text{Volume discharged} = 3000 \text{ litres/minute}$$
$$= 50 \text{ litres/second}$$
$$\Rightarrow \text{Mass discharged} = 50 \text{ kg/second.}$$

Power = work done in one second
$\qquad$ = KE given to the water in one second
$\qquad$ = $\frac{1}{2} \times 50(10)^2$ J/s
$\qquad$ = 2500 W = 2.5 kW

Example 13   Find the power needed to raise water 4 m from a well and discharge it through a hose of cross-sectional area 50 cm² at 3 m/s. ($1000 \text{ cm}^3 = 1$ litre)

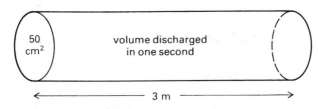

Volume discharged = 50×300 cm³/second
$\qquad$ = 15000 cm³/second
$\qquad$ = 15 litres/second
$\Rightarrow$ Mass discharged = 15 kg/second

Power = work done in one second
$\qquad$ = (increase in KE + increase in PE) per second
$\qquad$ = ($\frac{1}{2} \times 15 \times (3)^2 + 15 \times 9.8 \times 4$) J/s
$\qquad$ = 655.5 W

The Power of a Moving Vehicle

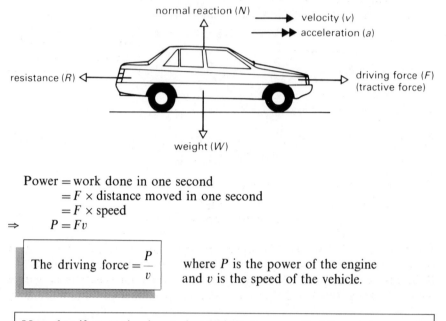

Power = work done in one second
$$= F \times \text{distance moved in one second}$$
$$= F \times \text{speed}$$
$$\Rightarrow \qquad P = Fv$$

> The driving force $= \dfrac{P}{v}$ where P is the power of the engine and v is the speed of the vehicle.

> **Note** that if an engine is rated at 20 kW, it means that its maximum rate of working is 20 kW, not that it always works at this rate.

Example 14   A cyclist racing along a level track against resistances of 50 N achieves a maximum constant speed of 10 m/s. What is the power of the cyclist?

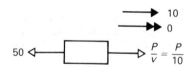

$F = ma$:

$$\frac{P}{10} - 50 = 0$$
$$\Rightarrow \text{power} = P = 500 \text{ W}$$

Example 15   A car of mass 1 tonne is driven along a level road against constant resistances totalling 250 N. If the engine is working at a steady rate of 5.5 kW, find the acceleration when the car's speed is 10 m/s and at its maximum speed.

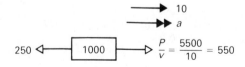

$F = ma$:
$$550 - 250 = 1000a$$
$$\Rightarrow \text{acceleration} = 0.3 \, \text{m/s}^2$$

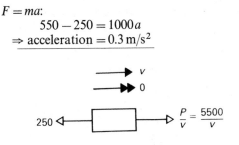

	At maximum speed, acceleration is zero.

$F = ma$:

$$\frac{5500}{v} - 250 = 0$$

$$\Rightarrow \qquad v = \frac{5500}{250} = 22$$

$$\Rightarrow \text{maximum speed} = 22 \, \text{m/s}$$

Example 16   What power is required to take a 200 tonne train at a constant speed of 10 m/s up an incline of $\sin^{-1} \frac{1}{100}$, if the resistances total $3000g$ N? If the engine works at the same rate coming *down* the same slope what would its maximum speed be?

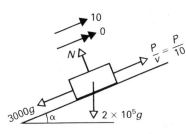

$F = ma$:

$$\frac{P}{10} - 3 \times 10^3 g - 2 \times 10^5 g \sin \alpha = 0$$

$$\Rightarrow \frac{P}{10} - 5 \times 10^3 g = 0 \leftarrow \boxed{\text{since } \sin \alpha = \frac{1}{100}}$$

$$\Rightarrow P = 5 \times 9.8 \times 10^4 \, \text{W}$$
$$\Rightarrow \text{power} = 490 \, \text{kW}$$

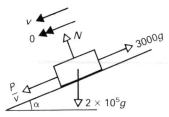

$F = ma$:

$$\frac{P}{v} + 2 \times 10^5 g \sin \alpha = 3 \times 10^3 g$$

$$\Rightarrow \frac{P}{v} = 10^3 g$$

$$\Rightarrow v = \frac{P}{10^3 g}$$

$$= \frac{5 \times 10^4 g}{10^3 g}$$

$$\Rightarrow \qquad \text{max. speed} = 50 \, \text{m/s}$$

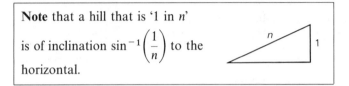

Note that a hill that is '1 in n' is of inclination $\sin^{-1}\left(\dfrac{1}{n}\right)$ to the horizontal.

•   Exercise 5.3   See page 148

5.4 Momentum and Impulse

> The **momentum** of a body of mass m, travelling with velocity **v**, is m**v**.

Since velocity is a vector quantity, momentum is also a vector quantity. The SI units of momentum are newton seconds (Ns).

> The **impulse** of a constant force **F**, acting for a time t, is **F**t.

Impulse $= \mathbf{F}t$ ⟶ | $\mathbf{F} = m\mathbf{a}$, where **a** is constant, since **F** is constant.
$\quad\quad\ = m\mathbf{a}t$ ⟵

$\quad\quad\ = m(\mathbf{v} - \mathbf{u})$ ↖
$\quad\quad\ = m\mathbf{v} - m\mathbf{u}$

| $\mathbf{v} = \mathbf{u} + \mathbf{a}t \Rightarrow \mathbf{a}t = \mathbf{v} - \mathbf{u}$

⇒ | Impulse $=$ change in momentum.

An impulsive force may act for any length of time, but, in many of the cases that we are going to consider, the impulse will be a very large force acting over a very short time, such as a hammer hitting a nail or a golf club striking a ball. The units of impulse are also newton seconds.

Example 17   A body of mass 2 kg is travelling is a straight line at 10 m/s. Find the magnitude of the change in momentum of the body when its speed changes to 15 m/s:
(a)   in the same direction   (b)   in the opposite direction.

(a)

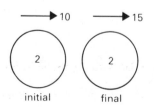

Change in momentum
$= |\text{final momentum} - \text{initial momentum}|$
$= |(2 \times 15) - (2 \times 10)|$
$= \underline{10\,\text{Ns}}$

(b)
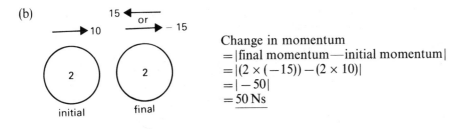

Change in momentum
= |final momentum—initial momentum|
= |(2 × (−15)) − (2 × 10)|
= |− 50|
= 50 Ns

Note that:
(i)   in the first case the change in momentum is in the initial direction of motion. Hence the impulse causing the change must also be in that direction.
(ii) in the second case the change in momentum is in the opposite direction to that of the initial motion. Hence the impulse must also be in the opposite direction.

Example 18   A golf ball of mass 45 grams is at rest on a tee. When the club strikes the ball it exerts a horizontal force on it of 6000 N for 0.0005 s. Find:
(a)   the magnitude of the impulse given to the ball
(b)   the momentum of the ball after the blow
(c)   the speed of the ball after the blow.

(a)   Impulse   = force × time
                   = 6000 × 0.0005
                   = 3 Ns

(b)   Impulse   = gain in momentum
                   = final momentum—initial momentum
$\Rightarrow$ 3   = final momentum—0
$\Rightarrow$ momentum = 3 Ns

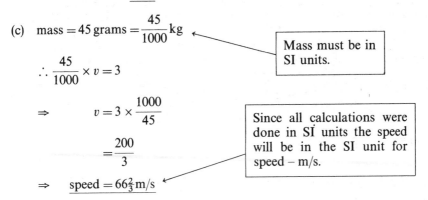

(c)   mass = 45 grams = $\frac{45}{1000}$ kg ← Mass must be in SI units.

$\therefore \frac{45}{1000} \times v = 3$

$\Rightarrow \quad v = 3 \times \frac{1000}{45}$

$\quad = \frac{200}{3}$

Since all calculations were done in SI units the speed will be in the SI unit for speed – m/s.

$\Rightarrow$ speed = $66\frac{2}{3}$ m/s

Example 19   A force of $(3\mathbf{i} + 6\mathbf{j})$ N acts on a body of mass 10 kg for 30 seconds, which was initially moving with a velocity of $(\mathbf{i} + \mathbf{j})$ m/s. Find the velocity of the body at the end of the 30 seconds and hence deduce its final speed.

Let final velocity $= (u\mathbf{i} + v\mathbf{j})$ m/s
Impulse $=$ gain in momentum

$$\Rightarrow \begin{pmatrix} 3 \\ 6 \end{pmatrix} \times 30 = 10\begin{pmatrix} u \\ v \end{pmatrix} - 10\begin{pmatrix} 1 \\ 1 \end{pmatrix}$$

$$\Rightarrow \quad 10\begin{pmatrix} u \\ v \end{pmatrix} = 30\begin{pmatrix} 3 \\ 6 \end{pmatrix} + 10\begin{pmatrix} 1 \\ 1 \end{pmatrix}$$

$$= \begin{pmatrix} 100 \\ 190 \end{pmatrix}$$

$$\Rightarrow \quad \begin{pmatrix} u \\ v \end{pmatrix} = \begin{pmatrix} 10 \\ 19 \end{pmatrix}$$

final velocity $= (10\mathbf{i} + 19\mathbf{j})$ m/s
and final speed $= \sqrt{10^2 + 19^2} \approx 21.5$ m/s

Force Exerted by Water on a Plane

When water strikes a plane without rebounding, the plane destroys the momentum of the water.

Momentum destroyed $=$ impulse $=$ force $\times$ time
$\Rightarrow$ Momentum destroyed/second $=$ force exerted by the plane

But, the force exerted by the plane on the water is equal and opposite to the force exerted by the water on the plane (Newton's Third Law).

$\Rightarrow$

> Force on plane $=$ momentum destroyed/second.

Example 20   Water is discharged from a hose at 200 litres per second with a speed of 8 m/s. The water strikes a wall at right angles and runs down the wall. Find the force exerted on the wall by the water. (1 litre of water has a mass of 1 kg.)

Volume discharged $\qquad = 200$ litres/s
$\Rightarrow$ Mass discharged $\qquad = 200$ kg/s
$\Rightarrow$ Momentum destroyed per second $= 200 \times 8$ Ns
$\Rightarrow$ Force $= 1600$ N

• Exercise 5.4   See page 149

5.5 Conservation of Momentum

Principle of Conservation of Linear Momentum

If there are no external forces acting on a system of bodies in a particular direction, the total momentum of the system in that direction remains constant.

Example 21 A bullet of mass 20 grams is fired with a speed of 400 m/s into a block of wood of mass 4 kg, resting on a smooth horizontal table. Find the common velocity of the bullet and block if the bullet becomes embedded into the block.

Momentum before impact $= 0.02 \times 400 = 8$
Momentum after impact $= 4.02 \times V$
By conservation of momentum:

$$4.02 \times V = 8$$

$$\Rightarrow \qquad V = \frac{8}{4.02} = 1.99$$

$\therefore$ Common velocity $= 1.99$ m/s.

Example 22 A pile driver of mass 5 tonnes falls through a vertical distance of 4 m onto a pile of mass 1 tonne. Find the common speed of the pile and its driver after the impact. If the pile is driven into the ground a distance of 4 cm with one blow, find the resistance of the ground.
(Let $g = 10$ m/s^2)

$u = 0$
$v = ?$
$a = 10$
$s = 4$

$v^2 = u^2 + 2as \Rightarrow v^2 = 0 + 2 \cdot 4 \cdot 10$
$\Rightarrow v = \sqrt{80} = 4\sqrt{5}$ m/s

$\therefore$ momentum of pile driver before impact $= 5000 \times 4\sqrt{5}$
momentum of pile and driver after impact $= 6000 \times V$
By conservation of momentum:

$$5000 \times 4\sqrt{5} = 6000 \times V$$
$$\Rightarrow \qquad V = 7.45 \text{ m/s}.$$

$\therefore$ Common speed of pile and driver $= 7.45$ m/s

After the blow:

$u = 7.45$
$v = 0$
$a = ?$
$s = 0.04$

$v^2 = u^2 + 2as \Rightarrow 0 = 7.45^2 + 2 \cdot a \cdot 0.04$

$$\Rightarrow a = -\frac{7.45^2}{2 \times 0.04} = -693.8 \text{ m/s}^2$$

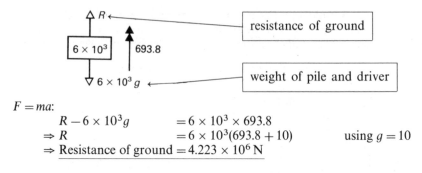

$F = ma$:

$$R - 6 \times 10^3 g = 6 \times 10^3 \times 693.8$$
$$\Rightarrow R = 6 \times 10^3 (693.8 + 10) \qquad \text{using } g = 10$$
$$\Rightarrow \text{Resistance of ground} = 4.223 \times 10^6 \,\text{N}$$

Recoil of a Gun

When a gun fires, an explosion occurs in the barrel of the gun. The explosion exerts a force on the shot and an equal and opposite force on the gun. Since both the shot and the gun are initially at rest the gain in momentum of the shot after the explosion will be equal and opposite to the gain in momentum of the gun.

Example 23   A bullet of mass 24 grams is fired from a gun with a horizontal velocity of 400 m/s. If the mass of the gun is 3 kg find the initial speed of recoil and the kinetic energy gained in the explosion.

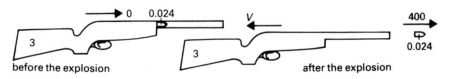

Momentum before explosion $= 0$
Momentum after explosion $= (400 \times 0.024) + 3(-V)$
By conservation of momentum:

$$400 \times 0.024 - 3V = 0$$

$$\Rightarrow \qquad V = \frac{400 \times 0.024}{3}$$

$$= 3.2$$

$\therefore$ Speed of recoil $= 3.2$ m/s

Gain in KE of gun $= \frac{1}{2} \times 3(3.2)^2 = 15.36$ J
Gain in KE of bullet $= \frac{1}{2} \times 0.024 \times 400^2 = 1920$ J
$\therefore$ total gain in KE $= 1935.36$ J

> $KE = \frac{1}{2}mv^2$ and its SI units are joules (J).

Impulsive Tensions in Strings

Example 24   Two masses of m and $3m$ are connected by a light inextensible string passing over a smooth pulley. Find the acceleration of the system. If the heavier mass hits the ground, without rebounding, after it has descended a distance 50 cm from rest find the time that elapses from this instant until the system is instantaneously at rest with the string taut.

First, think through what actually happens:
  (i)   the 3*m* mass descends a distance 0.5*m* with the acceleration of the system
  (ii) the string goes slack and the mass *m* rises and falls under gravity
  (iii) the string jerks the 3*m* mass back into motion
  (iv) the 3*m* mass rises until its velocity is zero.

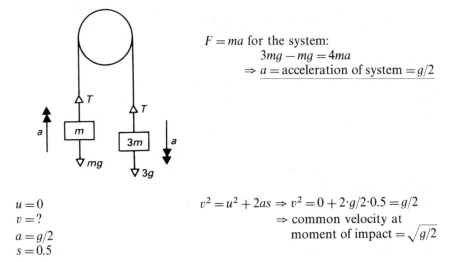

$F = ma$ for the system:
$$3mg - mg = 4ma$$
$$\Rightarrow a = \text{acceleration of system} = g/2$$

$u = 0$
$v = ?$
$a = g/2$
$s = 0.5$

$v^2 = u^2 + 2as \Rightarrow v^2 = 0 + 2 \cdot g/2 \cdot 0.5 = g/2$
$\Rightarrow$ common velocity at
moment of impact $= \sqrt{g/2}$

The 3*m* stays on the ground and the lighter mass moves upwards, under gravity with initial velocity $\sqrt{g/2}$. To find the time taken for it to return to this position, we put $s = 0$.

$u = \sqrt{g/2}$
$s = 0$
$a = -g$
$t = t_1$

$$s = ut + \tfrac{1}{2}at^2 \Rightarrow 0 = \sqrt{\frac{g}{2}}\, t_1 - \frac{1}{2}gt_1^2$$

$$\Rightarrow \sqrt{\frac{g}{2}}\, t_1 \left(1 - \sqrt{\frac{g}{2}}\, t_1 \right) = 0$$

Since $t_1 \neq 0$, $\quad t_1 = \sqrt{\dfrac{2}{g}} \approx 0.45\,\text{s}$.

When the lighter mass has completed its motion under gravity, the pulley system is jerked into motion.

Momentum before jerk $= m \times \sqrt{\dfrac{g}{2}}$

Momentum after jerk $= (m + 3m)V$

By conservation of momentum:

$$4mV = m\sqrt{\frac{g}{2}}$$

$$\Rightarrow \quad V = \frac{1}{4}\sqrt{\frac{g}{2}}$$

Both particles now move off with this velocity and under the acceleration of the system. That is, the $3m$ goes **up** but the acceleration is $g/2$ **down**.

$u = \frac{1}{4}\sqrt{g/2}$
$v = 0$
$a = -g/2$
$t = t_2$

$$v = u + at \Rightarrow 0 = \frac{1}{4}\sqrt{\frac{g}{2}} - \frac{g}{2}t_2$$

$$\Rightarrow t_2 = \frac{1}{4}\sqrt{\frac{2}{g}} \approx 0.11\,\text{s}$$

Total time to instantaneous rest $= t_1 + t_2 = 0.45\,\text{s} + 0.11\,\text{s}$
$$= 0.56\,\text{s}$$

- **Exercise 5.5**   See page 151

5.6 Direct Impact of Spheres

Direct impact occurs when the directions of the velocities of the spheres just before impact are along the line of centres on impact.

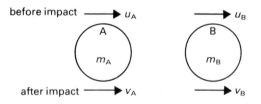

If A and B are travelling towards each other before impact then u_A and u_B will have opposite signs: one will be positive and the other negative.

Problems involving direct impact are solved using the following laws:

Principal of Conservation of Linear Momentum

Total momentum before impact = total momentum after impact.
Proof
When two bodies, A and B, collide the force exerted on A by B is equal and opposite to the force exerted by A on B, by Newton's third law. Since the period of contact is the same for the two bodies, the impulse on A is equal and opposite to the impulse on B. Thus the gain in momentum of one body is equal

to the loss in momentum of the other body and the total momentum of the two bodies is unchanged on impact.

$$m_A v_A + m_B v_B = m_A u_A + m_B u_B$$

Newton's Experimental Law of Impact

When two bodies impinge directly, the relative velocity after the impact bears a constant ratio to the relative velocity before the impact, and is in the opposite direction.

$$\Rightarrow \frac{v_A - v_B}{v_A - v_B} = -e$$

$$v_A - v_B = -e(u_A - u_B)$$

where e is the **coefficient of restitution**.
The constant e depends only on the material of which the bodies are made.

$$0 \leqslant e \leqslant 1$$

If $e = 1$ the bodies are said to be **perfectly elastic**
If $e = 0$ the bodies are said to be **inelastic**
For two glass balls $e \approx 0.9$
For two lead balls $e \approx 0.2$

Newton's Experimental Law is only approximately true and cannot be proved mathematically.

Example 25   A sphere of mass 6 kg, travelling at 4 m/s, collides directly with another sphere of mass 4 kg, travelling in the same direction at 2 m/s. If $e = \frac{1}{2}$, find the velocities after impact.

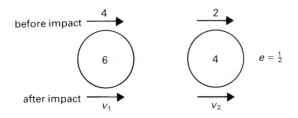

Conservation of Momentum:

$$6v_1 + 4v_2 = (6 \times 4) + (4 \times 2)$$
$$\Rightarrow 3v_1 + 2v_2 = 16 \tag{1}$$

Newton's Experimental Law:

$$v_1 - v_2 = -\tfrac{1}{2}(4 - 2)$$
$$\Rightarrow v_1 - v_2 = -1 \tag{2}$$

Solving (1) and (2) simultaneously $\Rightarrow v_1 = 2.8, v_2 = 3.8$
$\therefore$ Velocities after impact are 2.8 m/s and 3.8 m/s in the original direction

Example 26

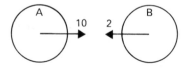

Two spheres, A and B, are moving in opposite directions in the same straight line when they collide. A has mass 4 kg and speed 10 m/s and B has mass 12 kg and speed 2 m/s. If $e = \frac{1}{3}$, find the velocities of A and B after impact.

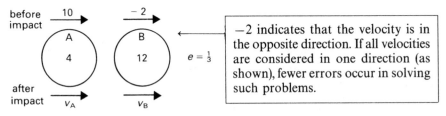

−2 indicates that the velocity is in the opposite direction. If all velocities are considered in one direction (as shown), fewer errors occur in solving such problems.

Conservation of momentum:

$$4v_A + 12v_B = (4 \times 10) + (12 \times -2)$$
$$\Rightarrow 4v_A + 12v_B = 16$$
$$\Rightarrow v_A + 3v_B = 4 \qquad (1)$$

Newton's experimental law:

$$v_A - v_B = -\tfrac{1}{3}[10 - (-2)]$$
$$\Rightarrow v_A - v_B = -4 \qquad (2)$$

Solving (1) and (2) simultaneously $\Rightarrow v_A = -2, v_B = 2$

Hence velocity of A is $\overset{2}{\leftarrow}$ and the velocity of B is $\overset{2}{\rightarrow}$
∴ after the impact both A and B move off at 2 m/s in the opposite directions to their original motion.

Example 27   Three smooth spheres, P, Q, R, of masses $3m, m, 2m$ respectively lie at rest in a straight line on a smooth horizontal table. P is projected directly at Q with a velocity u. If the coefficient of restitution is 0.5 show that Q is brought to rest after it hits R and show that, after Q collides with R for the second time, there will be no more impacts.

1st impact:

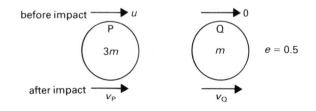

Conservation of momentum:

$$3mv_P + mv_Q = 3mu + m \cdot 0$$
$$\Rightarrow \quad 3v_P + v_Q = 3u \tag{1}$$

Newton's experimental law:

$$v_P - v_Q = -0.5(u - 0)$$
$$\Rightarrow \quad v_P - v_Q = -\frac{u}{2} \tag{2}$$

Solving (1) and (2) simultaneously $\Rightarrow v_P = \frac{5}{8}u$ and $v_Q = \frac{9}{8}u$

2nd impact:

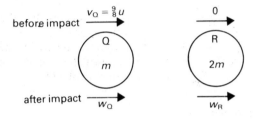

Conservation of momentum:

$$mw_Q + 2mw_R = \frac{9}{8}mv$$
$$\Rightarrow \quad w_Q + 2w_R = \frac{9}{8}u \tag{3}$$

Newton's experimental law:

$$w_Q - w_R = -0.5(\tfrac{9}{8}u - 0)$$
$$\Rightarrow \quad 2w_Q - 2w_R = -\frac{9}{8}u \tag{4}$$

Adding (3) and (4) $\Rightarrow 3w_Q = 0 \Rightarrow w_Q = 0$

$\therefore$ **Q is brought to rest after it hits R.**   QED

Also (3) $\Rightarrow w_R = \frac{9}{16}u$.

3rd impact:

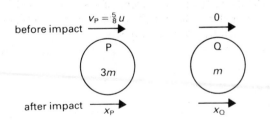

Conservation of momentum:

$$3mx_P + mx_Q = 3m \times \tfrac{5}{8}u$$
$$\Rightarrow \quad 3x_P + x_Q = \tfrac{15}{8}u \tag{5}$$

Newton's experimental law:

$$x_P - x_Q = -0.5(\tfrac{5}{8}u - 0)$$
$$= -\tfrac{5}{16}u \qquad (6)$$

(5) and (6) $\Rightarrow x_P = \tfrac{25}{64}u$, $x_Q = \tfrac{45}{64}u$

At this point, the velocity of Q is $x_Q = \tfrac{45}{64}u$ and the velocity of R is $w_R = \tfrac{9}{16}u = \tfrac{36}{64}u$. The velocity of Q is greater than the velocity of R so there will be another collision between Q and R.

4th impact:

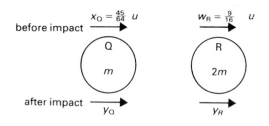

Conservation of momentum:

$$my_Q + 2my_R = m\tfrac{45}{64}u + 2m\tfrac{9}{16}u$$
$$\Rightarrow \quad y_Q + 2y_R = \tfrac{117}{64}u \qquad (7)$$

Newton's experimental law:

$$y_Q - y_R = -0.5(\tfrac{45}{64}u - \tfrac{9}{16}u)$$
$$\Rightarrow \quad y_Q - y_R = -\tfrac{9}{128}u \qquad (8)$$

(7) and (8) $\Rightarrow y_Q = \tfrac{36}{64}u$, $y_R = \tfrac{81}{128}u$

At this point, the velocity of $P = \tfrac{25}{64}u = \tfrac{50}{128}u$
the velocity of $Q = \tfrac{36}{64}u = \tfrac{72}{128}u$
and the velocity of $R = \tfrac{81}{128}u$
and, since R is moving faster than Q and Q is moving faster than P, there *will* be no more impacts.

- Exercise 5.6   See page 152

5.7 Further Impact Problems: Impact and Kinetic Energy

Kinetic Energy and Direct Impact

Unless the impact is perfectly elastic, $(e = 1)$, kinetic energy is lost because of permanent deformation of the spheres and the generation of vibrations and sound waves.

Example 28   A sphere of mass 5 kg is at rest on a smooth horizontal table when it is struck directly by a sphere of mass 3 kg travelling with a velocity of

7 m/s. After the impact the velocities of the spheres are in the ratio $3:2$. Find the velocities of the spheres after impact and the loss in kinetic energy.

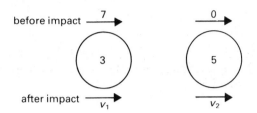

before impact

after impact v_1 v_2

Conservation of momentum:

$$3v_1 + 5v_2 = 21 \tag{1}$$

Also given:

> **Note** that, since e is not given or asked for, Newton's law cannot be used.

$$\frac{v_2}{v_1} = \frac{3}{2}$$

$$\Rightarrow \qquad v_2 = \tfrac{3}{2}v_1 \tag{2}$$

(1) and (2)

$$\Rightarrow 3v_1 + \tfrac{15}{2}v_1 = 21$$

$$\Rightarrow \qquad \tfrac{21}{2}v_1 = 21$$

$$\Rightarrow v_1 = 2 \text{ and } v_2 = 3$$

$\therefore$ velocities of the spheres after impact are $\underline{3 \text{ m/s and } 2 \text{ m/s}}$.

KE before impact $= \tfrac{1}{2} \cdot 3 \cdot 7^2 = \tfrac{147}{2}$

KE after impact $= \tfrac{1}{2} \cdot 3 \cdot 2^2 + \tfrac{1}{2} \cdot 5 \cdot 3^2 = \tfrac{57}{2}$

$\therefore$ loss in KE $= \tfrac{147}{2} - \tfrac{57}{2}$

$\phantom{\therefore \text{ loss in KE }} = \underline{45 \text{ J}}$

- **Exercise 5.7**   See page 153

5.8 Direct and Oblique Impacts of a Sphere on a Smooth Plane

Direct Impact of a Sphere on a Plane

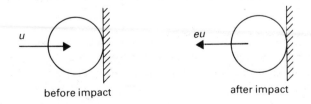

before impact   after impact

The velocity after impact is equal to $e \times$ (velocity before impact) and in the opposite direction.

Oblique Impact of a Sphere on a Plane

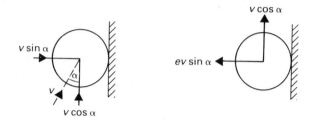

Parallel to the plane:
There is no impulse. The component of the velocity is unaltered in this direction.

Perpendicular to the plane:
The component of the velocity after impact is equal to $e \times$ (component of velocity before impact) and in the opposite direction.

Example 29   A ball falls from a height of 5 m onto a fixed horizontal plane, the coefficient of restitution being $\frac{1}{2}$. To what height does it rise after impact? (Let $g = 10 \, \text{m/s}^2$)

$u = 0$
$v = ?$
$a = g = 10$
$s = 5$

$v^2 = u^2 + 2as \Rightarrow \qquad v^2 = 2 \cdot 10 \cdot 5$
$\qquad\qquad\qquad \Rightarrow \qquad v = 10 \, \text{m/s}$
$\therefore$ velocity before impact $= 10 \, \text{m/s} \downarrow$.
$\Rightarrow$ velocity after impact $= \frac{1}{2} \cdot 10 \, \text{m/s} = 5 \, \text{m/s} \uparrow$

$u = 5$
$v = 0$
$a = -g = -10$
$s = ?$

$v^2 = u^2 + 2as \Rightarrow 0 = 25 - 2 \cdot 10 \cdot s$
$\qquad\qquad\qquad \Rightarrow s = \frac{25}{20} = \frac{5}{4}$
$\therefore$ it will rise to 1.25 m.

Example 30   A snooker ball moving with a speed of 8 m/s hits the cushion at an angle of $60°$ to the cushion. If the coefficient of restitution is $\frac{3}{4}$ find the magnitude and direction of its velocity after impact.

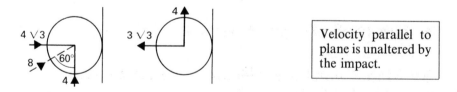

Velocity parallel to plane is unaltered by the impact.

Velocity after impact:

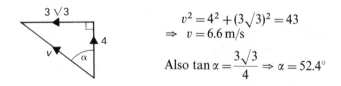

$v^2 = 4^2 + (3\sqrt{3})^2 = 43$
$\Rightarrow v = 6.6 \, \text{m/s}$

Also $\tan \alpha = \dfrac{3\sqrt{3}}{4} \Rightarrow \alpha = 52.4°$

$\therefore$ velocity after impact is 6.6 m/s at $52.4°$ to the cushion.

Example 31   A ball falls from a height of 10 m onto a plane inclined at 45°
to the horizontal. If the coefficient of restitution is $\frac{3}{7}$ find the velocity of the
ball immediately after impact.

$u = 0$

$v = ?$

$a = 9.8$

$s = 10$

$v^2 = u^2 + 2as$

$\Rightarrow v^2 = 2(9.8)\,10$

$\Rightarrow v = 14\,\text{m/s}$ just before impact.

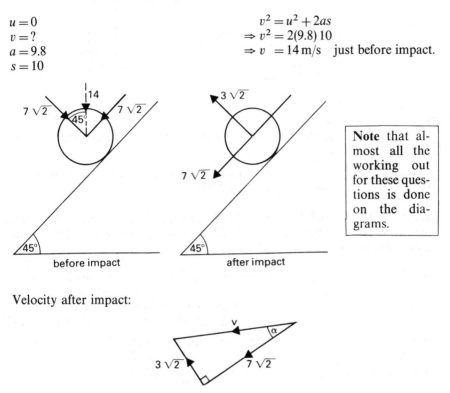

before impact   after impact

Note that al-
most all the
working out
for these ques-
tions is done
on the dia-
grams.

Velocity after impact:

$v^2 = (7\sqrt{2})^2 + (3\sqrt{2})^2$

$\Rightarrow v = 10.8\,\text{m/s}$

And $\tan\alpha = \frac{3}{7} \Rightarrow \alpha = 23.2°$

$\therefore$ velocity after impact is 10.8 m/s at 23.2° to the plane.

- Exercise 5.8   See page 154

5.9 More Complex Problems on Direct Impact

- Exercise 5.9   See page 154

5.10 + Further Impulsive Tensions: Revision of Unit:
A Level Questions

Example 32   Two small balls, A and B, of masses 3 kg and 2 kg respectively,
lie at rest on a smooth horizontal surface and are connected by a taut
inextensible string.

B is given an impulse **J** in a direction at 60° to $\overrightarrow{AB}$. Find the velocities with which A and B start to move and find the impulsive tension in AB.

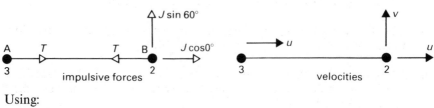

Using:
   Impulse = change in momentum for whole system

(i)   perpendicular to AB: $\dfrac{J\sqrt{3}}{2} = 2v \Rightarrow v = \dfrac{J\sqrt{3}}{4}$

(ii)   parallel to AB: $J\dfrac{1}{2} = 5u \Rightarrow u = \dfrac{J}{10}$

   $\Rightarrow$ Speed of A $= u = \dfrac{J}{10}$

   and speed of B $= \sqrt{u^2 + v^2} = J\sqrt{\dfrac{3}{16} + \dfrac{1}{100}} = \dfrac{J\sqrt{79}}{20}$

$\therefore$ velocity of A after the jerk $= \dfrac{J}{10}$ in the direction of AB

and velocity of B after the jerk $= \dfrac{J\sqrt{79}}{20}$ at $\tan^{-1} 5\sqrt{3}$ to AB

Using:
   Impulse = change in momentum of A
$\Rightarrow T = 3u$
$\Rightarrow$ Impulsive tension $= \dfrac{3J}{10}$

● **Miscellaneous Exercise 5B**   See page 156

Unit 5 Exercises

Exercise 5.1

1   Find the work done against gravity when a body of mass 3 kg is raised through a vertical distance of 5 m.

2 Find the work done when a body of mass 2 kg is lowered through a vertical distance of 3 m.

3 Find the work done when a woman of mass 60 kg climbs a ladder onto a scaffolding platform 10 m above the base of the ladder.

3 A bricklayer lifts 20 bricks a vertical distance of 1.5 m. If each brick has a mass of 4 kg, how much work is done against gravity?

5 A brick of mass 4 kg is moved upwards at a constant speed of 3 m/s. How much work is done against gravity per second?

6 A box is pulled with a constant velocity across a horizontal surface for a distance of 4 m. If the total resistance to motion is 9 N, find the work done against the resistance.

7 A box of mass 6 kg is pulled with a constant velocity across a rough horizontal surface for a distance of 4 m. If the only resisting force is the friction and the coefficient of friction is $\frac{1}{3}$, find the work done.

8 A smooth plane is inclined at 30° to the horizontal. A body of mass 5 kg slides 12 m down a line of greatest slope. Find the work done by gravity.

9 Find the kinetic energy of:
  (a) a body of mass 10 kg moving with a speed of 3 m/s
  (b) a cricket ball of mass 400 grams moving with a speed of 15 m/s
  (c) a lorry of mass 3 tonnes moving with a speed of 30 m/s
  (d) a particle of mass 0.1 gram moving with a speed of 100 m/s.

10 Find the potential energy gained when:
  (a) a child of mass 45 kg climbs from the ground up a ladder to the top of a slide 3 m above the ground
  (b) a golf ball of mass 100 grams is hit from the tree and reaches a height of 36 m.

11 Find the potential energy lost when:
  (a) a child of mass 40 kg slides down a slide to the ground from a height of 4 m
  (b) a diver of mass 65 kg dives from a platform 10 m above the water to a depth of 2 m.

12 Find the gain in kinetic energy when:
  (a) a car of mass 1.2 tonnes increases its speed from 10 m/s to 15 m/s
  (b) a ball of mass 120 grams increases its speed from 5 m/s to 20 m/s.

13 Find the loss of kinetic energy when a lorry of 3.5 tonnes brakes from 72 km/h to rest.

14 A body of mass 4 kg increases its kinetic energy by 54 J. Find its final speed if its initial speed is 3 m/s.

15 A car of mass 1 tonne accelerates at $1.5\,\text{m/s}^2$ for 10 seconds from rest. Find the gain in kinetic energy.

16 A body of mass 8 kg is dragged a distance of 6 m across a rough horizontal floor at constant speed. If the work done against friction is 47.04 J find the coefficient of friction.

17 A girl pulls a sledge 50 m along horizontal ground at a constant speed by a rope inclined at 60° to the horizontal. If the tension in the rope is 120 N, how much work does the girl do?

18 A father pushes a pram with a force of 50 N up a slope inclined at 60° to the horizontal. Find the work done after he has pushed it at a constant speed for half a kilometre.

19 A body of mass 20 kg is pulled 10 m up a line of greatest slope of a rough plane inclined at $\tan^{-1}\frac{3}{4}$ to the horizontal. If the coefficient of friction is $\frac{2}{7}$, find:

(a) the work done against gravity
(b) the work done against friction.

20 A body of mass m is pulled at a constant speed up a line of greatest slope of a rough plane inclined at θ to the horizontal. If the only resistances to motion are gravity and friction, show that the total work done on the body is

$$mg(\mu \cos \theta + \sin \theta)x$$

where x is the distance moved up the slope.

Exercise 5.2

1 A force of 10 N pushes a body of mass 6 kg in a straight line across a horizontal smooth surface from A to B, increasing its speed from 5 m/s to 10 m/s. Find:

(a) the gain in kinetic energy
(b) the work done by the force
(c) the distance from A to B.

2 A body of mass 10 kg travels from A to B in a horizontal straight line. Its motion is resisted so that it slows from 8 m/s at A to 3 m/s at B. Find:

(a) the loss in kinetic energy
(b) the work done against the resistance
(c) the magnitude of the resistance, given that it is constant and that A and B are 25 m apart.

3 P and Q are points 6 m apart on a horizontal smooth surface. A force of 100 N pushes a body of 12 kg from P to Q in a straight line. At P the speed of the body is 5 m/s. Find:

(a) the work done by the force

(b) the gain in kinetic energy

(c) the speed of the body at Q.

4   A constant force pushes a body of mass 4 kg up the line of greatest slope of a plane inclined at $\sin^{-1} \frac{3}{8}$ to the horizontal. P and Q are two points in its path. When it is at P its speed is 1 m/s and when it is at Q its speed is 5 m/s. P and Q are 4 m apart. Find:

(a) its increase in kinetic energy

(b) its increase in potential energy

(c) the work done by the force

(d) the magnitude of the force.

5   If a shot-putter raises a shot of mass 5 kg a distance of 1.5 m and then projects it with a speed of 15 m/s, calculate the total work done.

6   In a curling match, the stone, which has a mass of 18 kg, is given an initial speed of 10 m/s. It comes to rest when it has travelled 20 m across the ice. Find:

(a) the loss of kinetic energy

(b) the work done against the friction

(c) the coefficient of friction.

(Let $g = 10$ m/s^2)

7   A one kilogram bag of sugar falls from a shelf to the ground 90 cm below. Find the loss in potential energy and use the Principle of Conservation of Energy to calculate its speed when it hits the ground.

8   A bead of mass 200 grams is threaded onto a smooth wire in the shape of a circle of radius 35 cm. If the plane containing the wire is vertical and the bead is gently displaced from the highest point on the wire, find its maximum speed in the ensuing motion.

9   A pellet of mass 20 grams is projected vertically downwards into a viscous liquid with an initial speed of 12 m/s. When it has travelled a distance of 60 cm its speed has fallen to 2 m/s. What is the magnitude of the resistance of the liquid?

(Let $g = 10$ m/s^2)

10   A bullet of mass 12 grams is fired horizontally into a fixed wooden block at a speed of 400 m/s, and penetrates a distance of 20 cm. Find the loss of kinetic energy of the bullet and the resistance of the wood.

11   A bullet of mass 10 grams is fired horizontally into a fixed block which provides a constant resistance of 2000 N. If the speed of the bullet on impact is 360 m/s, find how far into the block the bullet will penetrate.

12   A body of mass 4 kg slides from rest down a smooth plane inclined at 30° to the horizontal. By considering the changes of energy find the speed of the body after it has travelled 6 m down a line of greatest slope.

13 A rough slope is inclined at $\tan^{-1}\frac{3}{4}$ to the horizontal. A box of mass 20 kg slides from rest a distance of 15 m down a line of greatest slope, at which point it has developed a speed of 12 m/s. Find the work done against friction and the coefficient of friction.

14 A bullet of mass 15 grams is fired horizontally into a vertical target of 20 cm thickness with a speed of 300 m/s. If the target offers a resistance of 3300 N, with what speed will the bullet emerge on the far side?

*15 A body of mass m is projected up the line of greatest slope of a rough plane which is inclined at α to the horizontal. The body travels a distance d before coming to rest. Show that the frictional force is $\mu mg \cos \alpha$ where μ is the coefficient of friction and find the work done by the friction. Show that the initial speed of the body is given by v where

$$v^2 = 2gd(\mu \cos \alpha + \sin \alpha)$$

Exercise 5.3

1 A monkey of mass 5 kg climbs 10 m in 5 seconds. At what rate is the monkey working?

2 An electrical pump discharges 1200 litres of water per minute at a speed of 5 m/s. Find the power required to do this.

3 Find the average rate of work of a mountaineer who climbs 240 m in an hour if the mass of the mountaineer and his equipment is 100 g.

4 A schoolboy designs a mechanical catapult that shoots pellets of mass 5 g at a rate of 36 per minute giving each pellet a speed of 10 m/s. What is the power of the catapult?

5 In one minute a pump raises 3 m³ of water from a well and ejects it 2 m above the water level with a speed of 4 m/s. Find the power required to do this.
($1000 \text{ cm}^3 = 1$ litre and 1 litre has a mass of 1 kg)

6 A train travels along a level track against constant resistances of 2000 N. Find the maximum speed possible if the engine works at a constant rate of 40 kW.

7 A car maintains a constant speed of 25 m/s on a horizontal road against constant resistances of 500 N. What is its tractive force? At what rate is its engine working?

8 A car of mass 1200 kg is driven along a level road against resistances totalling 300 N. If the engine works at a steady rate of 10 kW, find the acceleration of the car when its speed is 20 m/s and find its maximum speed.

9 A car of mass 1 tonne is travelling at a constant speed of 10 m/s up an incline of $\sin^{-1} \frac{1}{20}$. If its engine is working at 8 kW find the magnitude of the resistance to its motion.

10 A cyclist working at a rate of 400 W cycles up a hill of 1 in 40 against a resistance of 20 N. Find his acceleration when his speed is 5 m/s if the combined mass of the cyclist and his bike is 80 kg. What would his acceleration be if he cycled at the same rate *down* the hill?

11 The engine of a lorry of mass 5 tonnes can develop a constant 80 kW of power. If its maximum speed on a flat road is 120 km/h, find the magnitude of the resistance to its motion. If this resistance is constant find the acceleration of the lorry when it is travelling at a speed of 60 km/h, if the engine works at the same rate.

12 The combined mass of a cyclist and his machine is 75 kg and his maximum power is 400 W. If his greatest speed on a flat road is 8 m/s find his greatest speed up an incline of 1 in 10, assuming that the resistances remain the same.
(Let $g = 10 \text{ m/s}^2$)

13 A pump draws water from a tank, raises it a distance of 4 m and then ejects it from a nozzle with a cross-sectional area of 20 cm². Find the power required by the pump to raise the water and eject it with a speed of 5 m/s.

14 In every hour a pump draws 180 m³ of water from a well, raises it 4 m and issues it through a pipe of cross-section 50 cm².
  (a) Find the speed with which the water issues from the pipe.
  (b) Find the power required to do this.
  (c) The pump is inefficient and 20% of its power is wasted. Find the rate of work of the pump.

*15 The resistance to the motion of a lorry of mass M kg is proportional to the speed of the lorry. If the engine is working at P W, the lorry can reach a maximum speed V up an incline of 1 in n. Show that the resistance at the speed V is
$$\frac{P}{V} - \frac{Mg}{n}$$
and find the acceleration when the speed is $V/2$.

Exercise 5.4

1 Find the momentum, in newton seconds, of each of the following:
  (a) a car of mass 900 kg moving at 25 m/s
  (b) a bullet of mass 25 g moving at 500 m/s
  (c) an athlete of mass 75 kg running at 7 m/s
  (d) a ball of mass 50 g moving at 20 m/s
  (e) a ship of mass 800 tonnes moving at 0.2 m/s
  (f) a particle of mass 0.001 g moving at 1000 m/s
  (g) a car of mass 700 kg moving at 90 km/h.

2 A body of mass 4 kg receives an impulsive blow. Find the magnitude of the impulse if:

(a) the body continues to move in the same direction but its speed changes from 3 m/s to 5 m/s

(b) the body continues to move in the same direction but its speed changes from 5 m/s to 3 m/s

(c) the body's direction is reversed and its speed changes from 5 m/s to 3 m/s

(d) the body's direction is reversed and its speed changes from 1 m/s to 9 m/s.

3 A car of mass 900 kg is pushed along a level road for one minute and acquires a speed of 4 m/s. What constant force was applied?

4 A body of mass 5 kg is at rest when a force of 20 N is applied to it for 10 s.

Find: (a) the magnitude of the impulse

(b) the momentum of the body after the force has been apllied

(c) the final speed of the body.

5 A body of mass 5 kg is moving in the direction AB with a speed of 15 m/s when a force of 20 N is applied to it for 10 s. The force is also in the direction AB. Find the final speed of the body.

6 A cricket ball of mass 160 g strikes the bat at 5 m/s and leaves the bat in the opposite direction at 11 m/s. What is the impulse received by the ball?

7 A smooth sphere of mass 3 kg strikes a vertical wall in a direction normal to the wall. Before the impact the sphere was travelling at 5 m/s and after the impact at 4 m/s in the opposite direction. Find:

(a) the impulse of the wall on the sphere

(b) the impulse of the sphere on the wall.

8 A body of mass 8 kg is moving with a velocity of 3 m/s parallel to the positive direction of the x-axis. An impulse is applied to the body, after which its velocity is 7 m/s in the same direction. Find the magnitude and direction of the impulse.

9 A body of mass 10 kg is moving with a velocity of 4**i** m/s when it is given an impulse of 20**i** Ns. Find the final velocity of the body.

10 A body of mass 3 kg was travelling with a velocity of 5**i** m/s when it received a blow which changed the velocity to 3**i** m/s. What was the impulse of the blow?

11 A body of mass 8 kg is moving with a velocity of 5**i** m/s. A force of 2**i**/N acts on the body for t seconds, after which the body has a velocity of 12**i** m/s. Find the value of t.

12 A body of mass 3 kg is at rest when it receives an impulse $(12\mathbf{i} - 9\mathbf{j})$ Ns. Find the final velocity of the body and hence obtain its final speed.

13 A force of $(10\mathbf{i} + 25\mathbf{j})$N acts on a body of mass 5 kg for 10 s that was initially moving with a velocity of $(2\mathbf{i} + 3\mathbf{j})$ m/s. Find the final velocity of the body.

14 A body of mass 500 g is moving with a velocity of $(5\mathbf{i} - 2\mathbf{j})$ m/s. A force of $(p\mathbf{i} + q\mathbf{j})$N acts on the body for 10 seconds, after which the body has a velocity of $(25\mathbf{i} + 58\mathbf{j})$ m/s. Find p and q.

15 A water cannon emits water through a nozzle with cross-sectional area 200 cm^2 at a speed of 15 m/s. Find the force of the water on a rioter, if the water is aimed straight at him.

16 Water is squirted through the nozzle of a water pistol at a speed of 5 m/s and the nozzle has a cross-sectional area of 2 mm^2. If the water hits the face of the victim at right angles and does not rebound, find the magnitude of the force of the water on the face.

*17 In a factory, a pipe bursts and alcohol is forced out in a horizontal direction through an aperture of area 3 cm^2 with a speed of 12 m/s. Find the force with which it strikes a vertical wall adjacent to the aperture. The density of alcohol is 0.8 grams/cm^3.

Exercise 5.5

1 An empty goods truck of mass M is sitting in a siding, but is free to move. A full goods truck of mass $2M$ travelling at 3 m/s runs into the first truck and is automatically coupled to it. What is their common speed after the impact?

2 A bullet of mass 40 grams is fired with a velocity of 820 m/s into a block of wood of mass 6 kg resting on a smooth horizontal surface. If the bullet becomes embedded in the block, find the common velocity of the bullet and the block.

3 A pile driver of mass 3 tonnes falls through a height of 5 m onto a pile of mass 0.5 tonne. What is the momentum of the pile driver just before impact? After the impact the pile and the driver move together. Find their common speed.
(Take $g = 10$ m/s^2)

4 A bullet of mass 50 g is fired from a gun with a horizontal velocity of 500 m/s. If the gun has mass 2.5 kg find the initial speed of recoil of the gun.

5 A bullet is fired horizontally at 700 m/s into a block of wood of mass 0.275 kg which is resting on a smooth horizontal surface. The bullet becomes embedded in the block and the block starts to move with a speed of 50 m/s. Find the mass of the bullet.

6 A ball A of mass 0.4 kg is moving at 5 m/s when it collides directly with a stationary ball B of mass 0.1 kg. After the collision, A continues to

move in the same direction but with a speed 3.5 m/s. Calculate the speed of B after the collision and the loss of kinetic energy due to the collision.

7   Two marbles roll towards each other along a smooth horizontal groove. A has mass $2m$ and speed 4 m/s. B has mass $3m$ and speed 6 m/s. After the collision A's direction of motion is reversed and its speed is 3.5 m/s. Find B's velocity after the collision.

8   A hammer of mass 1.2 kg knocks a nail, whose mass is 20 grams, into a piece of wood. Just before the impact the speed of the hammer is 10 m/s. After the impact the hammer remains in contact with the nail. Find the speed with which the nail begins to penetrate the wood. If the wood offers a constant resistance of 7000 N, how far will the nail penetrate with each blow. (Take $g = 10\,\text{m/s}^2$.)

9   A shell of mass 10 kg is fired horizontally from a gun of mass 2 tonnes with a speed of 400 m/s. Find the velocity of recoil of the gun. The recoil is resisted by a constant force so that the gun moves back only 16 cm. Find the magnitude of this force.

*10   Two masses of 1.2 kg and 1 kg are connected by a light string passing over a smooth fixed pulley. The system is allowed to move from rest and the 1 kg mass, after it has risen 11 cm, passes through a fixed ring, without touching it, and picks up a mass of 500 grams which rises with the 1 kg mass. Show that, after the 500 g mass is jerked into motion, it will be carried to almost 6 cm above the ring. (Take $g = 10\,\text{m/s}^2$)

*11   A block of wood of mass 0.5 kg rests on a rough horizontal floor which has coefficient of friction 0.4. A bullet of mass 40 grams is fired into the block with a velocity of 150 m/s and becomes embedded in the block. Find the velocity with which the block and bullet begin to move after the impact and the distance which the block moves along the floor. Show that the ratio of the energy lost during the impact to that lost through friction is approximately 12.5:1.

Exercise 5.6

1   A sphere of mass 10 kg moving at 8 m/s collides directly with another sphere of mass 8 kg moving at 5 m/s in the same direction. If $e = \frac{1}{2}$ find the velocities after impact.

2   A sphere of mass 10 kg moving at 5 m/s collides directly with another sphere of mass 3 kg moving at 2 m/s in the same direction. If $e = \frac{1}{3}$, find the velocities after impact.

3   A sphere of mass 5 kg moving at 8 m/s collides directly with another sphere of mass 4 kg moving at 4 m/s in the opposite direction. Find the velocities after impact if: a) $e = \frac{2}{3}$ b) $e = \frac{1}{3}$.

4   A sphere of mass 1 kg moving at 7 m/s overtakes another sphere of mass 2 kg moving in the same direction at 1 m/s. If the first sphere is brought to rest by the impact what is the value of the coefficient of restitution?

5   A sphere of mass 4 kg, travelling at 20 m/s on a smooth horizontal surface, overtakes another sphere of mass 3 kg moving in the same direction at 15 m/s. Ten seconds after the impact the 3 kg sphere hits an obstacle and is brought to rest immediately. If the coefficient of restitution is $\frac{4}{5}$, show that the time that will elapse, after the impact of the spheres, until the 4 kg mass hits the 3 kg mass again, is approximately 12.5 seconds.

6   Three spheres, A, B, C, of masses $3m$, $2m$, $2m$ respectively lie in a straight line on a smooth horizontal table. If the coefficient of restitution is $\frac{1}{4}$ and if A is projected directly towards B with velocity u, show that there are only three impacts and that the final velocities of the spheres are

$$\frac{25u}{64}, \frac{57u}{128}, \frac{15u}{32}$$

*7   Three balls, A, B, C, of masses m_1, m_2, m_3, respectively, are lying at rest in a straight line on a smooth horizontal table. A is projected directly towards B with velocity v_1. After this impact, B collides with C. If the coefficient of restitution between A and B is e, and the coefficient of restitution between B and C is e', find the velocities of all three balls immediately after these two impacts.

Exercise 5.7

1   Two spheres of masses 2 kg and 3 kg are moving directly towards each other with velocities 8 m/s and 10 m/s respectively. If $e = \frac{3}{4}$, find their velocities after impact and the amount of kinetic energy lost in the collision.

2   Two trucks of masses 5 tonnes and 3 tonnes are free to move on a horizontal set of rails. The lighter truck is at rest when it is struck by the heavier truck travelling at 1.5 m/s. After the impact, the difference in the speeds of the trucks is 0.9 m/s. Find the speeds of the trucks after impact and calculate the kinetic energy lost in the collision.

3   A, B, C are three spheres of equal mass which are at rest in a straight smooth horizontal tube. A is projected towards B with speed u and, after the impact, B collides with C. Show that, after two impacts have taken place, the velocities of the spheres are

$$\tfrac{1}{2}(1 - e)u, \quad \tfrac{1}{4}(1 - e^2)u \text{ and } \tfrac{1}{4}(1 + e)^2 u$$

where e is the coefficient of restitution.

4   Three particles A, B, C with masses $10m$, $5m$, $3m$ respectively, lie at rest in a straight line on a smooth horizontal table. The particle A is projected horizontally with speed u and undergoes a direct perfectly elastic collision with B. Show that B acquires speed $4u/3$.
Subsequently B is in direct perfectly elastic collision with C. Show that immediately after this collision the velocities of A and B are equal.

[J]

5   A particle of mass m moving with speed u collides directly with an equal

particle at rest. The coefficient of restitution is e. Show that the overall loss of kinetic energy is

$$\frac{mu^2}{4}(1-e^2)$$ [J]

Exercise 5.8

1 A sphere strikes a fixed wall normally with a speed of 8 m/s. It rebounds with a speed of 6 m/s. Find the coefficient of restitution.

2 A ball falls from a height of 10 m onto a fixed horizontal plane. If it rebounds to a height of 3.6 m, find the coefficient of restitution.

3 A snooker ball moving with a speed of 10 m/s hits the cushion at an angle of 45° to the cushion. Find its velocity after the impact if the coefficient of restitution is 0.8.

4 A billiard ball of mass 200 grams moving at 4 m/s strikes the cushion at an angle of 30° to the cushion. If the coefficient of restitution is 0.9, find the loss of kinetic energy due to the impact.

5 A ball of mass 1 kg falls from a height of 2.5 m onto a plane inclined at 60° to the horizontal. If the coefficient of restitution is $\frac{2}{7}$, find the magnitude and direction of the velocity of the ball after impact and the loss of kinetic energy.

6 A ball falls from a height of 8 m onto horizontal ground. If the coefficient of restitution is $\frac{3}{4}$ find the height to which it rises after: (a) the first impact (b) the second impact.
How much time elapses between the two impacts?
(Use $g = 10 \, \text{m/s}^2$)

7 A billiard ball hits the edge of the table at an angle α to the edge. It rebounds in a direction making an angle β with the edge. Show that

$$\tan \beta = e \tan \alpha$$

where e is the coefficient of restitution.

*8 A ball falls vertically and strikes a fixed plane inclined at an angle θ ($\theta < 45°$) to the horizontal. The coefficient of restitution is $\frac{3}{5}$ and the ball rebounds horizontally. Show that $\theta = \tan^{-1} \sqrt{3/5}$ and that the fraction of kinetic energy lost is $\frac{2}{5}$.

Exercise 5.9

1 (a) Two particles A and B of masses $3m$ and $5m$ respectively are placed on a smooth horizontal table. The particle A is projected along the table with speed u directly towards B, which is at rest. After impact A continues to move in the same direction but with speed $u/6$.

Find:

   (i)   the speed of B after impact
   (ii) the coefficient of restitution between A and B
   (iii) the loss in kinetic energy due to the collision.

(b) The same particles A and B are now connected by a light inextensible string and are placed side by side on the smooth horizontal table. The particle A is projected horizontally, directly away from B, with a speed *v*. Calculate:

   (i)   the resulting common speed of the two particles after the string has tightened
   (ii) the magnitude of the impulse in the string when the string tightens
   (iii) the loss in kinetic energy due to the tightening of the string.
                                                                           [A]

2   Two particles, A of mass 2*m* and B of mass *m*, moving on a smooth horizontal table in opposite directions with speeds 5*u* and 3*u* respectively, collide directly. Find their velocities after the collision in terms of *u* and the coefficient of restitution *e*. Show that the magnitude of the impulse exerted by B on A is

$$\frac{16}{3}mu(1+e)$$

Find the value of *e* for which the speed of B after the collision is 3*u*. Moving at this speed B subsequently collides with a stationary particle C of mass *km*, and thereafter remains attached to C. Find the velocity of the combined particle and find the range of values of *k* for which a third collision will occur.
                                                                           [J]

3   A thin straight horizontal wire is smooth on one side of a point O of the wire and rough on the other side of O. Two spherical beads A and B, of masses 0.2 kg and 0.3 kg respectively, are threaded on the smooth part of the wire with bead B between bead A and O. Both beads are travelling towards O, A with speed 6 m/s and B with speed 1 m/s. Given that the beads collide whilst they are still on the smooth part of the wire, and that the coefficient of restitution is $\frac{1}{3}$, find the speed of each bead immediately after the impact and show that both beads continue moving towards O.
On the rough part of the wire the coefficient of friction between the wire and each bead is $\frac{1}{2}$. Find the distance between the beads when they have both come to rest. Find also the total work done by the frictional forces in reducing the beads to rest. (Take *g* as 10 m/s^2.)   [L]

4   State the law of conservation of momentum for two interacting particles which move in the same straight line, and derive the law from Newton's laws of motion.
Three identical smooth spheres A, B, C, each of mass *m*, lie on a smooth horizontal table in a straight line in the order A, B, C, with B closer to A than to C. The spheres A and C are simultaneously projected with speed *u* directly towards B. Given that the coefficient of restitution at each collision is *e*, find the velocity of each of the spheres just after the second collision.

Find also the magnitude and direction of the impulse exerted on B at the second collision.   [J]

5   Three smooth spheres A, B, C, of equal radii and masses m, λm, $\lambda^2 m$ respectively, where λ is a constant, are free to move along a straight horizontal groove with B between A and C. When any two spheres collide the impact is direct and the coefficient of restitution is e. Spheres B and C are initially at rest and sphere A is projected towards sphere B with speed u. Show that the velocities of A and B after the first impact are

$$\frac{1 - \lambda e}{1 + \lambda} u \text{ and } \frac{1 + e}{1 + \lambda} u \text{ respectively.}$$

Find the velocities of B and C after the second impact.
Given that $\lambda e < 1$ show that there is a third impact if $e < \lambda$.   [L]

6   A smooth hemispherical bowl is fixed with its circular rim horizontal and uppermost. Two particles A and B, of masses $2m$ and m respectively, are released from rest at opposite ends of a diameter of the rim. The particles collide at N, the lowest point of the bowl, when each is moving with speed u. The particle A is brought to rest by the impact. Show that the coefficient of restitution between the particles is $\frac{1}{2}$ and find the speed of B immediately after the collision.

Also find, in terms of m and u;
(a)   the magnitude of the impulse of the blow received by A
(b)   the kinetic energy lost in this collision.

Show that, immediately after the second collision of the particles, the speed of A is $\frac{1}{2}u$. By considering conservation of energy, find, in terms of u and g, the height above N to which A rises after this second collision.   [L]

Miscellaneous Exercise 5B

1   Two particles of equal mass lie at rest on a smooth horizontal table and are connected by a light taut inelastic string. An impulse **J** is applied to one of the particles in a direction that makes an angle α with the line of the particles and keeps the string taut. Show that the impulsive tension in the string is equal to $\frac{1}{2}J/\cos \alpha$.   [L: 5.10]

2   Three particles A, B and C of masses m, $2m$ and $3m$, respectively lie in that order in a straight line on a smooth horizontal plane. A light inextensible string of length $2a$ connects A and B and another such string connects B and C. Initially, the strings are slack and the particles are at rest with $AB = BC = a$. An impulse, of magnitude J, is then applied to A in the direction BA. Find, in terms of J, the impulse in the string AB when the particle B is jerked into motion. Show that the time taken for A to travel a distance $4a$ is $16ma/J$.   [J: 5.10]

3   A pump, working at an effective rate of 41 kW, raises 80 kg of water per second from a depth of 20 m. Calculate the speed with which the water is delivered. (Take g as 10 m/s².)   [L: 5.3]

4 Two particles of masses m and $2m$ are on a smooth horizontal table and are connected by a light inextensible string. Initially the string is slack and the particles are each moving with speed u along the same straight line and away from each other. When the string tightens, the particles begin to move in the same direction with speed v. Find v in terms of u and show that the loss of kinetic energy when the string tightens is

$$\tfrac{4}{3}mu^2 \qquad\qquad [\text{J}: 5.7, 5.9]$$

5 A particle of mass $4m$, which is at rest, explodes into two fragments, one of mass m and the other of mass $3m$. The explosion provides the fragments with total kinetic energy E. Find the velocities of the fragments just after the explosion. Hence find, in terms of m and E, the magnitude of the relative velocity of the fragments just after the explosion.

[J: 5.2, 5.5]

6 Two particles of masses $4m$ and $3m$ respectively are attached one to each end of a light inextensible string which passes over a small smooth pulley. The particles move in a vertical plane with both the hanging parts of the string vertical. Write down the equation of motion for each of the particles and hence determine, in terms of m and/or g as appropriate, the magnitude of the acceleration of the particles and of the tension in the string.
When the particle of mass $3m$ is moving upwards with a speed V it picks up from rest at a point A an additional mass $2m$ so as to form a composite particle Q of mass $5m$. Determine:

(a) the initial speed of the system
(b) the impulsive tension in the string immediately the additional particle has been picked up
(c) the height above A to which Q rises.   [A: 5.5]

7 A sphere of mass $3m$ is moving with speed $2u$ when it collides directly with another sphere, of the same radius but of mass m, which is moving in the opposite direction with speed u. The coefficient of restitution between the spheres is $\tfrac{1}{3}$. Calculate:

(a) the speed of each sphere immediately after impact
(b) the magnitude of the impulse received by each sphere on impact.

[L: 5.4, 5.6]

8 A motor truck of mass 15 tonnes travels up a straight slope of gradient $\sin^{-1}(0.05)$ against a resistance of 0.1 N per kilogram. Find the tractive force required to produce an acceleration of $0.1\,\mathrm{m\,s^{-2}}$ and the power in kilowatts which is then developed when the speed is 8 m/s.
Find the steady speed the truck can sustain on the same upward slope if the power exerted by the motor is 135 kW.

(1 tonne $= 1000\,\mathrm{kg}$. Take the acceleration due to gravity as $10\,\mathrm{m\,s^{-2}}$.)

[J: 2.8, 5.3]

9 Two particles, P and Q, of masses $2m$ and m, respectively, are connected by a light inextensible inelastic string of length $2a$ and placed on a smooth

horizontal table with P at the edge of the table and Q at a distance a from the edge. The height of the table is greater than $2a$ and the line joining P and Q is perpendicular to the edge. The particle P is then gently pushed over the edge. Show that the speed of the particles just after the string tightens is $\frac{2}{3}\sqrt{(2ga)}$. Find, in terms of m, g and a, the kinetic energy of the system when Q reaches the edge of the table.   [J: 5.1, 5.4, 1.4]

10 A smooth fixed plane is inclined at an angle of 30° to the horizontal. Two particles, P and Q, each of mass m, are connected by a light inextensible string of length $2l$. Initially P is held at a point A, at the top edge of the inclined plane, and Q rests on a horizontal platform at a point B, which is at a distance l vertically below A. Given that P is projected with speed $\sqrt{(gl)}$ down a line of greatest slope of the inclined plane, find the impulsive tension in the string when Q is jerked into motion.

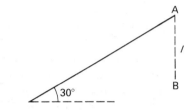

Find also the tension in the string in the subsequent upward motion of Q.   [J: 5.4]

11 A vehicle of mass 5000 kg is working at a constant rate of 40 kW while climbing a hill inclined at an angle $\sin^{-1}\frac{1}{25}$ to the horizontal. The resistance to motion is not constant. The speed increases from $2\,\text{m s}^{-1}$ to $5\,\text{m s}^{-1}$ in $4s$ whilst the vehicle covers a distance of 14m along the hill. Taking g to be 10m/s^2 and using energy considerations, find the work done against the resistance during this 4 second period.   [A: 5.1, 5.2]

12 Three identical particles A, B, C, lie at rest in a straight line on a smooth horizontal table, with B between A and C. The particle A is projected horizontally with speed u directly towards B. The coefficient of restitution at each collision is e $(0 < e < 1)$. Find the velocity of each of the particles just after C is set in motion.
Show that A strikes B a second time.   [J: 5.6]

13 The magnitude of the air and road resistance on a vehicle of mass 6000 kg is cv N where c is a constant and v m/s the speed of the vehicle. The maximum speed of the vehicle on a level road when working at a rate 40 kW is 10 m/s. Find c.
The vehicle now ascends a hill inclined at an angle $\sin^{-1}(\frac{1}{20})$ to the horizontal. It continues to work at a constant rate of 40 kW. Find the acceleration, in m/s^2, of the vehicle at the instant when the speed is 5 m/s. During this ascent the speed increased from 3 m/s to 5 m/s in a time 2.36s whilst the vehicle covers a distance of 9.87 m. Find, in kJ to one decimal place, the work done against the air and road resistance during this part of the ascent. (Take the acceleration due to gravity to be $10\,\text{m/s}^2$.)   [A: 5.1, 5.2, 5.3]

14 A train of mass 3×10^5 kg travels along a straight level track. The resistance to motion is 1.5×10^4 N. Find the tractive force required to produce an

acceleration of 0.1 m/s², and the power in kW which is then developed by the engine when the speed of the train is 10 m/s.
Find also the maximum speed attainable on the same track when the engine is working at a rate of 360 kW. [J: 5.3]

15 Two small smooth spheres A and B of equal radius but of masses $3m$ and $2m$ respectively are moving towards each other so that they collide directly. Immediately before the collision sphere A has speed $4u$ and sphere B has speed u. The collision is such that sphere B experiences an impulse of magnitude $6mcu$, where c is a constant. Find:

(a) in terms of u and c, the speeds of A and B immediately after collision
(b) the coefficient of restitution, in terms of c
(c) the range of values of c for which such a collision would be feasible
(d) the value of c such that $\frac{9}{16}$ of the total kinetic energy would be destroyed by the collision. [A: 5.4, 5.6, 5.7]

16 A ball is thrown vertically upwards with speed U from a point halfway between the floor and ceiling of a room of height h. After rebounding from the ceiling and then the floor it just reaches the ceiling a second time. The coefficient of restitution at both floor and ceiling is $\frac{1}{2}$.
Denoting the speed of the ball just before its first impact with the floor by V, show that

$$V = \sqrt{(8gh)}$$

and find U in terms of g and h.
Find the ratio of the magnitude of the impulse when the ball first reaches the ceiling to the magnitude of the impulse on the floor. [J: 1.4, 5.4, 5.8]

17 Three identical smooth spheres A, B, C are free to move in a straight smooth horizontal groove, with B between A and C. At a given instant sphere A and B are both moving with speed u towards sphere C, which is stationary. Show that after the first collision, which is between B and C, the speeds of B and C are

$$\tfrac{1}{2}u(1-e), \quad \tfrac{1}{2}u(1+e)$$

respectively, where e is the coefficient of restitution for all impacts between the spheres.
Find the speeds of the spheres after the second collision, which is between A and B.
Show that there will, in general, be a further collision stating the value of e for any exception to this. [L: 5.6]

18 A smooth peg, of negligible diameter, is fixed at a height $3l$ above a horizontal table and a light inextensible string of length $4l$ hangs over the peg. A particle of mass m is attached to one end of the string and a particle of mass $2m$ is attached to the other end. The system is held at rest with the particles hanging at the same level and at a distance l from the table. The parts of the string not in contact with the peg are vertical. The system is then released from rest. Find, in terms of g and l, the speed u with which the particle of mass $2m$ strikes the table. Given that this particle rebounds

from the table with speed $\frac{1}{2}u$, find, in terms of m, g and l, the magnitude of the impulse on the table.   [J: 2.3, 5.4]

19 Three smooth spheres A, B and C of equal radii and masses m, $3m$ and $5m$, respectively, lie at rest on a smooth horizontal table with their centres in a straight line. The sphere A is projected along the line of centres with speed u so as to strike B directly and subsequently B strikes C. The coefficient of restitution between A and B, and between B and C, is e. Given that B is brought to rest by its impact with C, find e and show that there will be no further impacts.   [J: 5.6]

20 When a car of mass 1 tonne is accelerating, the engine produces a constant tractive force of magnitude P N. The car pulls a caravan of mass 0.8 tonne up a hill of inclination arcsin $\frac{1}{20}$, and attains a speed of 54 km/h from rest, in a distance of 450 m. Show that the acceleration of the car and caravan is 0.25 m/s^2.
Given that the total non-gravitational resistances have a magnitude of 270 N, show that $P = 1620$.
Legal requirements prohibit the car and caravan from exceeding a speed of 54 km/h, so, once this speed is attained, the tractive force is adjusted so as to maintain a constant speed. Find the power exerted by the car's engine when it is pulling the caravan up the hill at a constant speed of 54 km/h.
The total non-gravitational resistances on the caravan have magnitude 120 N. Find the magnitude of the tension in the coupling between the car and the caravan whilst they are accelerating up the hill.
(Take g as 10 m/s^2.)   [L: 2.8, 5.3]

21 A particle moving along a smooth horizontal floor hits a smooth vertical wall and rebounds in a direction at right angles to its initial direction of motion. The coefficient of restitution is e. Find, in terms of e, the tangent of the angle between the initial direction of motion and the wall. Prove that the kinetic energy after rebound is e times the initial kinetic energy.
                                                                              [J: 5.8]

22 The resistance to the motion of a train of total mass 300 tonnes is 18 000 N. The greatest tractive force which the engine can exert is 30 000 N and the greatest power available is 450 kW. The train starts from rest and moves along a level line with the greatest possible acceleration. Show that the engine first develops its maximum power when the speed is 54 km/h. Show also that the speed of the train cannot exceed 90 km/h.
Find the acceleration of the train in m/s^2 when its speed is 72 km/h.
                                                                              [J: 5.3]

23 A car of mass 1500 kg tows a caravan of mass 500 kg. The car and caravan move on a straight level road with the engine of the car exerting a constant pull of 4100 N. Given that there are frictional resistances of 800 N on the car and 300 N on the caravan, find the magnitude of:
(a) the acceleration of the car
(b) the tension in the tow bar between the car and the caravan.

The car and caravan then go up a straight road which is inclined at an angle $\sin^{-1}\frac{1}{8}$ to the horizontal, the frictional resistances on the car and caravan remaining unchanged. The speed of the car and caravan increases from 10 m/s to 20 m/s in 16 seconds with the engine of the car now exerting a constant pull of P newtons. Find the value of P and the rate at which the engine is working when the speed is 15 m/s.
When the speed is 20 m/s, the tow-bar breaks. Find, to the nearest 10 m, the further distance travelled by the caravan before it comes momentarily to rest.
(Take g as 10 m/s².) [L: 2.2, 2.8, 5.3]

*24 A small sphere P of mass m is moving with speed u on a smooth horizontal plane along a straight line AO. At O the sphere collides with a smooth vertical wall which makes an acute angle θ with AO. Show that, if the coefficient of restitution at O is e, the sphere leaves O at an angle ϕ to the wall, where

$$\tan \phi = e \tan \theta$$

Find the value of θ for which the component, perpendicular to AO, of the velocity after collision is greatest.
A sphere Q of mass $2m$, with the same radius as P, is at rest on the plane and OQ is perpendicular at AO. Show that if $\cot \theta = \sqrt{e}$, P strikes Q directly and at speed $u\sqrt{e}$. Given that the spheres are perfectly elastic, find the impulse exerted by P on Q. [J: 5.8]

*25 State the law of conservation of momentum for two interacting particles which move in the same straight line, and derive the law from Newton's laws of motion.
Two particles, P of mass $4m$ and Q of mass $5m$, are subject to a mutual interaction which is variable and of unknown magnitude, and no other force acts on them. Initially Q is at rest and P is moving with speed u directly away from Q. Find the velocity of Q when P has velocity v.
At a certain instant P comes to rest. Find the total net work which has been done against the forces of the interaction since the initial instant.
When P comes to rest the force of interaction is zero, and it remains zero until Q collides with P. Immediately after the collision the velocity of P is $\frac{1}{2}u$. Find the coefficient of restitution between the particles. [J: 5.5, 5.1]

**26 A particle is free to move on a smooth horizontal floor on which there are two fixed smooth mutually perpendicular vertical walls that meet the floor in lines OX and OY. The particle is projected horizontally, from a point P on the floor, in a direction making an angle θ with XO. Show that it will rebound from OX at an angle ϕ to XO given by

$$\tan \phi = e \tan \theta,$$

where e is the coefficient of restitution between the particle and the wall OX. The particle returns to P (see diagram) after a subsequent perfectly elastic collision with OY. Prove that the coordinates $\{x, y\}$ of P, referred to OX and OY as axes, must satisfy

$$(1 + e)y = 2ex \tan \theta \qquad \text{[J: 5.8]}$$

**27 A particle of mass m, moving on a smooth horizontal surface, collides directly with a stationary particle B of mass $2m$, situated at a distance d from a wall which is at right angles to the direction of motion of A. The coefficient of restitution between A and B and between B and the wall is e $(e < \frac{1}{2})$. Show that after B has rebounded from the wall, the particles collide again at a distance

$$\frac{3e^2 d}{1 - e + e^2}$$

from the wall. (Assume that the entire motion takes place in a straight line perpendicular to the wall.)   [J: 5.8]

CENTRES OF GRAVITY

6.1 Systems of Coplanar Particles: Uniform Bodies

The **weight of a body** is the resultant of the weights of the particles that make up the body. It acts vertically downwards through the **centre of gravity** which is a fixed point that is independent of the orientation of the body.

> **Note that** (i) the weights of the particles may be considered as parallel forces and their resultant found using the techniques in 4.3.
> (ii) the centre of gravity (CG) is a convenient mathematical notion and may not even be on the body itself. For example the centre of gravity of a hollow sphere is at the centre of the sphere.
> (iii) the centre of gravity will usually be labelled G.

Example 1 A straight rod AB, of negligible mass and of length 10 cm has three masses fixed to it: 5 kg at A, 2 kg at 4 cm from A and 3 kg at B. Where is the centre of gravity of the three masses?

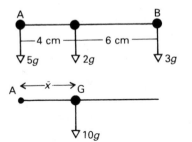

By symmetry we expect the centre of gravity to be on the rod. The total weight $= 5g + 2g + 3g = 10g$ N. Let the rod AB be horizontal and consider an equivalent system composed only of the resultant.

Taking moments about A for both systems:

$$\bar{x}\, 10g = 2g \times 4 + 3g \times 10$$
$$= 38g$$
$$\Rightarrow \bar{x} \quad = 3.8 \,\text{cm}$$

∴ the centre of gravity is 3.8 cm from A.

Centre of Gravity of a System of Coplanar Particles

Moments about Oy:

$$(m_1 + m_2 + m_3 + \cdots)g\bar{x} = m_1 gx_1 + m_2 gx_2 + m_3 gx_3 + \cdots$$
$$\Rightarrow \bar{x} = \frac{m_1 x_1 + m_2 x_2 + m_3 x_3 + \cdots}{m_1 + m_2 + m_3 + \cdots}$$

Moments about Ox:

$$(m_1 + m_2 + m_3 + \cdots)g\bar{y} = m_1 gy_1 + m_2 gy_2 + m_3 gy_3 + \cdots$$
$$\Rightarrow \bar{y} = \frac{m_1 y_1 + m_2 y_2 + m_3 y_3 + \cdots}{m_1 + m_2 + m_3 + \cdots}$$

$$\boxed{\bar{x} = \frac{\sum m_r x_r}{\sum m_r} \qquad \bar{y} = \frac{\sum m_r y_r}{\sum m_r}}$$

It is **strongly recommended** that the technique of taking moments about two axes (as shown above) is used rather than quoting these formulae.

Example 2   The rectangle PQRS is such that PQ $= 20$ cm and QR $= 10$ cm. Particles of mass 3 kg, 2 kg, 4 kg, 1 kg we placed at P, Q, R, S respectively. Find the position of the centre of gravity of the system.

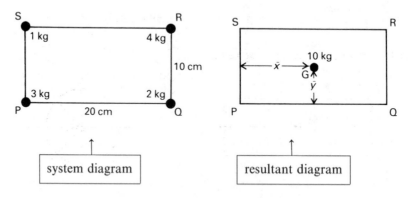

Taking moments about PS:

$$(4g \times 20) + (2g \times 20) = 10g \times \bar{x}$$
$$\Rightarrow 120g = 10g\bar{x}$$
$$\Rightarrow \quad \bar{x} = 12 \text{ cm}$$

Taking moments about PQ:

$$(1g \times 10) + (4g \times 10) = 10g \times \bar{y}$$
$$\Rightarrow 50g = 10g\bar{y}$$
$$\Rightarrow \quad y = 5 \text{ cm}$$

$\therefore$ the CG is 12 cm from PS and 5 cm from PQ.

Example 3   Find the position vector of the centre of gravity of the system
of particles with the following position vectors:

$$2\,\text{kg} \quad \text{at} \quad 3\mathbf{i} + 2\mathbf{j}$$
$$4.2\,\text{kg at} \quad -\mathbf{i} + 5\mathbf{j}$$
$$1.8\,\text{kg at} \quad -2\mathbf{i} - 4\mathbf{j}$$

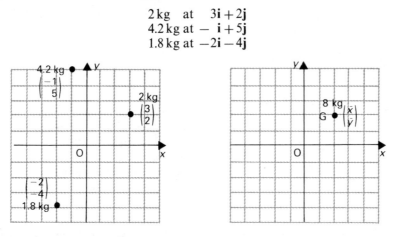

Taking moments about Oy:

$$(2g \times 3) - (4.2g \times 1) - (1\cdot8g \times 2) = 8g \times \bar{x}$$
$$\Rightarrow \quad -1.8g = 8g\bar{x}$$
$$\Rightarrow \qquad \bar{x} = -0.225$$

Taking moments about Ox:

$$(2g \times 2) + (4.2g \times 5) - (1.8g \times 4) = 8g \times \bar{y}$$
$$\Rightarrow \quad 17.8g = 8g\bar{y}$$
$$\Rightarrow \qquad \bar{y} = 2.225$$

$\therefore$ position vector of centre of gravity is $-0.225\mathbf{i} + 2.225\mathbf{j}$

Centre of Gravity of a Uniform Body

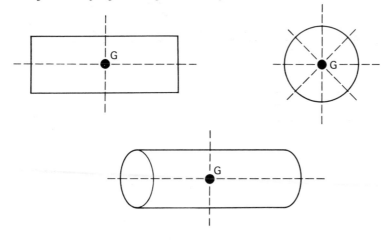

The centre of gravity of a uniform body lies on every line of symmetry of
the body.

Centre of gravity of a Uniform Triangle

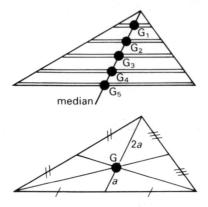

median

> This does not necessarily have symmetry.

Consider a series of strips parallel to one side. The centres of gravity of these strips lie at the middle of each strip and these points lie on a median of the triangle. Hence the centre of gravity of the triangle must lie on this median. Similarly the centre of gravity must lie on all three medians of the triangle.

> The centre of gravity of a triangle lies at the intersection of the medians— that is, two thirds of the distance from each vertex to the mid point of the opposite side.

Example 4   Find the coordinates of the centre of gravity of a uniform triangular lamina whose vertices are O(0,0), A(6,4) and B(8,0).

The midpoint of OB is M(4,0).

> Midpoint of line connecting (x_1, y_1) and (x_2, y_2) is
> $$\left(\frac{x_1 + x_2}{2}, \frac{y_1 + y_2}{2}\right).$$

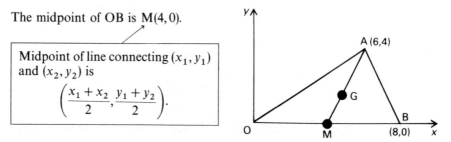

We now need to find the coordinates of G, which is one third of the way up MA from M.

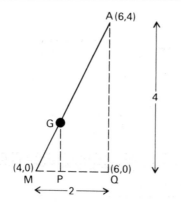

> $\Delta s\ \text{MGP} \brace \text{MAQ}$ are similar
>
> $\therefore$ MP is $\frac{1}{3} \times$ MQ
> and PG is $\frac{1}{3} \times$ QA

$$\text{MP} = \tfrac{1}{3} \times \text{MQ} = \tfrac{1}{3} \times 2 = \tfrac{2}{3}$$
$$\text{PG} = \tfrac{1}{3} \times \text{QA} = \tfrac{1}{3} \times 4 = 1\tfrac{1}{3}$$

$$\Rightarrow x\text{-coordinate of } G = 4 + \tfrac{2}{3} = 4\tfrac{2}{3}$$
$$\text{and } y\text{-coordinate of } G = 0 + 1\tfrac{1}{3} = 1\tfrac{1}{3}$$
$$\therefore \underline{G \text{ is } (4\tfrac{2}{3}, 1\tfrac{2}{3})}$$

● Exercise 6.1   See page 180

6.2 Centres of Gravity of Composite Bodies and Remainders

Standard Centres of Gravity of Some Uniform Bodies	
Triangular Lamina:	Centre of gravity is two thirds of the distance from each vertex to the midpoint of the opposite side.
Semicircular lamina of radius r:	Centre of gravity is on the axis of symmetry, a distance $\dfrac{4r}{3\pi}$ from the straight edge.
Solid right circular cone of height h:	Centre of gravity is on the axis of symmetry, a distance $\dfrac{h}{4}$ from the plane base.
Solid hemisphere of radius r:	Centre of gravity is on the axis of symmetry, a distance $\dfrac{3r}{8}$ from the plane surface.

These will be used in this section but proved later in the unit.

One Dimensional Problems

Example 5   A shape is cut out of uniform sheet metal. It is composed of a square of side 8 cm with a 5, 5, 8 triangle attached to one side. Find the position of its centre of gravity.

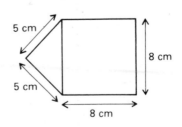

The lamina is symmetrical and the centres of gravity of the two parts and the composite body lie on the line of symmetry.

Method 1

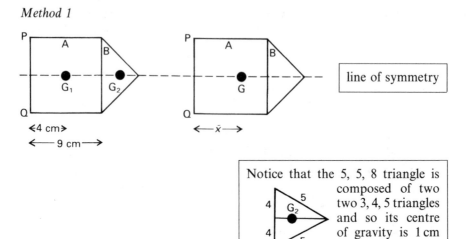

<4 cm>

← 9 cm →

←$\bar{x}$→

| line of symmetry |

> Notice that the 5, 5, 8 triangle is composed of two two 3, 4, 5 triangles and so its centre of gravity is 1 cm from the 8 cm side.

Body	Weight	Distance of CG from PQ
A	64	4
B	12	9
A+B	76	$\bar{x}$

> These are, in fact, the areas as the weights are proportional to the areas in this case.

Taking moments about PQ:

$$76 \times \bar{x} = (64 \times 4) + (12 \times 9)$$
$$\Rightarrow \bar{x} = 4.79 \text{ cm}.$$

∴ the CG is on the line of symmetry, 4.79 cm from the edge opposite the triangle.

Method 2

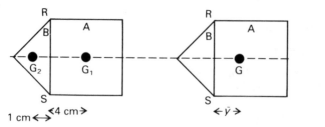

1 cm ↔ <4 cm→

←$\bar{y}$→

| line of symmetry |

Body	Weight	Distance of CG from RS
A	64	4
B	12	1
A+B	76	$\bar{y}$

Taking moments about RS:

$$76 \times \bar{y} = (64 \times 4) - (12 \times 1)$$
$$\Rightarrow \bar{y} = 3.21 \, \text{cm}$$

$\therefore$ CG is on the line of symmetry, 3.21 cm from the common edge of the two shapes.

Simplifying the data in the 'weight' column

1 For a wire framework, weight is proportional to **length** of wire.
   For a lamina,   weight is proportional to **area** of lamina.
   For a solid,   weight is proportional to **volume** of solid.

2 Since, in the moments equations, we are only interested in the ratio of the weights to each other the questions become much more straightforward if we simplify the ratios as much as possible.
For example, $\frac{1}{3}a^2\pi g : \frac{2}{3}a^2\pi g : \frac{4}{3}a^2\pi g$
can be used as   1 : 2 : 4.

Example 6   A hole of diameter 2 cm is cut from a uniform circular disc of diameter 10 cm. Find the position of the centre of gravity of the remainder if the centre of the hole lies 2 cm from the centre of the disc.

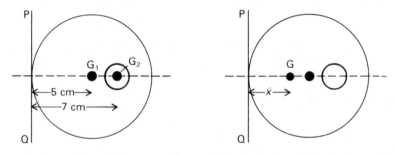

Note that, by symmetry, the centre of gravity of the remainder lies on the line through the centres of the disc and the hole.

Body	Weight	Distance from PQ
◯	25	5
∘	1	7
◎	24	$\bar{x}$

Note that the weights are $25\pi g$, πg, $24\pi g$ but their ratios simplify to $25:1:24$.

Taking moments about PQ:

$$24 \times \bar{x} = (25 \times 5) - (1 \times 7)$$
$$\Rightarrow \bar{x} = \tfrac{59}{12}\text{cm} = 4\tfrac{11}{12}\text{cm}.$$

$\therefore$ CG is $\tfrac{1}{12}$ cm from the centre of the disc on the other side of the centre from the hole.

Two-Dimensional Problems

Example 7   A uniform length of wire is bent to form the shape shown in the diagram. Find the position of the centre of gravity.

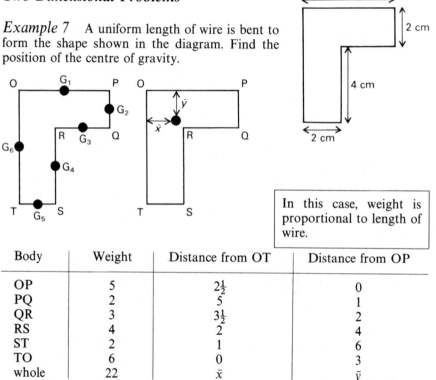

In this case, weight is proportional to length of wire.

Body	Weight	Distance from OT	Distance from OP
OP	5	$2\tfrac{1}{2}$	0
PQ	2	5	1
QR	3	$3\tfrac{1}{2}$	2
RS	4	2	4
ST	2	1	6
TO	6	0	3
whole	22	$\bar{x}$	$\bar{y}$

Taking moments about OP:

$$22\bar{x} = (5 \times 2\tfrac{1}{2}) + (2 \times 5) + (3 \times 3\tfrac{1}{2}) + (4 \times 2) + (2 \times 1) + (6 \times 0)$$
$$\bar{x} = \tfrac{43}{22}\text{cm}$$

Taking moments about OQ:

$$22\bar{y} = (5 \times 0) + (2 \times 1) + (3 \times 2) + (4 \times 4) + (2 \times 6) + (6 \times 3)$$
$$\bar{y} = \tfrac{27}{11}\text{cm}.$$

$\therefore$ CG is $\tfrac{43}{22}$ cm from the 6 cm wire and $\tfrac{27}{11}$ cm from the 5 cm wire.

Example 8   A uniform lamina is composed of a square with an equilateral triangle and a semi-circle attached to two adjacent edges. If the square is of side 2a, find the position of the centre of gravity of the composite lamina.

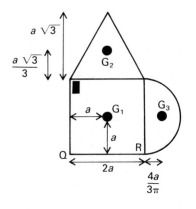

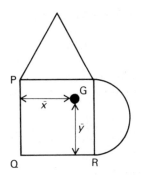

Weight is proportional to area.

Body	Weight	Distance of CG from PQ	Distance of CG from QR
□	$4a^2$	a	a
△	$\sqrt{3}a^2$	a	$a + \dfrac{a\sqrt{3}}{3}$
○	$\frac{1}{2}a^2\pi$	$2a + \dfrac{4a}{3\pi}$	a
whole	$\left(\dfrac{\pi}{2}+4+\sqrt{3}\right)a^2$	$\bar{x}$	$\bar{y}$

Taking moments about PQ:

$$\left(\frac{\pi}{2}+4+\sqrt{3}\right)a^2\bar{x}=(4a^2\times a)+(\sqrt{3}a^2\times a)+\tfrac{1}{2}a^2\pi\left(2a+\frac{4a}{3\pi}\right)$$

$$\boxed{\div\ a^2}\ \rightarrow\Rightarrow\left(\frac{\pi}{2}+4+\sqrt{3}\right)\bar{x}=\left(4+\sqrt{3}+\pi+\frac{2}{3}\right)a$$

$$\Rightarrow\qquad\qquad \bar{x}\approx 1.31a$$

Taking moments about QR

$$\left(\frac{\pi}{2}+4+\sqrt{3}\right)a^2\bar{y}=(4a^2\times a)+\sqrt{3}a^2\times\left(a+\frac{a\sqrt{3}}{3}\right)+\tfrac{1}{2}a^2\pi\times a$$

$$\Rightarrow\left(\frac{\pi}{2}+4+\sqrt{3}\right)\bar{y}=\left(4+\sqrt{3}+1+\frac{\pi}{2}\right)a$$

$$\Rightarrow\qquad\qquad \bar{y}\approx 1.14a$$

∴ CG is $1.31a$ from the edge opposite the semicircle and $1.14a$ from the edge opposite the triangle.

- **Exercise 6.2**   See page 182

6.3 Centres of Gravity of Laminae by Integration

When a body cannot be divided into a finite number of parts with known centres of gravity, it may be divided up into as in finite number of very small parts all with known centres of gravity. The summing of the moments of the parts is done by integration.

One-Dimensional Problems

Example 9   A uniform lamina is bounded by the curve $y^2 = x$ and the double ordinate $x = 6$. Find the position of its centre of gravity.

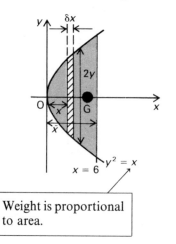

By symmetry, the centre of gravity lies on Ox
The lamina is divided into thin strips parallel to Oy, each of thickness δx.

Weight is proportional to area.

Body	Weight	Distances from Oy
Elemental strip	$2y\delta x$	x
Lamina	$2 \times \displaystyle\int_0^6 y \, dx$	$\bar{x}$

Note that $\int_0^6 y \, dx$ only gives the area above the x-axis.

Taking moments about Oy:

$$\sum_{x=0}^{x=6} (2y\delta x)x = \left[2 \times \int_0^6 y \, dx \right] \bar{x}$$

The sum of the moment of each of the strips about Oy.

$$\Rightarrow \int_0^6 2xy\,\mathrm{d}x = 2\bar{x}\int_0^6 y\,\mathrm{d}x$$

$\mathrm{d}x$ tells us that we are integrating a function w.r.t. x, so it is necessary to express y as a function of x.

But $y^2 = x \Rightarrow y = x^{1/2}$

$$\therefore 2\int_0^6 x\cdot x^{1/2}\,\mathrm{d}x = 2\bar{x}\int_0^6 x^{1/2}\,\mathrm{d}x$$

$$\Rightarrow \int_0^6 x^{3/2}\,\mathrm{d}x = \bar{x}\int_0^6 x^{1/2}\,\mathrm{d}x$$

$$\Rightarrow \left[\frac{2x^{5/2}}{5}\right]_0^6 = \bar{x}\left[\frac{2x^{3/2}}{3}\right]_0^6$$

$$\Rightarrow \bar{x} = \frac{2\cdot 6^{5/2}}{5}\times\frac{3}{2\cdot 6^{3/2}} = \frac{18}{5} = 3\cdot 6$$

$\therefore$ CG is at $(3\cdot 6, 0)$

Two-Dimensional Problems

Example 10   Find the centre of gravity of a uniform lamina bounded by the curve $y = x^2 + 2$, the x-axis and the ordinates $x = 1$ and $x = 3$.

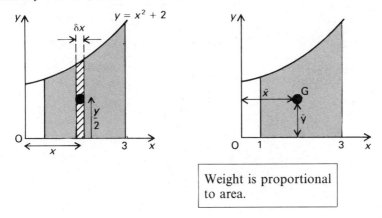

Weight is proportional to area.

The lamina is divided into this strips parallel to Oy, each of thickness δx.

Body	Weight	Distance from Oy	Distance from Ox
Elemental strip	$y\delta x$	x	$\dfrac{y}{2}$
Lamina	$\displaystyle\int_1^3 y\,\mathrm{d}x$	$\bar{x}$	$\bar{y}$

Taking moments about Oy:

$$\sum_{x=1}^{x=3} [(y\delta x)x] = \bar{x} \int_1^3 y\,dx$$

$$\Rightarrow \quad \int_1^3 xy\,dx = \bar{x} \int_1^3 y\,dx$$

But $y = x^2 + 2$

$$\therefore \int_1^3 x(x^2 + 2)\,dx = \bar{x} \int_1^3 (x^2 + 2)\,dx$$

$$\Rightarrow \quad \left[\frac{x^4}{4} + x^2\right]_1^3 = \bar{x}\left[\frac{x^3}{3} + 2x\right]_1^3$$

$$\Rightarrow \quad 28 = \bar{x}\,\frac{38}{3}$$

$$\Rightarrow \quad \bar{x} = \frac{42}{19}$$

Taking moments about Ox:

$$\sum_{x=1}^{x=3} \left[(y\delta x)\frac{y}{2}\right] = \bar{y} \int_1^3 y\,dx$$

$$\Rightarrow \quad \int_1^3 \frac{y^2}{2}\,dx = \bar{y} \int_1^3 y\,dx$$

Using $y = x^2 + 2$

$$\Rightarrow \int_1^3 \tfrac{1}{2}(x^2 + 2)^2\,dx = \bar{y}\,\frac{38}{3}$$

Evaluated in first part of problem.

$$\Rightarrow \frac{38}{3}\bar{y} = \frac{1}{2} \int_1^3 (x^4 + 4x^2 + 4)\,dx$$

$$= \frac{1}{2}\left[\frac{x^5}{5} + \frac{4x^2}{3} + 4x\right]_1^3 = \frac{683}{15}$$

$$\Rightarrow \quad \bar{y} = \frac{683}{15} \times \frac{3}{38} = \frac{683}{190}$$

$$\therefore \text{CG is at } \left(\frac{42}{19}, \frac{683}{190}\right).$$

- Exercise 6.3   See page 185

6.4 Centres of Gravity of Solids by Integration

Example 11   The curve $y = x^2$ between $x = 0$ and $x = 3$ is rotated around the x-axis to form a solid of revolution. Find the coordinates of the centre of gravity.

By symmetry, the centre of gravity lies on O*x*.

The solid is divided into thin discs parallel to O*y*, each of thickness δx.

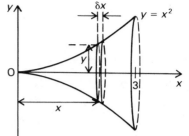

> Weight is proportional to volume.

Body	Weight	Distance from O*y*
Elemental disc	$\pi y^2 \, \delta x$	x
Solid of revolution	$\pi \displaystyle\int_0^3 y^2 \, dx$	$\bar{x}$

Taking moments about O*y*:

$$\sum_{x=0}^{x=3} [(\pi y^2 \delta x)x] = \left[\pi \int_0^3 y^2 \, dx \right] \bar{x}$$

$$\Rightarrow \quad \pi \int_0^3 y^2 x \, dx = \pi \bar{x} \int_0^3 y^2 \, dx$$

volume of revolution

But $y = x^2$

$$\therefore \quad \int_0^3 x^5 \, dx = \bar{x} \int_0^3 x^4 \, dx$$

$$\Rightarrow \quad \left[\frac{x^6}{6} \right]_0^3 = \bar{x} \left[\frac{x^5}{5} \right]_0^3$$

$$\Rightarrow \quad \bar{x} = \frac{3^6}{6} \times \frac{5}{3^5} = \tfrac{5}{2}$$

∴ CG is at $(2\tfrac{1}{2}, 0)$.

Example 12   Find the position of the centre of gravity of a cone of radius 4 cm and height 12 cm.

By symmetry the centre of gravity lies on O*x*.

The cone is divided into thin discs parallel to O*y*, each of thickness δx.

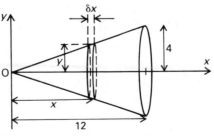

> Weight is proportional to volume.

Body	Weight	Distance from Oy
Elemental disc	$\pi y^2 \delta x$	x
Cone	$\frac{1}{3}\pi \cdot 4^2 \cdot 12$	$\bar{x}$

Taking moments about Oy:

$$\sum_{x=0}^{x=12} \left[(\pi y^2 \delta x)x \right] = \tfrac{1}{3}\pi \cdot 4^2 \cdot 12\bar{x}$$

$$\Rightarrow \quad \pi \int_0^{12} y^2 x \, dx = 64\pi\bar{x}$$

> But we need to express y in terms of x.
>
> Using similar triangles, $\dfrac{y}{4} = \dfrac{x}{12} \Rightarrow y = \tfrac{1}{3}x.$

Using $y = \tfrac{1}{3}x$

$$\Rightarrow 64\bar{x} = \int_0^{12} \tfrac{1}{9}x^3 \, dx$$

$$= \left[\frac{x^4}{36} \right]_0^{12} = 576$$

$$\Rightarrow \quad \bar{x} = \frac{576}{64} = 9$$

∴ CG is on the axis of the cone 9 cm from the vertex.

- **Exercise 6.4**   See page 186

6.5 Centres of Gravity of Shells and Non-Uniform Bodies by Integration

Example 13   A piece of uniform wire is bent to form an arc of a circle of radius 10 cm, subtending an angle $\pi/3$ at the centre of the circle. Find the position of the centre of gravity of the wire.

> Weight is proportional to length.

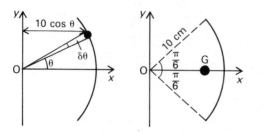

The centre of gravity lies on the line of symmetry Ox.

The wire is divided into elemental pieces each subtending an angle $\delta\theta$ at the centre of the circle. The length of each piece is $10\delta\theta$.

Body	Weight	Distance
Elemental piece	$10\delta\theta$	$10\cos\theta$
Arc	$10 \times \dfrac{\pi}{3}$	$\bar{x}$

Taking moments about Oy:

$$\sum_{\theta=-\pi/6}^{\theta=\pi/6} [(10\delta\theta)10\cos\theta] = \bar{x}\frac{10\pi}{3}$$

$$\Rightarrow \qquad \int_{-\pi/6}^{\pi/6} 100\cos\theta\,\mathrm{d}\theta = \bar{x}\frac{10\pi}{3}$$

$$\Rightarrow \qquad [100\sin\theta]_{-\pi/6}^{\pi/6} = \bar{x}\frac{10\pi}{3}$$

$$\Rightarrow \qquad 100[\tfrac{1}{2}-(-\tfrac{1}{2})] = \bar{x}\frac{10\pi}{3}$$

$$\Rightarrow \qquad \bar{x} = \frac{30}{\pi}$$

$\therefore$ CG is on the line of symmetry at $30/\pi$ cm from the centre of the circle.

Example 14   Find the centre of gravity of a non-uniform rod AB of length l, given that its weight per unit length is wx at a point distant x from A.

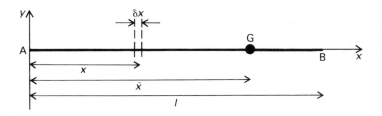

The body is divided into small elements each of length δx.

Body	Weight	Distance from Oy
Elemental	$(wx)\delta x$	x
Rod	$\displaystyle\int_{0}^{l} (wx)\,\mathrm{d}x$	$\bar{x}$

Taking moments about Ay:

$$\sum_{x=0}^{x=l} [(wx\,\delta x)x] = \bar{x} \int_0^l wx\,\mathrm{d}x$$

$$\Rightarrow \quad \int_0^l wx^2\,\mathrm{d}x = \bar{x} \int_0^l wx\,\mathrm{d}x$$

$$\Rightarrow \quad \left[\frac{wx^3}{3}\right]_0^l = \bar{x}\left[\frac{wx^2}{2}\right]_0^l$$

$$\Rightarrow \quad \frac{wl^3}{3} = \bar{x}\,\frac{wl^2}{2}$$

$$\Rightarrow \quad \bar{x} = \frac{2l}{3}$$

$\therefore$ CG is $2l/3$ from A.

- Exercise 6.5   See page 187

6.6 Equilibrium of Hanging Bodies: Condition For Toppling

Uniform Body	Position of CG on axis of symmetry
Solid cone Solid pyramid	$\dfrac{h}{4}$ from base
Hollow cone Hollow pyramid	$\dfrac{h}{3}$ from base
Solid hemisphere	$\dfrac{3r}{8}$ from plane face
Hollow hemisphere	$\dfrac{r}{2}$ from the plane face
Arc subtending 2α at the centre	$\dfrac{r\sin\alpha}{\alpha}$ from centre
Semicircular arc	$\dfrac{2r}{\pi}$ from centre
Sector subtending 2α at the centre	$\dfrac{2r\sin\alpha}{3\alpha}$ from centre
Semi-circular lamina	$\dfrac{4r}{3\pi}$ from centre

Hanging Bodies

Since there are only two forces acting on the body

AG is vertical.

Condition For Toppling

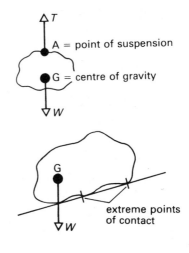

A = point of suspension

G = centre of gravity

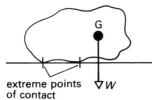

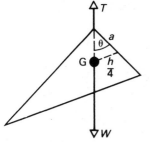

extreme points of contact

If the vertical through the centre of gravity lies outside the extreme points of contact of the body with the plane, the body will topple.

Example 15 A solid cone of radius a and height h is suspended from a point on the rim of the base of the cone. Find the angle that the base makes with the vertical in the position of equilibrium.

G is directly below the point of suspension.

$$\therefore \tan\theta = \frac{h}{4} \div a$$

$$\Rightarrow \quad \theta = \tan^{-1}\left(\frac{h}{4a}\right)$$

Example 16 A cylinder, of radius r and height r and weight W, lies with one generator along a line of greater slope of a plane of inclination α. If the plane is rough enough to prevent slipping, show that the cylinder will not topple if $\alpha \leqslant \tan^{-1}(0.5)$. For values of $\alpha > \tan^{-1}(0.5)$, find the magnitude of the least force applied to the lower plane face of the cylinder that will maintain equilibrium.

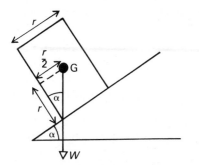

When the cylinder is at the point of toppling, the weight acts through the lowest point of contact of the cylinder and the plane.
In this position

$$\tan\alpha = \frac{r/2}{r} = \tfrac{1}{2}$$

For equilibrium, $\alpha \leqslant \tan^{-1}(0.5)$.   [QED]

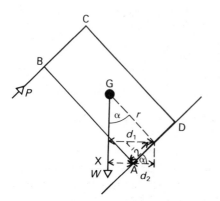

To stop the cylinder toppling about A, a force must be applied to counteract the moment of W about A. The minimum possible force is that with its line of action at a maximum distance from A. This force acts at B, perpendicular to the diameter AB.

Taking moments about A:

$$W \times \text{AX} = P \times \text{AB}$$

$$\Rightarrow W(d_1 - d_2) = P \times 2r$$

But $d_1 = r \sin \alpha$

and $d_2 = \dfrac{r}{2} \cos \alpha$

$$\therefore \qquad \text{AX} = r \sin \alpha - \frac{r}{2} \cos \alpha$$

$$\therefore \text{Minimum value of } P = \frac{W}{2r}\left(r \sin \alpha - \frac{r}{2}\cos \alpha\right)$$

$$\Rightarrow \text{Minimum value of } P = \frac{W}{4}(2 \sin \alpha - \cos \alpha).$$

- Exercise 6.6   See page 188

6.7 + Revision of Unit: A Level Questions

- Miscellaneous Exercise 6B   See page 190

Unit 6 Exercises

Exercise 6.1

In each of the questions 1–3 find the coordinates of the centre of gravity of the system of particles:

1   **2**   **3**

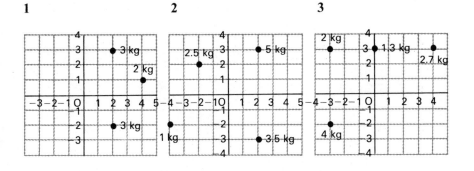

4   ABCD is a square of side 8 cm. Particles of mass 1 kg, 2 kg, 3 kg and 4 kg are placed at A, B, C, D respectively. Find the distances of the centre of gravity of the system of particles from AB and from AD.

5   Find the position vectors of the centre of gravity of particles of mass 1 kg, 5 kg, 2 kg, 2 kg which are at points with position vectors $5\mathbf{i} + 2\mathbf{j}$, $2\mathbf{i} + 5\mathbf{j}$, $2\mathbf{i} - 2\mathbf{j}$, $-2\mathbf{i} + 2\mathbf{j}$ respectively.

In each of questions 6–13 find the coordinates of the centre of gravity of the uniform lamina:

6   **7**   **8**

9

11

10

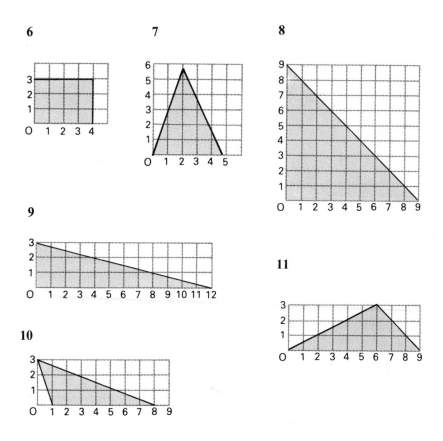

12

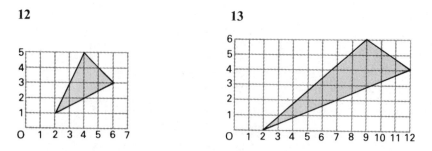

13

14 The rectangle ABCD is such that AB = 6 cm and BC = 2 cm. Particles of mass 3 kg, 5 kg, 7 kg, 1 kg, 4 kg lie at A, B, C, D and the centre of the rectangle, respectively. Find the distances of the centre of gravity from AB and from AD.

15 Four particles of mass 2 kg, 3 kg, 4 kg, 1 kg are situated at points with position vectors $\begin{pmatrix} 2 \\ 4 \end{pmatrix}$, $\begin{pmatrix} 5 \\ 2 \end{pmatrix}$, $\begin{pmatrix} -4 \\ -2 \end{pmatrix}$, $\begin{pmatrix} p \\ q \end{pmatrix}$ respectively.

If the centre of gravity of this system has position vector $\begin{pmatrix} 1 \\ 3 \end{pmatrix}$ find p and q.

16 Particles of mass $6m$, $7m$, $3m$ are placed at the points $(-3, 4)$, $(2, 1)$, $(-2, -\frac{1}{3})$ respectively. Where must a particle of $4m$ be placed so that the centre of gravity is at the origin?

17 Show that the centre of gravity of a uniform lamina in the shape of a rhombus is at the intersection of the diagonals.

18 Show that the centre of gravity of a uniform lamina in the shape of a parallelogram is at the intersection of the diagonals.

*19 Three particles of equal mass are placed at the vertices A, B, C of a triangle. By finding the resultant weight of two particles at a time, show that the centre of gravity of the three particles is at the same position as the centre of gravity of a uniform lamina in the shape of the triangle ABC.

*20

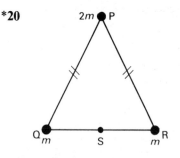

Triangle PQR is isosceles. S is the midpoint of QR. Particles of mass $2m$, m, m are placed at P, Q, R respectively. Show that their centre of gravity is at the midpoint of PS.

Exercise 6.2

In each of questions 1–4 find the coordinates of the centre of gravity of the uniform lamina:

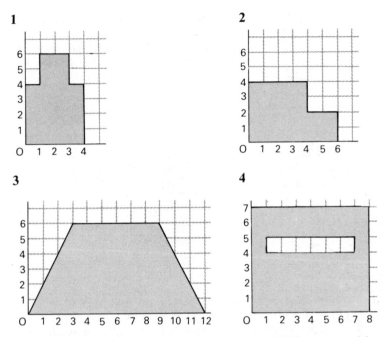

1

2

3

4

5   The handle of a tennis racket has a mass of 260 grams and its centre of gravity is 15 cm from the end of the handle. The oval frame weighs 160 grams and its centre of gravity is 50 cm from the end of the handle. Find the distance of the centre of gravity of the racket from the end of the handle.

6

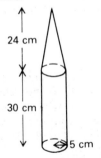

A solid cylinder of radius 5 cm and height 30 cm is surmounted by a solid right cone, made from the same material and with the same radius. If the cone has height 24 cm, show that the distance of the centre of gravity of the composite body from the base of the cylinder is approximately 19.4 cm. (Volume of a cone $= \frac{1}{3}\pi r^2 h$)

7   A square of side 10 cm is cut from one corner of a square of side 20 cm. Show that the centre of gravity of the remainder is $25\sqrt{2}/3$ cm from the corner opposite the one that has been removed.

8

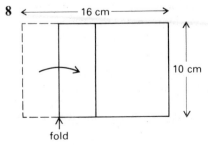

A uniform rectangular metal lamina is 16 cm long and 10 cm wide. One quarter of the lamina is folded back onto the rest of the lamina, as shown in the diagram. Find the distance of the centre of gravity from the fold.

9

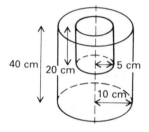

A uniform cylinder of radius 10 cm and height 40 cm has a cylindrical hole cut in one end. The hole has radius 5 cm and depth 20 cm. Find the position of the centre of gravity.

10 A square ABCD is made up of four pieces of wire each 4 cm long. AB and AD are twice the density of BC and CD. Find the distance of the centre of gravity from AB and from AD.

11

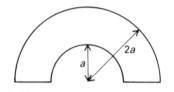

A semicircular annulus is formed by cutting a semicircle of radius a from a semicircle of radius $2a$. Find the position of the centre of gravity of the annulus.

12 A disc of radius 10 cm has its centre at the origin of a system of coordinates marked in cm. Two holes are cut out of the disc: one with its centre at $(5, 0)$ of radius 4, and one with its centre at $(-3, -3)$ of radius 3. Find the coordinates of the centre of gravity of the remainder.

*13

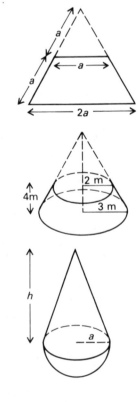

An equilateral triangle of side $2a$ has a piece removed, as shown in the diagram. Find the distance of the centre of gravity of the remainder from the base.

*14

A **frustum** is formed when the top of a cone is cut off by a plane parallel to the base. The radii of the circular faces of a particular frustum are 2 m and 3 m and its height is 4 m. Find the distance of the centre of gravity of the frustum from its base.

*15

A child's toy consists of a solid hemisphere of radius a joined to a solid right cone of radius a and height h. The hemispherical base is made of a material that is three times as dense as the conical top. Show that the centre of gravity of the body lies at a distance

$$\frac{h^2 - 9a^2}{4(6a + h)}$$

from the common face of the two solids.

Exercise 6.3

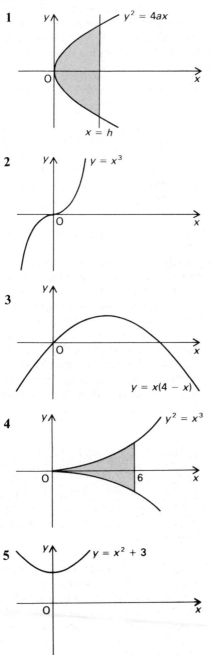

1 Find the coordinates of the centre of gravity of a uniform lamina bounded by the parabola $y^2 = 4ax$ and the double ordinate $x = h$.

2 Find the coordinates of the centre of gravity, the uniform lamina bounded by the curve $y = x^3$, the x-axis and the ordinate $x = 4$.

3 Find the coordinates of the centre of gravity, the uniform lamina bounded by the curve $y = x(4 - x)$ and the x-axis.

4 Find the coordinates of the centre of gravity of the uniform lamina bounded by the curve $y^2 = x^3$ and the double ordinate $x = 6$.

5 Find the coordinates of the centre of gravity of the uniform lamina bounded by the curve $y = x^2 + 3$ and the lines $x = 1$ and $x = 4$.

6 Find the coordinates of the centre of gravity of the uniform lamina bounded by the curve $y = x^2$ and the line $y = 3x$.

*7 Find the coordinates of the centre of gravity of the uniform lamina bounded by the curves $y = x^2$ and $x = y^2$.

*8

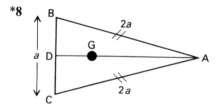

An isosceles triangle ABC has AB = AC = 2a, BC = a and the midpoint of BC is D. Show, by integration, that the centre of gravity of the triangle is one third of the way along AD from D.

Exercise 6.4

1

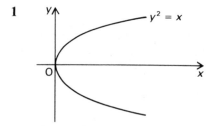

The area between the curve $y^2 = x$, the x-axis and the ordinate $x = 8$ is rotated through 360° about the x-axis. Find the coordinates of the centre of gravity of this solid of revolution.

2   Find the coordinates of the centre of gravity of the solid formed by rotating the area between the curve $y^2 = x^3$ and the line $x = 4$ about the x-axis.

3

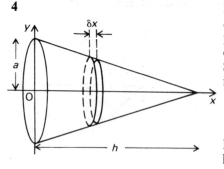

The diagram shows a uniform solid hemisphere of radius a. By dividing the hemisphere into thin elemental discs parallel to Oy and of thickness δx, show that the centre of gravity of the hemisphere lies on its axis of symmetry at a distance $3a/8$ from its plane base.

4

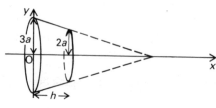

Divide a right solid cone of radius a and height h into elemental discs parallel to its base, each disc being distance x from the base of the cone. Show that the radius of such a disc, y, is given by

$$y = \frac{a(h - x)}{h}$$

Hence show that its centre of gravity lies at a distance $h/4$ from its base. (Volume of cone $= \frac{1}{3}\pi r^2 h$)

*5

The plane circular faces of the frustum of a right cone have radii $2a$ and $3a$ and the height of the frustum is h. Find the distance of its centre of gravity from its base.

*6

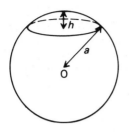

A spherical cap is cut from a uniform solid sphere of radius a by a plane. The maximum depth of the cap is h. Show that the distance of the centre of gravity of the cap from its plane face is

$$\frac{4ah - h^2}{4(3a - h)}$$

Exercise 6.5

1

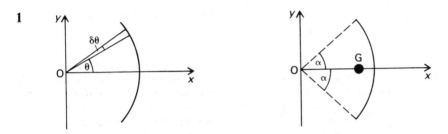

(a)   A piece of uniform wire is bent into the form of an arc of a circle of radius r, subtending an angle of 2α at the centre of the circle. Show that the centre of gravity of the wire is at a distance

$$\frac{r \sin \alpha}{\alpha}$$

from the centre of the circle.

(b)   Find the distance of the centre of gravity of a wire, in the form of a semicircular arc of radius r, from the centre of the circle.

2

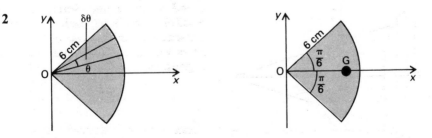

A lamina is in the form of the sector of a circle of radius 6 cm which subtends an angle $\pi/3$ at the centre of the circle. The sector may be divided into elemental sectors, each subtending an angle $\delta\theta$ at the centre of the circle, at an angle θ to the x-axis. These elemental sectors may be considered as triangles. Show that:

(a)   the area of each elemental sector is approximately $18\delta\theta$ cm² (note that for a small angle ϕ $\sin \phi \approx \phi$)

(b)   the distance of the centre of gravity of an element of an angle θ to the x-axis is $4 \cos \theta$

(c) the area of the whole sector is $6\pi\,\mathrm{cm}^2$
(d) the distance of the centre of gravity of the sector from Oy is $12/\pi\,\mathrm{cm}$.

3   Use the techniques of the last question to show that the centre of gravity of a uniform lamina in the form of a sector of a circle of radius r, subtending an angle 2α at the centre, is a distance

$$\frac{2r\sin\alpha}{3\alpha}$$

from the centre of the circle.

4   Use the result of question 3 to show that the centre of gravity of a uniform semicircular lamina of radius r is a distance $4r/3\pi$ from the centre of the circle.

5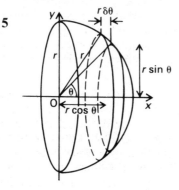

The centre of gravity of a uniform hemispherical shell is found by dividing it up into elemental rings of thickness $r\delta\theta$ parallel to the plane face of the hemisphere and at a distance $r\cos\theta$ from it. By considering values of θ from 0 to $\pi/2$, show that the centre of gravity of the shell lies on its axis of symmetry at a distance $r/2$ from the centre of the plane face.

6   AB is a non-uniform rod of length l whose weight per unit length is $(2+x)w$ at a point distant x from A. Show that the distance of its centre of gravity from A is

$$\frac{2(3+l)l}{3(4+l)}$$

*7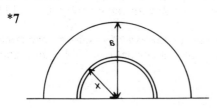

The weight per unit area of a non-uniform semi-circular lamina of radius a is wx, where x is the distance from the centre of the straight edge. Divide the lamina into semicircular rings and show that its centre of gravity is $3a/2\pi$ from the centre of the straight edge.

*8   A hollow cone of radius r and height h has no base. Find the distance of its centre of gravity from its vertex.

Exercise 6.6

. **1**   A semi-circular lamina is suspended in a vertical plane from the point at one end of its straight edge. Show that, in equilibrium, the angle between the straight edge and the vertical is $\tan^{-1}\left(\frac{4}{3}\pi\right)$.

2 A uniform solid hemisphere of weigh W is placed with its spherical surface on a smooth horizontal plane. A particle P of weight $\frac{1}{2}W$ is attached to its rim. In the position of equilibrium the plane face of the hemisphere is inclined at an angle θ to the horizontal, as shown in the diagram. Find $\tan\theta$.   [J]

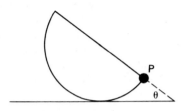

3

D ─────── C

2 m

A   X ┈┈┈┈ B
    ⟨1.25 m⟩

A uniform square lamina ABCD has side of length 2 m. A point X is marked on the side AB a distance 1.25 m from B. The piece BCX is removed. If the remaining lamina is placed in a vertical plane, with AX resting on a rough horizontal plane, show that the lamina will not topple over.

4

Y

O

⟵─ r ─⟶

A sphere of radius $(r/2)$ is cut from a uniform solid hemisphere of radius r. O is the centre of the plane base and Y is the other end of the diameter of the spherical hole from O. Show that the centre gravity of the remaining solid is on OY at a distance $r/3$ from O.

This solid is placed with its plane face on a rough plane of inclination α. A gradually increasing force P is applied at Y in a direction parallel to the line of greatest slope and *up the plane*. If the solid is of weight W find the value of P when the body is about to tilt, in terms of W and α.

5 A uniform wire of weight W, bent into a semicircular arc, has a particle of weight W attached to one end and is suspended freely from the other end. Show that when the system is in equilibrium the straight line through the ends of the wire makes an angle $\tan^{-1}(3\pi/2)$ with the horizontal.   [J]

6 The diagram shows a plane section through the axis of a uniform solid cone of height h and base radius $2h$ from which a coaxial cone of height a and base radius h has been removed.

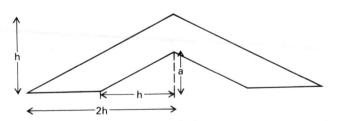

Given that the centre of gravity of the resulting solid S is at a distance $4h/15$ from its plane base, find the possible values of a.

The solid S, which is of weight W, is now placed with its flat base on a plane inclined at an angle θ to the horizontal, the plane being sufficiently rough to prevent slipping. Find the minimum value of $\tan\theta$ such that S cannot stay unsupported on the plane and, for such values of θ, obtain the magnitude of the least force, in terms of W and θ, applied at the vertex V, that will maintain equilibrium.   [A]

7 A uniform solid is made by joining together two solid right circular cones, C_1 of height h_1 and C_2 of height h_2 (where $h_2 > h_1$), each with base radius a, so that their bases coincide. Show that the centre of mass of the solid lies in the cone C_2 at a distance $\frac{1}{4}(h_2 - h_1)$ from the common base.
Show that the solid can rest in equilibrium on a horizontal table with the curved surface of C_1 touching the table if

$$h_2 < h_1 + \frac{4a^2}{h_1} \qquad \text{[J]}$$

8 A uniform solid hemispherical bowl has internal radius $(a/2)$ and external radius a and the axes of the inner and outer surfaces coincide. The point O is the centre of a diameter of the outer rim of the bowl. Show that the centre of mass of the bowl is at a distance $45a/112$ from O.
The bowl is suspended by a light inextensible string, one end of which is attached at the point A, the other end of which is attached to a fixed point C. Find the tangent of the angle which AB makes with the vertical when the bowl hangs in equilibrium.   [L]

Miscellaneous Exercise 6B

1 A uniform solid hemisphere is of base radius $2a$. Find, by integration, the perpendicular distance from the base to the centre of gravity of the hemisphere.   [A: 6.4]

2 In a uniform circular disc of centre O and diameter 18 cm a circular hole of diameter 6 cm is cut. The centre of the hole is 4 cm from O. How far from O is the centre of mass of the remainder of the disc? What would the diameter of the hole have had to be if the centre of mass of the remainder was 1 cm from O?   [A: 6.2]

3 Show that the centre of mass of a uniform solid hemisphere of radius a is at a distance $3a/8$ from the centre of its plane face.
The hemisphere rests in equilibrium with its plane face vertical and its curved surface on a rough inclined plane. Find, correct to the nearest degree, the angle between the plane and the horizontal.   [J: 6.4]

4 The diagram shows a square OABC of side a. The mid-point of BC is D. Show that, with respect to OA and OC as axes, the coordinates of the centroid F of the triangular region ABD are $(\frac{5}{6}a, \frac{2}{3}a)$.

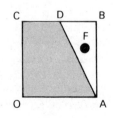

Find the coordinates of the centre of mass of a uniform lamina in the form of the figure OADC.   [J: 6.1, 6.2]

5   Show that the centre of mass of a uniform solid right circular cone of height h and base-radius a is at a distance $\frac{3}{4}h$ from the vertex.

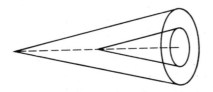

A hole in the shape of a right circular cone of base-radius $\frac{1}{2}a$ and height $\frac{2}{3}h$ is bored out of this cone; the axis of the hole coincides with that of the cone. The resulting solid is shown in the diagram. Find the distance of the centre of mass of this solid from the vertex of the original cone.   [J: 6.4]

6   Use integration to show that the centre of mass of a uniform semi-circular lamina, of radius a, is at a distance $4a/3\pi$ from O the mid-point of its straight edge.
A semi-circular lamina, of radius b and with O as the mid-point of its straight edge, is removed from the first lamina. Show that the centre of mass of the resulting lamina is at a distance $\bar{x}$ from O, where

$$\bar{x} = \frac{4}{3\pi} \frac{(a^2 + ab + b^2)}{(a + b)}$$

Hence find the position of the centre of mass of a uniform semi-circular arc of radius a.   [L: 6.2]

7   Prove that the centre of mass of a uniform semicircular lamina of radius r is at a distance $(4r)/(3\pi)$ from the centre of the semicircle.
From a uniform lamina in the form of a semicircle of radius $2a$ a concentric semicircular portion of radius a is removed. Find the distance from the centre of the semicircles to the centre of mass of the remaining material.   [J: 6.5]

8   Show that the centre of gravity of a uniform triangular lamina lies at the point of intersection of the medians.
A lamina ABCD, in the form of a rhombus, consists of two uniform triangular laminae, ABC and ADC, joined along AC. Both triangles are equilateral of side $2a$, ABC is of mass $2m$ and ADC is of mass m. Determine the perpendicular distance of the centre of gravity G of the rhombic lamina from AC and from AB.
The rhombic lamina is then suspended by a string attached to A. Determine:
(a)   the tangent of the angle between AC and the vertical
(b)   the tension in the string when a mass is attached at D so that AC is vertical.   [A: 6.2, 6.6]

9 A uniform wire of length πa is bent to form a semicircle of radius a. Prove that the distance of the centre of mass of the wire from the centre of the circle is $2a/\pi$.

Another uniform wire of length $(2 + \pi)a$ is bent into the form of a semicircular arc together with the diameter AB joining the ends of the arc, and hangs in equilibrium from the point A. Show that AB makes an angle θ with the vertical, where

$$\tan \theta = \frac{2}{2 + \pi}$$

[J: 6.5, 6.6]

10 A uniform square lamina ABCD of side $2a$ has a square PQRS of side $2b$ $(b < a)$ cut from it, where the point P and Q lie on BC. Find the distance of G, the centre of mass of the remaining lamina, from BC. Show that G will lie on the straight line through R and S if

$$2b = (\sqrt{5} - 1)a$$

[J: 6.2]

*11 Show that the centre of mass of a uniform solid hemisphere of radius a is at a distance $3a/8$ from the centre of its plane face.

Such a hemisphere, of weight W, stands on a rough horizontal table with its plane face vertical. Equilibrium is maintained by a uniform rod AB of length $a\sqrt{8}$ and weight W which is hinged to the highest point A of the hemisphere and to a point B on the table. The rod lies in the vertical plane containing the axis of symmetry of the hemisphere. Find the force of friction at the point of contact between the hemisphere and the table and show that equilibrium is possible only if the coefficient of friction is at least $\frac{1}{9}$.

Evaluate the horizontal and vertical components of the force exerted on the hemisphere by the rod at A, indicating clearly the directions in which they act.
[J:6.4, 6.6]

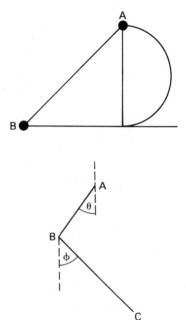

*12 Two thin uniform rods AB and BC, of lengths $2a$ and $2b$ respectively and of the same mass per unit length, are rigidly connected at B and suspended freely from A. In the equilibrium position AB and BC make angles θ and φ, respectively, with the downward vertical (as shown in the diagram). Express the ratio a/b in terms of $\sin \theta$ and $\sin \varphi$.   [J: 6.6]

13 A composite body B is constructed by joining, at their rims, the bases of a uniform solid hemisphere of base radius a and a solid cylinder, constructed from different uniform material, and whose base is also of radius a. The mass of the cylinder is twice that of the hemisphere. When B is suspended from a point on the rim of the base of the hemisphere the axis of the cylinder is inclined at an angle $\sin^{-1}(1/\sqrt{5})$ to the vertical. Show that the centre of mass of B is at a distance $2a$ from the base of the hemisphere and find the height of the cylinder. [A: 6.2]

*14 Prove by integration that the centre of mass of a uniform solid right circular cone of height h and base radius r is at a distance $\frac{3}{4}h$ from the vertex. Such a cone is joined to a uniform solid right circular cylinder, of the same material, with base radius r and height l, so that the plane base of the cone coincides with a plane face of the cylinder. Find the centre of mass of the solid thus formed. Show that, if $6l^2 \geqslant h^2$, this solid can rest in equilibrium on a horizontal plane, with the curved surface of the cylinder touching the plane.
Given that $l = h$, show that the solid can rest in equilibrium with its conical surface touching the plane provided that $r \geqslant \frac{1}{4}h\sqrt{5}$. [J: 6.4, 6.6]

15 Show that the centre of mass of a uniform solid hemisphere of radius a is at a distance $3a/8$ from its plane face.
A solid S is formed by joining the hemisphere to a uniform solid cylinder of the same material and of radius a and height $3a/2$ so that the plane face of the hemisphere coincides with one end of the cylinder. Find the distance of the centre of mass of S from its plane face.
Show that S could rest with its plane face on a sufficiently rough plane inclined to the horizontal at an angle β provided that

$$\tan \beta < \frac{52}{57}$$

The solid S, of weight W, rests with a point of its hemispherical surface on a horizontal plane and with its axis inclined at an angle θ to the vertical, equilibrium being maintained by a couple. Calculate the moment of this couple and indicate its sense in a diagram. [J: 6.2, 6.4, 6.6]

16 Show by integration that the centre of mass of a uniform solid right circular cone of height h is at a distance $\frac{3}{4}h$ from the vertex.
Two uniform solid right circular cones, each with the same base radius a and the same density, have heights h and λh, where $\lambda > 1$. These cones are joined together, with their circular bases coinciding, to form a spindle. Show that the centre of mass of this spindle is at a distance $\frac{1}{4}h(3\lambda + 1)$ from the vertex of the larger cone.
Given that $a = h$, show that the spindle can rest in equilibrium with the curved surface of the smaller cone in contact with a horizontal plane provided that $\lambda \leqslant 5$. [L: 6.2, 6.4, 6.6]

17 Prove, by integration, that the position of the centre of mass of a uniform solid right circular cone is one quarter of the way up the axis from the base. From a uniform solid right circular cone of height H is removed a cone with the same base and of height h, the two axes coinciding. Show that the centre of mass of the remaining solid S is a distance

$$\tfrac{1}{4}(3H - h)$$

from the vertex of the original cone.

The solid S is suspended by two vertical strings, one attached to the vertex and the other attached to a point on the bounding circular base. Given that S is in equilibrium, with its axes of symmetry horizontal, find, in terms of H and h, the ratio of the magnitude of the tension in the string attached to the vertex to that in the other string.   [L: 6.2, 6.4, 6.6]

*18 The diagram represents a vertical central cross-section ABCD of a uniform cube of side 2a and weight W resting on a rough horizontal plane. The coefficient of friction between the cube and the plane is μ. A force of magnitude P and inclined at an acute angle θ ($>45°$) to the downward vertical, is applied at A, in the plane ABCD, as shown in the diagram. Find the value of P when:

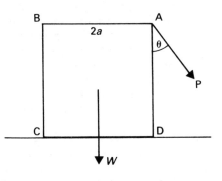

(a) the cube is about to tilt
(b) the cube is about to slide.

Hence determine the range of values of μ such that, as P is gradually increased, the cube slips before it tilts.

The direction of the force at A is now changed so that while remaining at an angle θ to the downward vertical, its positive horizontal component is towards the inside of the cube, the positive vertical component still being downward. Determine the value of P when the cube is about to tilt, provided μ is sufficiently large.   [A: 6.6]

*19 Show by integration that the centre of mass of a uniform triangular lamina PQR is at a distance $\tfrac{1}{3}h$ from QR, where h is the length of the altitude PS. A uniform piece of cardboard is in the form of a rectangle ABCD, in which AB = 10a, BC = 6a, and E is the point in AB such that AE = 4a. The cardboard is folded along CE so that the edge CB lies along the edge CD to form a trapezium shaped lamina AECD and the triangular part CEB is of double thickness. Find the distance of the centre of mass of the lamina AECD from (a) AD, (b) AE. Show that, when this lamina is freely suspended from the vertex A, the edge AD makes an angle $\tan^{-1}(11/9)$ with the vertical.

The lamina AECD, which is of weight W, is next suspended by two vertical strings attached at A and D and is at rest with AD horizontal. Find the tension in each string.   [L: 6.5, 6.2, 6.6]

**20 Two uniform spheres of equal density but of radii na and a respectively are in contact. Show that the distance from the centre of the sphere of radius na to the centre of gravity of the two spheres regarded as one combined body is

$$\frac{a(n + 1)}{n^3 + 1}$$

Two such spheres are in equilibrium in contact with each other on the inner surface of a smooth fixed hemispherical bowl. The circular rim of the bowl is horizontal and of radius $5a$.

(a)   When the spheres are of equal radius a and weight W, determine the reaction between the spheres and the reactions between the spheres and the bowl.

(b)   When the spheres are of radii $2a$ and a respectively and the line of centres is horizontal, find the ratio of the magnitude of the reaction of the bowl on the sphere of radius $2a$ to that of the bowl on the sphere of radius a.   [A:6.2, 4.5]

CIRCULAR MOTION

7.1 Speed and Acceleration

Linear Speed and Angular Speed

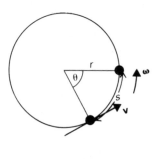

When a body is moving in a circular path.

$$\text{linear speed} = v = \frac{ds}{dt}$$

where $s = $ distance travelled along the arc

and

$$\text{angular speed} = \omega = \frac{d\theta}{dt}$$

where θ is the angle termed through in radians, ω is read as omega.

Also $s = r\theta$ where $r = $ radius $= $ constant
Differentiating w.r.t. time:

$$\frac{ds}{dt} = \frac{r\,d\theta}{dt}$$

$$\Rightarrow \boxed{v = r\omega}$$ ← **Note** that this relationship only holds if ω is measured in radians/ unit time.

Units of Angular Speed

Angular speed is usually measured in either rad/s or rev/min. Engineers use rev/min because it is simple to measure, but for calculations it is necessary to use a unit that involves radians.

Example 1   Find the speed of a particle on the edge of a disc of radius 40 cm which is rotating with an angular speed of (a) 10 rad/s (b) 10 rev/min.

(a) $r = 40\,\text{cm} = 0.4\,\text{m}$ $\omega = 10\,\text{rad/s}$.
    $v = r\omega \Rightarrow v = 0.4 \times 10 = 4\,\text{m/s}$
        $\therefore$ speed of particle $= 4\,\text{m/s}$

(b) $\omega = 10\,\text{rev/min}$ $r = 0.4\,\text{m}.$
$\quad\quad = 10 \times 2\pi\,\text{rad/min}$
$$= \frac{10 \times 2\pi}{60}\,\text{rad/s}$$

$$v = r\omega \Rightarrow v = 0.4 \times \frac{10 \times 2\pi}{60} = \frac{2\pi}{15}$$

$$\therefore \text{speed of particle} = \frac{2\pi}{15}\,\text{m/s} \; (\approx 0.419\,\text{m/s})$$

Example 2   A body is fastened onto the end of a piece of string and swung round in a circle of radius 2 m with a speed of 10 m/s. Find the angular speed of the body in (a) rad/s (b) rev/min.

(a) $v = 10\,\text{m/s}\quad r = 2\,\text{m}$
$$v = r\omega \Rightarrow \omega = \frac{v}{r} \Rightarrow \omega = \frac{10}{2} = 5\,\text{rad/s}.$$
$$\therefore \text{angular speed} = 5\,\text{rad/s}.$$

(b) $1\,\text{rev} = 2\pi\,\text{rad}\quad \Rightarrow 1\,\text{rad} = \frac{1}{2\pi}\,\text{rev}.$

$$\therefore 5\,\text{rad/s} = \frac{5}{2\pi}\,\text{rev/s} = \frac{5}{2\pi} \times 60\,\text{rev/min} = \frac{150}{\pi}\,\text{rev/min}$$

$$\therefore \text{angular speed} = \frac{150}{\pi}\,\text{rev/min} \; (\approx 48\,\text{rev/min})$$

Linear Velocity and Angular Velocity

When a body is moving in a circular path its linear velocity has magnitude v and its direction is always tangential to the circle. Also its angular velocity has magnitude ω and its direction is either clockwise or anticlockwise.

Angular Acceleration

The angular acceleration of a body is $\dfrac{d^2\theta}{dt^2}$ or $\ddot{\theta}$

Acceleration

$$\text{Acceleration} = \frac{\text{change in velocity}}{\text{time}}$$
and is therefore a vector.

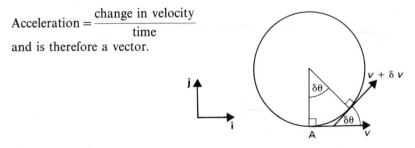

Acceleration at $A = \underset{\delta t \to 0}{\text{Lim}} \dfrac{1}{\delta t} [(v + \delta v)\cos \delta \theta \mathbf{i} + (v + \delta v)\sin \delta \theta \mathbf{j} - v\mathbf{i}]$

$$= \underset{\delta t \to 0}{\text{Lim}} \left[\frac{(v\cos \delta \theta - v)\mathbf{i}}{\delta t} + \frac{\delta v}{\delta t}\cos \delta \theta \mathbf{i} + \frac{v \sin \delta \theta}{\delta t}\mathbf{j} + \frac{\delta v}{\delta t}\sin \delta \theta \mathbf{j} \right]$$

$$= 0\mathbf{i} + \frac{dv}{dt}\mathbf{i} + \frac{v\, d\theta}{dt}j + 0\mathbf{j}$$

$$\qquad\qquad \uparrow \qquad\quad \uparrow$$

As $\delta t \to 0$, $\delta \theta \to 0$, $\sin \delta \theta \to \delta \theta$, $\cos \delta \theta \to 1$, $\dfrac{\delta v}{\delta t} \to \dfrac{dv}{dt}$, $\dfrac{\delta \theta}{\delta t} \to \dfrac{d\theta}{dt}$.

$\therefore$ Acceleration $= \dfrac{dv}{dt}\mathbf{i} + v\omega\mathbf{j}$

Since $v = r\omega$, $v\omega = r\omega^2$ or $\dfrac{v^2}{r}$

Acceleration $= r\omega^2$ or $\dfrac{v^2}{r}$ towards the centre of the circle

and $\dfrac{dv}{dt}$ tangentially.

Motion in a Circle at Constant Speed

In this case, $\dfrac{dv}{dt} = 0$

$\therefore$ acceleration is purely $r\omega^2 \left(\text{or } \dfrac{v^2}{r} \right)$ towards the centre of the circle.

learn both forms

Example 3   A body moves in a horizontal circle of radius 2 m at a speed of 10 m/s. Find the magnitude and direction of its acceleration.

$v = \text{constant} = 10 \Rightarrow \dfrac{dv}{dt} = 0$

$\therefore$ acceleration $= \dfrac{v^2}{r} = \dfrac{10^2}{2}$

$\Rightarrow$ acceleration $= 50 \text{ m/s}^2$ towards the centre of the circle

- Exercise 7.1   See page 215

7.2 Circular Motion with Constant Speed

Force Required to Keep a Body Moving in a Circle at Constant Speed

$$F = \frac{mv^2}{r}$$

or $F = mr\omega^2$ towards the centre of the circle.

Examples of central forces:

(i) Particle on a string moving on a smooth horizontal surface.

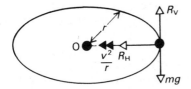

The tension in the string causes the radial acceleration.

$$T = \frac{mv^2}{r}$$

If the string breaks, the particle moves off tangentially.

(ii) Bead on a smooth horizontal circular wire.

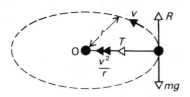

The normal reaction between the bead and the wire has two components: R_V vertically upwards to balance mg and R_H horizontally, where

$$R_H = \frac{mv^2}{r}$$

(iii) Car rounding a bend on a level road.

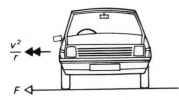

The central force is provided by the friction between tyres and road,

$$F = \frac{mv^2}{r}$$

(iv) Particle resting on a rotating disc.

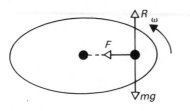

The central force is provided by the friction between the particle and the rotating disc.

Note that, if ω varied, a second frictional force, normal to the radial friction, would be induced to keep the particle in place on the disc.

Example 4   A particle of mass 5 kg is attached to a point on a smooth horizontal table by a rope 1.2 m long. If it describes circles at 150 rev/min, find the tension in the rope.

angular velocity $= \omega = 150$ rev/min $= 5\pi$ rad/s

$$T = mr\omega^2 = 5 \times 1.2 \times (5\pi)^2$$
$$= 150\pi^2 \text{ N}$$
$$\Rightarrow T = 1480 \text{ N}$$

Example 5   A particle of mass 200 grams lies on a horizontal turntable at a distance of 30 cm from the axis of revolution. The particle is on the point of slipping when travelling at 1.5 m/s. Find the coefficient of friction.

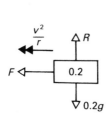

Resolving vertically:

$$R = 0.2g$$

$$F = \mu R:$$

$$F = \mu 0.2g$$

$$F = ma \rightarrow:$$

$$F = \frac{mv^2}{r}$$

$$\Rightarrow \mu 0.2g = 0.2 \frac{2.25}{0.3}$$

$$\Rightarrow \quad \mu = \frac{0.2 \times 2.25}{9.8 \times 0.3}$$

$$\Rightarrow \quad \mu = 0.15$$

Example 6   Two particles, X and Y, both of mass 2 kg are connected by a light inextensible string which passes through a small smooth hole in a smooth horizontal table. The strip is 1 m long and Y hangs freely 20 cm below the table when X makes horizontal circles on top of the table. How many revolutions per minute does X make for Y to hang in this position?

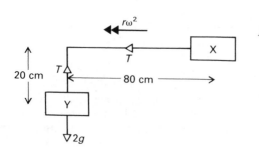

$$F = ma \uparrow \text{for Y:}$$

$$T = 2g$$

$$F = ma \leftarrow \text{for X:}$$

$$T = mr\omega^2$$

$$\Rightarrow 2g = 2 \times 0.8 \times \omega^2$$

$$\Rightarrow \omega^2 = \frac{9.8}{0.8}$$

$$\Rightarrow \quad \omega = 3.5 \text{ rad/s}$$

$$\Rightarrow \quad \omega = \frac{3.5}{2\pi} \times 60 \, \text{rev/min}$$

$$\omega = 33.4 \, \text{rev/min}$$

● Exercise 7.2   See page 217

7.3 The Conical Pendulum

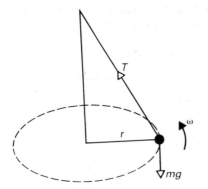

A light inextensible string has one end attached to a fixed point. The other end is fastened to a bob which describes a horizontal circle. Since the moving string describes the curved surface of a cone, this system is known as a **conical pendulum.**

Example 7   A bob of mass 5 kg moves as a conical pendulum at the end of a string of length 90 cm. If the bob describes a circle with a constant angular velocity of 4 rad/s, find the tension in the string and the angle the string makes with the vertical.

$F = ma\uparrow$:

$$T \cos \theta = 5g \qquad\qquad (1)$$

$F = ma\leftarrow$:

$$T \sin \theta = mr\omega^2$$
$$= 5 \cdot r \cdot 16$$

but $r = 0.9 \sin \theta$

$$\therefore \quad T \sin \theta = 80 \times (0.9 \sin \theta)$$
$$\Rightarrow \qquad T = 80 \times 0.9 = 72$$
$$\therefore \text{ tension} = 72 \, \text{N}$$

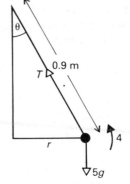

$$(1) \Rightarrow \cos \theta = \frac{5g}{72} = 0.680\dot{5}$$
$$\Rightarrow \qquad \theta = 47.1°$$

∴ the string makes an angle of 47.1° to the vertical.

● Exercise 7.3   See page 218

7.4 Further Problems on Conical Pendulums

Example 8   Two light inextensible strings, each of length l, are each attached at one end to a particle of mass m. The other ends of the strings are attached to two fixed points A and B, where B is a distance $\sqrt{2}l$ vertically below A. When both strings are taut, the particle is given an angular velocity ω and then moves in a horizontal circle with constant speed. Find the tensions in the two strings and show that this motion is only possible if $\omega^2 \geqslant (g\sqrt{2}/l)$

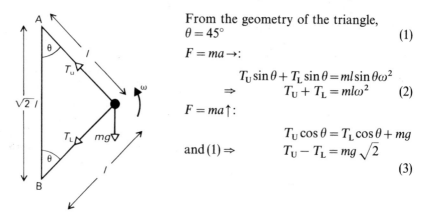

From the geometry of the triangle,
$$\theta = 45° \tag{1}$$

$F = ma \rightarrow$:
$$T_U \sin\theta + T_L \sin\theta = ml\sin\theta\omega^2$$
$$\Rightarrow \qquad T_U + T_L = ml\omega^2 \tag{2}$$

$F = ma \uparrow$:
$$T_U \cos\theta = T_L \cos\theta + mg$$
and (1) $\Rightarrow \qquad T_U - T_L = mg\sqrt{2} \tag{3}$$

Solving (2) and (3) simultaneously:
$$T_U = \tfrac{1}{2}m[l\omega^2 + g\sqrt{2}]$$
$$\underline{T_L = \tfrac{1}{2}m[l\omega^2 - g\sqrt{2}]}$$

This motion is only possible if $T_L \geqslant 0 \longleftarrow$ | That is, T_L cannot be negative.

$$\Rightarrow l\omega^2 - g\sqrt{2} \geqslant 0$$
$$\Rightarrow \omega^2 \geqslant \frac{g\sqrt{2}}{l}$$ QED

Remember that if the particle is attached to two separate strings or fastened to a point on one string, then the tensions in both posts of the string are different. But, if the particle is free to move on one string then the tension is the same throughout the string—see the next example.

Example 9   A bead of mass m is threaded onto a string whose ends are fastened to two points A and B with B a distance $2a$ vertically below A. The bead rotates in a horizontal circle with centre B and radius a. Show that the angular velocity of the bead is $[(g/2a)(\sqrt{5}+1)]^{1/2}$

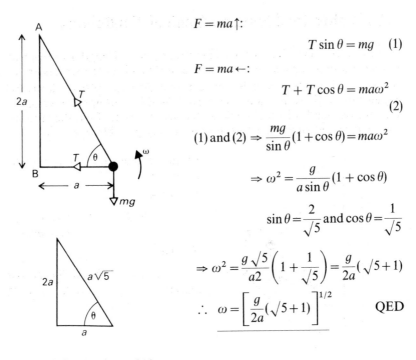

$F = ma\uparrow$:

$$T \sin \theta = mg \quad (1)$$

$F = ma\leftarrow$:

$$T + T \cos \theta = ma\omega^2 \quad (2)$$

(1) and (2) $\Rightarrow \dfrac{mg}{\sin \theta}(1 + \cos \theta) = ma\omega^2$

$$\Rightarrow \omega^2 = \frac{g}{a \sin \theta}(1 + \cos \theta)$$

$$\sin \theta = \frac{2}{\sqrt{5}} \text{ and } \cos \theta = \frac{1}{\sqrt{5}}$$

$$\Rightarrow \omega^2 = \frac{g\sqrt{5}}{a2}\left(1 + \frac{1}{\sqrt{5}}\right) = \frac{g}{2a}(\sqrt{5} + 1)$$

$$\therefore \quad \omega = \left[\frac{g}{2a}(\sqrt{5} + 1)\right]^{1/2} \qquad \text{QED}$$

• Exercise 7.4   See page 219

7.5 Motion of a Vehicle Round a Curve: Banking

Constant Speed Round a Bend on a Horizontal Road

Case 1:   Cars

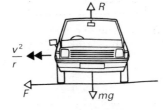

The central force required is $\dfrac{mv^2}{r}$ and this is provided by friction between the tyres and the road. If, however, $F_{\text{max}} < \dfrac{mv^2}{r}$ the car will side slip.

Case 2:   Cycles

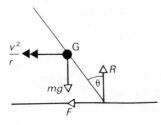

The central force is provided by the friction.

Note also that, since there are only three forces acting on the cycle, the resultant of R and F must pass through G.

Constant Speed Round Banked Curves

Case 1:   Car travelling at Optimum Speed

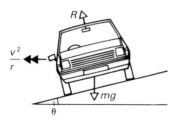

At the optimum speed, the car has no tendency to slip up or down the banked slope and

$$R \sin \theta = \frac{mv^2}{r}$$

Case 2:   Car travelling at higher than Optimum Speed

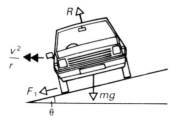

When the speed is higher than the optimum, $R \sin \theta$ is not sufficient to provide the central force. Hence the car tends to sideslip **up** the slope and

$$R \sin \theta + F_1 \cos \theta = \frac{mv^2}{r}$$

Case 3:   Car travelling at lower than Optimum Speed

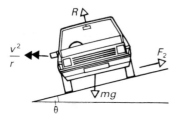

When the speed is lower than optimum, $R \sin \theta > (v^2/r)$. Hence the car tends to sideslip **down** the slope and

$$R \sin \theta - F_2 \cos \theta = \frac{mv^2}{r}$$

Example 10   A car travels along a road round a bend of radius 250 m which is banked at an angle θ. If the car has no tendency to sideslip at 20 m/s find θ.

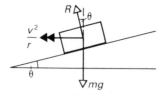

Resolving vertically:

$$R \cos \theta = mg \qquad (1)$$

$F = ma \rightarrow$:

$$R \sin \theta = \frac{mv^2}{r} = \frac{m20^2}{250} \qquad (2)$$

$$(1) \text{ and } (2) \Rightarrow \frac{R \sin \theta}{R \cos \theta} = \frac{m400}{250} \times \frac{1}{mg}$$

$$\Rightarrow \quad \tan \theta = \frac{400}{250g}$$

$$\Rightarrow \quad \theta = 9.3°$$

Example 11   A car rounds a bend of radius 100 m which is banked at $\tan^{-1}\frac{1}{2}$ to the horizontal. Find the greatest and least speeds at which the car can be driven without slipping occurring if the coefficient of friction between the tyres and the road is 0.4.

At maximum speed, the car is on the point of slipping up the slope and

$$F_1 = \mu R \qquad\qquad (1)$$

Resolving vertically:

$$R\cos\alpha = F_1 \sin\alpha + mg \qquad (2)$$

$F = ma \rightarrow$:

$$F_1 \cos\alpha + R\sin\alpha = \frac{mv_1^2}{100} \qquad\qquad (3)$$

$$(1) \text{ and } (2) \Rightarrow R(\cos\alpha - \mu\sin\alpha) = mg \qquad\qquad (4)$$

$$(1) \text{ and } (3) \Rightarrow R(\mu\cos\alpha + \sin\alpha) = \frac{mv_1^2}{100} \qquad\qquad (5)$$

$$(4) \text{ and } (5) \Rightarrow \frac{v_1^2}{100g} = \frac{\mu\cos\alpha + \sin\alpha}{\cos\alpha - \mu\sin\alpha}$$

$$\Rightarrow \frac{v_1^2}{100g} = \frac{\mu + \tan\alpha}{1 - \mu\tan\alpha}$$

but $\mu = 0.4$ and $\tan\alpha = \frac{1}{2}$.

$$\therefore \frac{v_1^2}{100g} = \frac{0.9}{0.8} \Rightarrow v_1 = v_{\max} = 33.2 \text{ m/s}.$$

At minimum speed, the car is on the point of slipping down the slope and

$$F_2 = \mu R_2 \qquad (6)$$

Resolving vertically:

$$R\cos\alpha + F_2 \sin\alpha = mg \qquad (7)$$

$F = ma \rightarrow$:

$$R\sin\alpha - F_2\cos\alpha = \frac{mv_2^2}{100} \qquad (8)$$

$$(6) \text{ and } (7) \quad\Rightarrow\quad R(\cos\alpha + \mu\sin\alpha) = mg \qquad\qquad (9)$$

$$(6) \text{ and } (8) \quad\Rightarrow\quad R(\sin\alpha - \mu\cos\alpha) = \frac{mv_2^2}{100} \qquad\qquad (10)$$

$$(9) \text{ and } (10) \Rightarrow \frac{v_2^2}{100g} = \frac{\sin\alpha - \mu\cos\alpha}{\cos\alpha + \mu\sin\alpha} = \frac{\tan\alpha - \mu}{1 + \mu\tan\alpha}$$

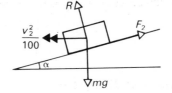

but $\mu = 0.4$ and $\tan \alpha = \frac{1}{2}$

$$\therefore \quad \frac{v_2^2}{100g} = \frac{0.1}{1.2} \Rightarrow v_2 = v_{min} = 9.04\,\text{m/s}$$

Example 12   In Japan the gauge of railway tracks is 1.1 m. A particular stretch of track has a radius of 320 m and trains round this bend at an average speed of 144 km/h. How much should the outer rail be raised above the inner rail in order that trains travelling at this speed should exert no lateral force on the rails?

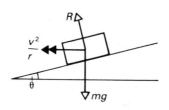

$$v = 144\,\text{km/h} = \frac{144 \times 1000}{3600}\,\text{m/s}$$

$$\Rightarrow v = 40\,\text{m/s}$$
Also $r = 320$ m.

Resolving vertically:

$$R\cos\theta = mg \qquad (1)$$

$$F = ma \rightarrow:$$

$$R\sin\theta = \frac{m40^2}{320} = 5m \qquad (2)$$

$(2) \div (1) \Rightarrow \tan\theta = \dfrac{5m}{mg} = \dfrac{5}{9.8}$ where θ = angle of bank

$$\Rightarrow \quad \theta = 27°$$

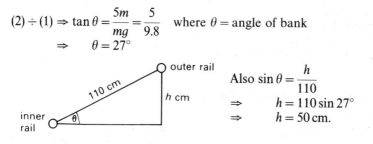

Also $\sin\theta = \dfrac{h}{110}$

$$\Rightarrow \quad h = 110\sin 27°$$
$$\Rightarrow \quad h = 50\,\text{cm}.$$

$\therefore$ outer rail should be raised 50 cm above the inner rail.

Angle of Bank

For a railway track the angle is chosen for the average speed at which trains take the bend. At higher speeds than this, there is a sideways thrust on the outer rail and, at lower speeds, there is a sideways thrust on the inner rail. For a racing track, the banking gets gradually steeper towards the outside of the track. For higher speeds, the car is steered, or skids, towards the outer part of the track, where it is steeper.

• Exercise 7.5   See page 220

7.6 Motion in a Vertical Circle I: Circular Motion on a String

The only external force acting on the body is the tension. As the tension is perpendicular to the direction of motion, it does no work. Thus this system of forces is conservative and the Law of Conservation of Energy can be applied.

To solve problems like these, we apply $F = ma$ radially and the Law of Conservation of Energy.

Case 1:   The string stays below the horizontal
Conservation of energy:

$$\tfrac{1}{2}mv_A^2 - \tfrac{1}{2}mv_P^2 = mgr(1 - \cos\theta)$$

$F = ma$, radially:

$$T - mg\cos\theta = \frac{mv_P^2}{r}$$

$$\Rightarrow T = mg\cos\theta + \frac{mv_P^2}{r}$$

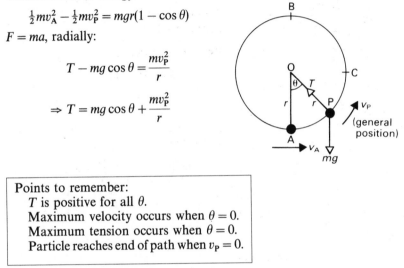

Points to remember:
  T is positive for all θ.
  Maximum velocity occurs when $\theta = 0$.
  Maximum tension occurs when $\theta = 0$.
  Particle reaches end of path when $v_P = 0$.

Case 2:   The string moves above the horizontal
Conservation of energy:

$$\tfrac{1}{2}mv_A^2 - \tfrac{1}{2}mv_P^2 = mgr(1 + \cos\alpha)$$

$F = ma$ radially:

$$T + mg\cos\alpha = \frac{mv_P^2}{r}$$

$$\Rightarrow T = \frac{mv_P^2}{r} - mg\cos\alpha$$

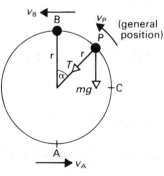

Points to remember:
  String goes slack when $T = 0$.
  Particle leaves circular path when $T = 0$.
  Condition for complete revolution is that
  the tension at B must not be negative $\Rightarrow T_B \geqslant 0$.

Example 13   A particle of mass m is suspended from a fixed point O by a light inextensible string of length l. It is given a horizontal velocity of $3\sqrt{(gl/2)}$. Find the tension in the string when the string is (a)   30° below the horizontal (b)   30° above the horizontal. Also find the distance of the particle above O when the string goes slack.

(a)

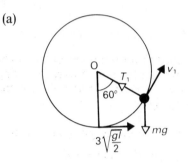

Conservation of energy:
$$\tfrac{1}{2}m(\tfrac{9}{2}gl - v_1^2) = mgl(1 - \cos 60°)$$
$$\Rightarrow \qquad \tfrac{9}{2}gl - v_1^2 = 2gl(1 - \tfrac{1}{2})$$
$$\Rightarrow \qquad v_1^2 = \tfrac{9}{2}gl - gl = \tfrac{7}{2}gl \qquad (1)$$

$F = ma$ radially:
$$T_1 - mg\cos 60° = \frac{mv_1^2}{l}$$

$$(1) \Rightarrow \qquad T_1 = \tfrac{7}{2}mg + \frac{mg}{2}$$

$$\Rightarrow \qquad \underline{T_1 = 4mg}$$

(b)

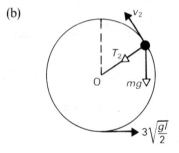

Conservation of energy:
$$\tfrac{1}{2}m(\tfrac{9}{2}gl - v_2^2) = mgl(1 + \cos 60°)$$
$$\Rightarrow \qquad v_2^2 = \tfrac{9}{2}gl - 3gl = \frac{3gl}{2} \qquad (2)$$

$F = ma$ radially:
$$T_2 + mg\cos 60° = \frac{mv_2^2}{l}$$

$$(2) \Rightarrow \qquad T_2 = \frac{3mg}{2} - \frac{mg}{2}$$

$$\Rightarrow \qquad \underline{T_2 = mg}$$

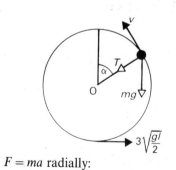

Conservation of energy:
$$\tfrac{1}{2}m(\tfrac{9}{2}gl - v^2) = mgl(1 + \cos \alpha)$$
$$\Rightarrow \qquad v^2 = \tfrac{9}{2}gl - 2gl - 2gl\cos \alpha$$
$$\Rightarrow \qquad v^2 = \tfrac{5}{2}gl - 2gl\cos \alpha \qquad (3)$$

$F = ma$ radially:
$$T + mg\cos \alpha = \frac{mv^2}{l}$$

$(3) \Rightarrow \qquad\qquad T = \tfrac{5}{2}mg - 2mg \cos \alpha - mg \cos \alpha$

String becomes slack when $T = 0$

$$\Rightarrow 3mg \cos \alpha = \tfrac{5}{2}mg$$

$$\Rightarrow \qquad \cos \alpha = \tfrac{5}{6}$$

Distance above $O = l \cos \alpha = \dfrac{5l}{6}$

Example 14   A particle of mass m is attached to a fixed point O by a light string of length l. It is held on the same horizontal level as O with the string taut and is projected vertically downwards with speed V. Show that, if the particle makes complete revolutions then $V^2 \geqslant 3gl$.

Conservation of energy from A to B:

$$\tfrac{1}{2}m(V^2 - v_T^2) = mgl$$

$$\therefore v_T^2 = V^2 - 2gl \qquad (1)$$

$F = ma$ radially at B:

$$T + mg = \frac{mv_T^2}{l}$$

$(1) \Rightarrow \qquad T = \dfrac{m}{l}(V^2 - 2gl) - mg$

$\Rightarrow \qquad T = \dfrac{mV^2}{l} - 3mg \qquad (2)$

For complete revolutions, the tension at the top of the circle must be $\geqslant 0$.

$$\therefore (2) \Rightarrow \frac{mV^2}{l} - 3mg \geqslant 0$$

$$\Rightarrow \underline{V^2 \geqslant 3gl} \qquad\qquad [\text{QED}]$$

- Exercise 7.6   See page 221

7.7 Motion in a Vertical Circle II: Various Configurations

Problems involving motion in a vertical circle can be classified into two types:

Type 1: Where the body could leave the circle under certain circumstances
Examples of this motion are:
 (i) particle on the end of a string
 (ii) particle on the inside of a circular arc
 (iii) particle on the outside of a circular arc.
In each case, circular motion ceases when the tension, or normal reaction, becomes zero. The condition for complete revolutions is that, at the highest point of the motion, the tension, or the normal reaction, is greater than, or equal to zero.

Type 2: Where the body connot leave the circle
Examples of this motion are:
  (i)   particle on the end of a light rod
 (ii)   bead threaded onto a circular, vertical, wire
(iii)   particle moving in a circular, vertical, tube.
In these cases the condition for complete revolutions is that $v_T^2 \geqslant 0$ where v_T is
the velocity at the highest point.

Example 15   A particle of mass m is free to move in a smooth narrow tube
in the form of a circle of centre O and radius 60 cm. It is projected from the
lowest point with a velocity u. For what values of u does it make complete
revolutions?

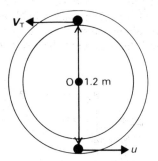

Conservation of energy:

$$\tfrac{1}{2}m(u^2 - v_T^2) = mg\,1.2$$
$$\Rightarrow v_T^2 = u^2 - 2.4g$$

Condition for complete revolutions
is that $v_T^2 \geqslant 0$
$\Rightarrow u^2 - 2.4g \geqslant 0$
$\Rightarrow \qquad u^2 \geqslant 2.4g$
$\Rightarrow \qquad \underline{u \geqslant \sqrt{2.4g}}$

Example 16   A small smooth body is at rest on top of a smooth sphere of
radius 2 m. The small body is projected horizontally with a speed of 50 cm/s.
Find its distance below the point of projection when it leaves the surface.

Conservation of energy:

$$\tfrac{1}{2}mv^2 - \tfrac{1}{2}m(0.5)^2 = mg(2 - 2\cos\theta)$$
$$\Rightarrow v^2 = 4g(1 - \cos\theta) + 0.25 \qquad (1)$$

$F = ma$ radially:

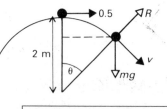

$$mg\cos\theta - R = m\frac{v^2}{2}$$

$$\Rightarrow R = m\left(g\cos\theta - \frac{v^2}{2}\right) \quad (2)$$

Note that the acceleration
is *towards* the centre of
the circle so it must
be $mg\cos\theta - R$ not
$R - mg\cos\theta$.

(1) and (2) $\Rightarrow R = m(g\cos\theta - 2g + 2g\cos\theta - \tfrac{1}{8})$

When the body leaves the surface, $R = 0$

$$\Rightarrow 3\cos\theta = \tfrac{17}{8}$$
$$\Rightarrow \quad \cos\theta = \tfrac{17}{24}$$

Distance below point of projection $= 2(1 - \cos\theta)\,\text{m}$
$$= 2(1 - \tfrac{17}{24})\,\text{m}$$
$$= \tfrac{7}{12}\,\text{m}$$

● **Exercise 7.7** See page 222

7.8 Motion in a Vertical Circle III: More Difficult Problems

Example 17   A mass of 1.5 kg is attached to a fixed point O by a light string of length 1 m. The mass is held on the same, horizontal level as O and allowed to fall. When the string is vertical it catches on a smooth peg, *P*, 60 cm below O. Find the tension in the string when the lower part next becomes horizontal.

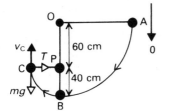

The first part of the motion, from A to B is part of a circle, centre O, radius 1 m. The second part of the motion, from B to C, is part of a circle, centre *P*, radius 40 cm.

Conservation of energy from A to C:
$$\tfrac{1}{2}mv_c^2 = mg\,0.6$$
$$\Rightarrow \quad v_c^2 = 1.2g \tag{1}$$

$F = ma$ radially at C:
$$T = 1.5\frac{v_c^2}{0.4} = \frac{1.5 \times 1.2g}{0.4} \qquad \text{using (1)}$$

$$\Rightarrow \text{tension} = 44.1\,\text{N}$$

Example 18

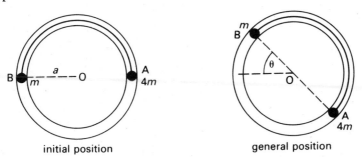

initial position   general position

A smooth narrow tube is in the form of a circle centre O, radius *a*, which is fixed in a vertical plane. The tube contains two particles, A of mass 4*m* and B of mass m which are connected by a light inextensible string. Initially A and B are on the same horizontal level as O and the system is released from rest. If, after a time t, the line AOB has turned through an angle θ, show that

$$5a\left(\frac{d\theta}{dt}\right)^2 = 6g \sin \theta.$$

Find the reaction between B and the tube in terms of m, g and θ. Find $a\dfrac{d^2\theta}{dt^2}$ in terms of g and θ and hence find the tension in the string.

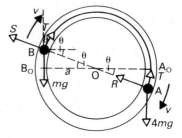

> The particles move with the same speed as long as the string is taut.

loss in PE $= 4mga \sin \theta - mga \sin \theta$
$\qquad = 3mga \sin \theta$
gain in KE $= \frac{1}{2}4mv^2 + \frac{1}{2}mv^2 = \frac{1}{2}5mv^2$

Conservation of energy:

$$\frac{1}{2}5mv^2 = 3mga \sin \theta$$
$$\Rightarrow \qquad v^2 = \frac{6}{5}ga \sin \theta$$

But
$$v = r\omega \Rightarrow \qquad v = a\frac{d\theta}{dt}$$

$$\therefore \ a^2\left(\frac{d\theta}{dt}\right)^2 = \frac{6}{5}ga \sin \theta$$

$$\Rightarrow 5a\left(\frac{d\theta}{dt}\right)^2 = 6g \sin \theta \qquad (1) \qquad \qquad \text{[QED]}$$

$F = ma$ radially for B:

$$mg \sin \theta - S = ma\left(\frac{d\theta}{dt}\right)^2$$

$$\Rightarrow \qquad\qquad S = mg \sin \theta - ma\left(\frac{d\theta}{dt}\right)^2$$

and (1) $\Rightarrow \qquad\qquad S = mg \sin \theta - m[\frac{6}{5}g \sin \theta]$

$$\Rightarrow \qquad\qquad S = -\frac{mg}{5} \sin \theta$$

$\therefore \ S = \frac{1}{5}mg \sin \theta$ towards the centre

Differentiating (1) w.r.t. t

$$\Rightarrow 10a\left(\frac{d\theta}{dt}\right)\frac{d^2\theta}{dt^2} = 6g\cos\theta\frac{d\theta}{dt}$$

$$\Rightarrow \qquad a\frac{d^2\theta}{dt^2} = \tfrac{3}{5}g\cos\theta \qquad (1)$$

Using the chain rule

$$\frac{d}{dt}(\sin\theta) = \frac{d}{d\theta}(\sin\theta) \times \frac{d\theta}{dt}$$

$$= \cos\theta\frac{d\theta}{dt}$$

and $\dfrac{d}{dt}\left[\left(\dfrac{d\theta}{dt}\right)^2\right] = 2\left(\dfrac{d\theta}{dt}\right)\dfrac{d^2\theta}{dt^2}$

$F = ma$ tangentially for B:

$$T - mg\cos\theta = m\frac{dv}{dt} = ma\frac{d^2\theta}{dt^2}$$

and (2) $\Rightarrow$ $\qquad T = mg\cos\theta + \tfrac{3}{5}mg\cos\theta$

$\Rightarrow$ $\qquad\qquad T = \tfrac{8}{5}mg\cos\theta^5$

$$v = a\frac{d\theta}{dt}$$

$$\Rightarrow \frac{dv}{dt} = a\frac{d^2\theta}{dt}$$

- **Exercise 7.8**   See page 224

7.9 Motion in a Vertical Circle IV: Problems Involving Impact

Example 19   A mass of 400 grams is attached to a fixed point O by a light string of length 1.5 m. The mass is allowed to fall from a point on the same horizontal level as O, with the string taut. Find the maximum velocity of the particle in the subsequent motion.
When it passes through the lowest point, the mass picks up a ring of mass 100 grams and takes it with it. How high does the combined mass rise?

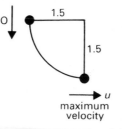

The particle has its maximum velocity at the lowest point of the swing.

Conservation of energy:

$$\tfrac{1}{2}0.4u^2 = 0.4g\,1.5$$

$$\Rightarrow \qquad u^2 = 3g$$

$$\Rightarrow \text{maximum velocity} = \sqrt{3g}\ \text{m/s}.$$

Momentum before impact $= 0.4u$
Momentum after impact $\ = 0.5u'$ $\longleftarrow$ After impact, the combined mass is $(0.4 + 0.1)\,\text{kg}$.

Conservation of momentum:

$$0.4u = 0.5u'$$

$$\Rightarrow \qquad u' = 0.8u = 0.8\sqrt{3g}$$

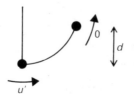

Conservation of energy:

$$\tfrac{1}{2}0.5u'^2 = 0.5gd$$

$$\Rightarrow d = \frac{u'^2}{2g} = \frac{0.64 \times 3g}{2g}$$

$$\Rightarrow d \qquad = 0.96\,\text{m}$$

∴ combined mass rises 96 cm.

• **Exercise 7.9** See page 225

7.10 + Orbits: Revision of Unit: A Level Questions

The Period of a Circular Path

$$\omega \text{ rad/s} = \frac{\omega}{2\pi} \text{ rev/sec}$$

$$\Rightarrow \frac{\omega}{2\pi} \text{ rev are performed in one second.}$$

$$\Rightarrow 1 \text{ rev is performed in } \frac{2\pi}{\omega} \text{ seconds.}$$

$$\Rightarrow \text{time for one revolution} = \boxed{\text{Period} = \frac{2\pi}{\omega}}$$

Orbits

Example 20   The gravitational force of the earth on an object of mass m is $\dfrac{km}{r^2}$ where r is the distance of the object from the the centre of the earth. A body of mass m on the surface of the earth is attracted towards the earth's centre with a force mg where $g = 9.81\,\text{m/s}^2$. If the mean radius of the earth is 6371 km, show that the value of the constant k is approximately 3.98×10^{14}. A satellite is in orbit 500 km above the earth. Find:

(a) its gravitational acceleration
(b) its angular speed
(c) the period of its orbit.

gravitational force $= \dfrac{km}{r^2} \Rightarrow$ gravitational acceleration $= \dfrac{k}{r^2}$.

When $r = 6.371 \times 10^6\,\text{m}$, acceleration $= 9.81\,\text{m/s}^2$

$$\frac{k}{(6.371 \times 10^6)^2} = 9.81$$

$$\Rightarrow \qquad k = 9.81 \times 6.371^2 \times 10^{12}$$

$$\Rightarrow \qquad k \approx 3.98 \times 10^{14} \qquad\qquad \text{[Q.E.D]}$$

(a)   distance of satellite from centre of earth $= 6371 + 500\,\text{km}$
$$= 6871\,\text{km}$$
$$= 6.871 \times 10^6\,\text{m}.$$

$$\therefore \text{ gravitational acceleration} = \frac{k}{r^2} = \frac{3.98 \times 10^{14}}{6.871^2 \times 10^{12}} = \underline{8.43\,\text{m/s}^2}$$

(b)   Also, gravitational acceleration $= r\omega^2$

$$\therefore 6.871 \times 10^6 \omega^2 = 8.43$$

$$\Rightarrow \qquad \omega^2 = \frac{8.43}{6.871 \times 10^6}$$

$$\Rightarrow \qquad \omega = 1.108 \times 10^{-3}\,\text{rad/s}$$

(c)   Period $= \dfrac{2\pi}{\omega} = \dfrac{2\pi}{1.108 \times 10^{-3}}$ seconds

$$= 5.67 \times 10^3 \text{ seconds}$$
$$= 94.51 \text{ minutes}$$
$$\approx \underline{1 \text{ hour } 35 \text{ minutes}}$$

- **Miscellaneous Exercise 7B**   See page 226

Unit 7 Exercises

Exercise 7.1

1   Express the following angular speeds in the form stated:
(a)   20 rev/min in rad/min
(b)   30 rev/min in rad/s
(c)   25 rad/s in rev/min.

2   A wheel completes one revolution every 5 seconds. What is the angular speed of the wheel in   (a) rev/min   (b) rad/s?

3   An LP record has radius 15 cm and it spins at $33\frac{1}{3}$ rev/min. Find:
(a)   its speed in rad/s
(b)   the speed of a point on the edge of the record.

4   If a body is moving in a circular path of radius 2 m, find the magnitude and direction of its acceleration if it has:
(a)   a constant speed of 3 m/s
(b)   a constant angular speed of 3 rad/s.

5   The speed v of a body moving in a circular path of radius 2 m is given by $v = t^2/100$ where t is the time elapsed in seconds. Find the tangential and normal components of its acceleration after 20 seconds.

6   A car takes a bend of radius 25 m at 12 m/s. What is its acceleration?

7   The rim of a water wheel is pushed round by a fast flowing stream which gives the rim a speed of 0.5 m/s. If the wheel has diameter 4 m, how many revolutions does it make in one minute?

8   A fairground roundabout is turning at 4 rev/min. A small boy is sitting in a car 3 m from the centre of the roundabout. What speed is he travelling at? If the roundabout is speeded up so that the boy travels at 2 m/s, how long would one revolution of the roundabout take?

9   A drill of diameter 7 mm turns at 2500 rev/min. What is the speed of its cutting edge?

10   A playground roundabout revolves at 10 rev/min. Mary hangs onto the edge of the roundabout at a distance of 1.8 m from its centre. Sue, more daring, climbs up onto the roundabout and stands at a distance of 140 cm from the centre. With what speeds are Mary and Sue moving?

11   The minute hand of a townhall clock is 2.4 m long. Find the speed of its tip.

12   A pilot knows that if he accelerates faster than $4g$ the wings of his plane will rip off. If he wants to fly a circular loop of radius 800 m, what is the fastest speed at which he can fly in:
(a) m/s   (b) km/h.
If 1 km $\approx \frac{5}{8}$ miles what is his maximum speed in miles/h?

13   The orbit of the moon around the earth is nearly circular and with a period of 27.32166 days. The mean earth-moon distance is 384,400 km. Calculate the mean orbital speed round the earth in km/s to 4 significant figures.

*14

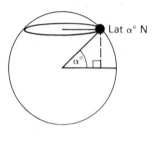

The earth is almost spherical with a mean radius of 6371 km and it spins on an axis through its two poles. Calculate its angular speed in rad/h to 4 significant figures.
Hence find the velocities (in km/h), due to the rotation of the earth, of Entebbe, on the equator, Cairo, on latitude 30°N, and Edinburgh, on latitude 56°N.
Show that the acceleration of Edinburgh is approximately 244 km/h².

*15   A bend on a level track has radius 100 m. If a car of mass 1500 kg can go round the bend without slipping sideways at 180 km/h, calculate the

acceleration in m/s² of the car and the frictional force required to prevent it slipping. If it is on the point of slipping, calculate the coefficient of friction between the tyres and the ground.
(Take $g = 10\,\text{m/s}^2$).

Exercise 7.2

1 A string, 1.5 m long has one end attached to a point on a smooth horizontal table and the other end attached to a mass of 2 kg. If the 2 kg mass describes a horizontal circle with a speed of 3 m/s, find the tension in the string.

2 A particle of mass 200 grams is attached to one end of a string 60 cm long whose other end is attached to a point on a smooth horizontal table. If the particle describes a circle and the tension in the string is a constant 1.47 N, find the angular velocity of the particle.

3 A railway engine of mass 30 tonnes moves along a level curved track in the form of an arc of a circle of radius 360 m, at 108 km/h. Find the horizontal force exerted by the rails on the flanges of the wheels.

4 A string 1.2 m long can just support a weight of mass 18 kg without breaking. One end of the string is attached to a point on a smooth horizontal table and the other end is attached to a particle of mass 3 kg. What is the greatest number of revolutions that the mass can make per minute without the string breaking?

5 A car of mass 1200 kg rounds a curve of radius 600 m on level road at 72 km/h. What is the frictional force required between the tyres and the road to prevent slipping?

6 A block of mass 2 kg is placed onto a disc revolving at a constant angular speed of 5 rad/s. If the coefficient of friction between the block and the disc is 0.4, find the maximum distance that the block can be placed from the axis of revolution without slipping.

7 Two particles, P of mass 2 kg and Q of mass 3 kg, are connected by a light inextensible string of length 50 cm. Q lies on a smooth horizontal table and the string passes through a small smooth hole in the table to P hanging freely below. Q is made to perform horizontal circles on the table and P hangs at rest 10 cm below the table. How many revolutions per minute does Q make?

8 A parcel of mass 20 kg rests on the floor of a van. When the van rounds a bend of radius 50 m on level road at a uniform speed of 18 m/s, the parcel is on the point of slipping. Find the coefficient of friction between the parcel and the floor of the van to 2 decimal places.

9 A bead of mass 100 grams is threaded on a horizontal smooth circular wire of radius 40 cm. If it moves along the wire with a constant speed of

3 m/s, find the magnitude and direction of the vertical and horizontal components of the force exerted on the bead by the wire.

10 If the coefficient of friction between the tyres of a car and the road is 0.6, find the maximum speed, in m/s to 2 d.p., at which it can go round a curve of radius 120 m on level road without sideslipping

11 Two particles, A of mass 200 grams and B of mass 150 grams are connected by a light inextensible string. A lies on a smooth horizontal table and the string passes through a small smooth hole in the table to B hanging freely below. When A describes circles at 50 rev/min, B hangs at rest 15 cm below the table. How long is the string?

12 A string of length l m has one end attached to a point on a smooth horizontal table and the other end attached to a particle of mass m. If the particle describes circles with a speed of $\sqrt{2gl}$ m/s, find the tension in the string.

13 Two particles, A of mass m and B of mass M are connected by a light inextensible string of length l m. A lies on a smooth horizontal table and the string passes through a small smooth hole in the table to B hanging freely below. If A describes horizontal circles with angular speed ω, B hangs at rest at a distance x below the table. Find x in terms of m, M, l, g and ω.

Exercise 7.3

1 A bob of mass 4 kg moves as a conical pendulum at the end of a string of length 50 cm, which is inclined at $30°$ to the vertical. Show that the tension in the string is $8\sqrt{3g}/3$ N and find the angular velocity of the bob.

2 A bob of mass 3 kg moves as a conical pendulum at the end of a string of length 80 cm. If the bob makes 90 revolutions per minute find the tension and the angle of the string to the vertical.

3 A particle of mass 200 grams is tied to the end of a string 1.2 m long whose other end is fixed to a point A. The particle describes horizontal circles below A at an angular speed of 4 rad/s. Find the tension in the string and the radius of the circles.

4 A bob of mass m is fastened to the end B of a string AB of length l. The end A is fixed and the bob describes horizontal circles at a distance d below A with constant angular speed ω. Show that $T = ml\omega^2$ and find d in terms of ω and g.

5 One end of a light inextensible string of length 60 cm is attached to a fixed point O which is at a height of 40 cm above a smooth horizontal table. At the other end of the string, which is taut, is a bob of mass 2 kg which describes a horizontal circle on the table with angular velocity ω. The centre of the circle is directly below O. Find the tension in the string and the magnitude of the reaction of the table in the case when $\omega = 3$ rad/s. What is the greatest value of ω^2 for which the particle remains on the table?

6 Two weights, of masses 6 kg and 3 kg, are connected by a light inextensible string, 90 cm long which passes through a smooth fixed ring. The 6 kg weight hangs 60 cm below the ring and the 3 kg weight describes a horizontal circle. Show that the centre of this circle is 15 cm below the ring and find how many revolutions the 3 kg weight makes per minute.

*7 Two weights, of masses m and $2m$, are connected by a light inextensible string passing through a smooth ring, fixed at a height h above a smooth horizontal table. The smaller mass hangs at rest below the ring and above the table whilst the larger mass describes a horizontal circle of radius $h/3$. The circle has its centre on the table directly below the ring. Show that the time taken to describe the circle once is

$$2\pi \left(\frac{2h \sqrt{10}}{3g} \right)^{1/2}$$

8 A particle P is attached by a light inextensible string of length l to a fixed point O. The particle is held with the string taut and OP at an acute angle α to the downward vertical, and is then projected horizontally at right angles to the string with speed u chosen so that it describes a circle in the horizontal plane. Show that

$$u^2 = gl \sin \alpha \tan \alpha$$

The string will break when the tension exceeds twice the weight of P. Find the greatest possible value of α. [J]

Exercise 7.4

1 A light inextensible string AB of length 80 cm has a particle of mass 500 grams fastened to a point 50 cm from A. A and B are attached to two fixed points with A 40 cm vertically above B. The particle describes a horizontal circle about the line AB with angular velocity 6 rad/s and both parts of the string are taut. Show that the lower part of the string is horizontal and find the tensions in both parts of the string.

2 A light inextensible string of length 1.40 m has its ends fastened to two fixed points A and B where B is 1 m vertically below A. A smooth bead of mass 200 grammes is threaded on the string and moves in a horizontal circle, with constant angular speed and the string taut, about the line AB. If the bead rotates so that its distance from A is 80 cm, find its angular speed and the tension in the string.

3 A particle of mass m is attached to one end of a light inextensible string of length l, the other end of which is attached to a fixed point O at a height h above a smooth horizontal table. With the string taut the particle describes a circle on the table, at a constant angular speed ω. The centre of the circle is vertically below O. Find the tension in the string and the magnitude of the reaction of the table. Determine the greatest value of ω^2 for which such a motion is possible. [A]

4 A particle of mass m is attached to the midpoint of a light inextensible string of length $2l$. The ends of the string are fastened to two fixed points A and B where B is a distance l vertically below A. The particle describes a horizontal circle about AB with constant angular speed ω. If the tension in the upper string is twice that in the lower string, show that

$$\omega = \sqrt{(6g/l)}$$

5 A particle of mass m is attached by two strings of equal length to two fixed points A and B where B is a distance d vertically below A. The particle describes a horizontal circle with constant angular speed. Show that, if both strings are to remain taut, then the angular speed must exceed $\sqrt{(2g/d)}$. If the speed is $2\sqrt{(2g/d)}$ find the ratio of the tensions in the strings.

6 A small bead of mass m is threaded on a light smooth inextensible string of length $2l$. The ends of the string are attached to two fixed points A and B such that A is at a distance l vertically above B. The bead describes a horizontal circle about B as centre with constant angular velocity ω. Show that the radius of the circle is $3l/4$. Find, in terms of m and g, the tension in the string, and, in terms of l and g, the time taken for the bead to describe the circle once.   [J]

7 A light inextensible string of length $5a$ has one end fixed at a point A and the other end fixed at a point B which is vertically below A and at a distance $4a$ from it. A particle P of mass m is fastened to the midpoint of the string and moves with speed u, and with the parts AP and BP of the string both taut, in a horizontal circular path whose centre is the midpoint of AB. Find, in terms of m, u, a and g, the tensions in the two parts of the string and show that the motion described can take place only if $8u^2 \geqslant 9ga$.   [J]

Exercise 7.5

1 A bend of a racetrack has a radius of 60 m and is sloping downwards towards the inside of the curve at an angle of $\tan^{-1}\frac{1}{3}$. At what speed must a car take the bend so there is no tendency to sideslip?

2 The radius of a bend on a railway track is 500 m and the gauge of the track is 1.4 m. The average speed of trains rounding this bend is 90 km/h. If there is to be no sideways thrust on the rails at this speed, how much higher must the outer rail be above the inner?

3

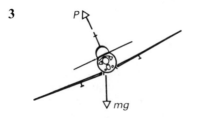

An aeroplane banks as it describes a horizontal circle of radius 250 m at a speed of 270 km/h. If the air pressure can be assumed to act through the centre of gravity of the aeroplane, in a direction perpendicular to the plane containing the wings and the body, find the angle to the horizontal at which the plane is banked.

4 A curve of a railway track has a radius of 180 m. The track is banked so that a train travelling at 108 km/h exerts no lateral force on the rails. A train of mass 80 tonnes is at rest on this bend. What lateral force does it exert on the track?

5 A car travels round a bend of radius 100 m which is banked at $\tan^{-1}\frac{5}{12}$ to the horizontal. If the coefficient of friction between the tyres and the road is $\frac{2}{5}$, find the maximum speed at which the car can take the bend without slipping.

6 A bend of radius 80 m is banked at 30° to the horizontal. It is found that the minimum speed at which a car can travel round the bend, without slipping down, is 10 m/s. What is the coefficient of friction between the tyres and the road?

7 At a certain height on the 'Wall of Death' the track is banked at 45° to the horizontal and has radius 15 m. If the coefficient of friction between the motorcycle and the track is 0.5, find the greatest and least speeds with which the cycle can travel at this height without slipping.

8 A car goes round a bend of radius 50 m, whose road surface is horizontal, at 14 m/s without sideslipping. Calculate the least value of the coefficient of friction between the car and the road.

At what angle to the horizontal should the bend be banked so that the cars can go round it at 14 m/s without any tendency to sideslip?

9 A cyclist travelling on level road takes a bend of radius 20 m at a speed of 18 km/h. What angle does the plane containing the bike and rider make with the vertical? What is the minimum value of the coefficient of friction so that the machine will not sideslip?

10 A bend of a motorcycle racetrack has a radius of 100 m. At what angle should the track be banked in order that a rider travelling at 72 km/h can take the bend without any sideways frictional reaction between his tyres and the track?

11 A racetrack bend is banked at an angle α to the horizontal so that, at a speed U, cars do not tend to sideslip. A car takes the bend at a speed V, where $V > U$. Show that the minimum value of the coefficient of friction needed to prevent sideslipping is

$$\frac{(V^2 - U^2)\sin \alpha \cos \alpha}{U^2 \cos^2 \alpha + V^2 \sin^2 \alpha} \quad .$$

Exercise 7.6

1 A mass of 2 kg is suspended from a fixed point O by a light string of length 1 m. It is given a horizontal velocity of 4 m/s. Find the tension in the string when it makes an angle of 60° with the downward vertical and, when it is at the highest point of its swing, find its distance below O.

2   A mass of 50 grams is suspended from a fixed point O by a light string of length 90 cm. It is given a horizontal velocity u. Find the least value of u in order that:

(a)   the string becomes horizontal
(b)   the mass makes complete revolutions.

3   A mass of 250 grams is attached to a fixed point O by a string of length 1.2 m. The mass is held at a point on the same horizontal level as O, with the string taut, and allowed to fall. Find the tension in the string when the string makes an angle of 30° with the horizontal and show that the maximum tension is $3g/4$ N.

4   A mass m is attached to a fixed point O by a string of length l. The mass is held at a point on the same horizontal level as O, with the string taut, and is projected vertically downwards with velocity u. Find the tension in the string when it is at an angle θ below the horizontal.

5   If a particle is projected in the same way as in question 4, show that, when the string goes slack, the particle is at a distance $u^2/3g$ above O.

6   A mass m is suspended from a fixed point O by a string of length $2a$. It is given a horizontal velocity u. Show that the condition that it makes complete revolutions is that $u^2 \geqslant 10ga$.
       If $u = 2\sqrt{3ga}$, find the ratio of the greatest and least tensions.

*7   A mass m is attached to a fixed point O by a light string of length l. The mass is projected horizontally from a point at a distance l vertically above O with a velocity $\sqrt{gl}$. Show that the tension in the string at the lowest point is $6mg$.
       If, however, the string breaks when the tension is $5mg$, what will be the distance of the mass below O when the string breaks?

Exercise 7.7

1   In order to go under a low bridge, a road dips in the form of an arc of a circle of radius 25 m. A car of mass 1.2 tonnes travels along the road and its speed at the bottom of the dip is 54 km/h. What is the reaction between the car and the road at this point?

· 2   The road surface of a humpback bridge is in the form of a circular arc of radius 16 m. What is the greatest speed at which a car can cross the bridge without leaving the ground at the highest point?

· 3   A particle slides, from rest at the highest point, down the outside of a smooth sphere of radius r. Show that it leaves the surface at a distance $2r/3$ above the centre of the sphere. If, instead, it slid from rest from a point that is a distance $r/2$ above the centre of the sphere, find its height above the centre when it leaves the surface.

4   A light rigid rod AB of length l is pivoted about A and has a particle of mass m at B. When hanging vertically the rod is given a horizontal velocity of $2\sqrt{gl}$. When the rod makes an angle θ with the downward vertical, show that its angular velocity, $\dfrac{d\theta}{dt}$, is given by

$$l\left(\frac{d\theta}{dt}\right)^2 = 2g(1 + \cos\theta)$$

and find the tension in the rod at this instant.

5   A bead of mass m is threaded on a smooth wire fixed in a vertical plane in the form of a circle of radius a. The bead is projected from the lowest point with speed u. For what values of u will the particle make complete revolutions?
If θ is the angle that the radius to the bead makes with the upward vertical, show that there is no reaction between the bead and the wire when $\theta = \alpha$, where

$$\cos\alpha = \frac{u^2}{3ag} - \frac{2}{3}$$

State the direction of the reaction of the wire on the bead when θ is   (a) less than α (b) more than α.

6   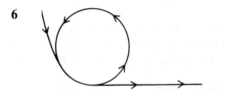   A toy racetrack has a section on which cars loop the loop. If the cars start their descent from rest, at a point 60 cm above the lowest point of the loop, show that the greatest possible radius of the loop is 24 cm.

7   A small smooth marble of mass m is placed inside a smooth circular tube of radius a, fixed in a vertical plane. The marble is slightly displaced from rest at the highest point of the tube. If T_1 is the pressure on the marble at one end of a horizontal diameter and T_2 is the pressure on the marble at the lowest point, find the ratio $T_1 : T_2$.

8   A particle attached to a light string makes vertical circles of radius 50 cm. If the ratio of the greatest and least tensions in the string is 10:1, show that the least velocity of the particle is $5g/6$.

9   A particle P is projected horizontally with speed u from the lowest point A of the smooth inside surface of a fixed hollow sphere of internal radius a.
(i)   In the case when $u^2 = ga$ show that P does not leave the surface of the sphere. Show also that, when P has moved halfway along its path from A to the point at which it first comes to rest, its speed is

$$ga(\sqrt{3} - 1)$$

(ii)   Find u^2 in terms of ga in the case when P leaves the surface at a height $3a/2$ above A, and find, in terms of a and g, the speed of P as it leaves the surface.   [J]

Exercise 7.8

1 A mass of 500 g is attached to a fixed point O by a light string of length 3 m. The mass is held at the same horizontal level as O, with the string taut, and allowed to fall. Calculate the tension in the string when it is vertical. When the string is vertical it encounters a small smooth peg, C, a distance 2 m below O. The mass than starts to describe a circular path about C. Find the tension in the string immediately after it hits the peg and find whether the mass makes a complete revolution about C.

2 A particle attached to one end of a string, whose other end is fixed, is allowed to fall when the string is taut and horizontal. The string is of length 2*l* and it catches on a peg a distance *l* vertically below the fixed end of the string. The particle starts to move in a circular path round the peg. Find how high above the peg the particle will go before the string goes slack.

3 A mass *m* is attached to a fixed point O by a light string of length *l*. The mass is held at the same horizontal level as O with the string taut and allowed to fall. It encounters a fixed peg at a distance $(l - d)$ below O and then starts to move in a circle around this peg. Show that the condition that it completes a full revolution about this peg is

$$d \leqslant \frac{2l}{5}$$

4 The diagram shows a smooth narrow tube in the form of a circle of centre O and radius *a*, which is fixed in a vertical plane. The tube contains two particles, P of mass *m* and Q of mass 3*m*, which are connected by a light inextensible string of length $\pi a/2$.

The system is released from rest with P at the level of O and with Q at the highest point of the tube.

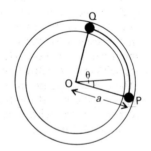

Show that, if after time *t* the line OP has turned through an angle θ (as shown in the diagram), then

$$2a\left(\frac{d\theta}{dt}\right)^2 = g(3 - 3\cos\theta + \sin\theta),$$

provided that the string remains taut.

Find the reaction between P and the tube in terms of *m*, *g* and θ.

Obtain $a\dfrac{d^2\theta}{dt^2}$ in terms of *g* and θ, and find the tension in the string in terms of *m*, *g* and θ. Deduce that the string becomes slack when $\theta = \pi/4$.   [J]

5. A particle P of mass *m* is at rest at a point A on the smooth outer surface of a fixed sphere of centre O and radius *a*, OA being horizontal. The particle is attached to one end of a light inextensible string which is taut and passes over the sphere, its other end carrying a particle Q of mass 2*m* which hangs freely. The string lies in the vertical plane containing OA.

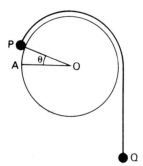

The system is released from rest, and after time *t* the angle POA is θ, as shown in the diagram. Show that, while P remains in contact with the sphere.

$$3a\left(\frac{d\theta}{dt}\right)^2 = 2g(2\theta - \sin\theta)$$

Find the tension in the string and the reaction of the sphere on P in terms of *m*, *g* and θ. [J]

Exercise 7.9

1 A small bead P of mass *m* is threaded on a smooth wire fixed in a vertical plane and in the form of a circle of radius *a* and with its centre at O. The bead is projected from the lowest point of the wire with speed *U*. Obtain the condition necessary for the bead to make complete revolutions and, if θ denotes the angle the radius to the bead makes with the upward vertical, find an expression for the reaction on the bead in terms of *m*, *U*, *g*, *a* and θ. A small bead Q is now fixed to the wire at the same horizontal level as O. The coefficient of restitution between P and Q is *e*. The bead P is projected from the lowest point of the wire towards Q with speed *V*, where $V > 2(ga)^{1/2}$. Show that the bead P will not subsequently pass through the highest point of the wire unless

$$V^2 > 2ga(1 + e^{-2})$$

2 The rim of a smooth hemispherical bowl is a circle of centre O and radius *a*. The bowl is fixed with its rim horizontal and uppermost. A particle P of mass *m* is released from rest at a point A on the rim, and during the subsequent descent of the particle into the bowl the angle AOP is denoted by φ. Find, in terms of *a*, *g*, *m* and φ:

(a) the speed of the particle
(b) the components of acceleration along and perpendicular to PO
(c) the reaction exerted by the bowl on the particle

indicating clearly, where appropriate, the direction associated with each quantity.

When P reaches the lowest point of the bowl it collides directly with a stationary particle Q of mass $\frac{1}{2}m$. After the collision Q just reaches the rim of the bowl. Find the coefficient of restitution between P and Q. [J]

3   A rigid wire ABC is fixed in a vertical plane. The portion AB, of length b, is straight and horizontal and BC is a smooth circular arc of centre O (vertically above B), radius a and length $\frac{1}{2}\pi a$. A bead P of mass m is threaded on the wire and projected from B with speed u towards C. Denoting by θ the angle BOP when P is between B and C, write down an equation of motion relating $d^2\theta/dt^2$ to θ.

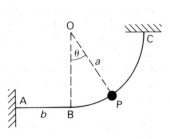

Show that, while P is moving from B to C,

$$\left(\frac{d\theta}{dt}\right)^2 = \frac{u^2}{a^2} - \frac{2g}{a}(1 - \cos\theta)$$

Find the reaction of the wire on the bead in terms of m, g, a, u and θ, indicating clearly its direction.

The bead collides with a fixed stop at C at which the coefficient of restitution is e. Find its speed when it returns to B.

On the straight portion BA the motion is resisted by a constant force F. Show that the bead will reach A if

$$Fb \leqslant \tfrac{1}{2}me^2u^2 + m(1 - e^2)ga \qquad\qquad \text{[J]}$$

4   A smooth wire in the form of a circle of centre O and radius a is fixed in a vertical plane. Two beads, P of mass $2m$ and Q of mass m, are threaded on to the wire. Initially Q is at rest at A, the lowest point of the wire, and P is held at a point B on the wire, where the angle AOB is $\frac{1}{3}\pi$.

The particle P is then released. Find the speed of P when the angle AOP is θ and P has not yet reached A, and find the reaction then exerted on P by the wire. Show that, when P reaches A, its speed is $\sqrt{(ga)}$. Show that, when the particles collide, the impulse of the reaction between them is of magnitude $\frac{2}{3}m(1 + e)\sqrt{(ga)}$, where e is the coefficient of restitution.

In the case when $e = \frac{1}{5}$ find the height above A which Q subsequently reaches before coming instantaneously to rest. [J]

Miscellaneous Exercise 7B

1   A particle P of mass 0.1 kg moving on a smooth horizontal table with constant speed v m/s describes a circle centre O such that $OP = r$ m. The particle is attracted towards O by a force of magnitude $4v$ newtons and repelled from O by a force of magnitude k/r newtons where k is a constant.

(a)   Given that $v = 40$ and the time of one revolution is $\pi/10$ s find r and k.
(b)   Given that $k = 30$ and $r = 1$ find the possible values of v.
(c)   Find the range of values of k if $r = 1$. [A: 7.2, 7.9]

2   An artificial satellite of mass m moves under the action of a gravitational force which is directed towards the centre, O, of the earth and is of magnitude F. The orbit of the satellite is a circle of radius a and centre O. Obtain an expression for T, the period of the satellite, in terms of m, a and F.

Show that, if the gravitational force acting on a body of mass m at a distance r from O is $m\mu/r^2$, where μ is a constant, then $T^2\mu = 4\pi^2 a^3$.

Assuming that the radius of the earth is $6\,400\,\mathrm{km}$ and that the acceleration due to gravity at the surface of the earth is $10\,\mathrm{m/s^2}$, show that $\mu = (6.4)^2 10^{13}\,\mathrm{m^3/s^2}$.

Hence, or otherwise, find the period of revolution, in hours to 2 decimal places, of the satellite when it travels in a circular orbit $600\,\mathrm{km}$ above the surface of the earth. [L: 7.9]

3   At any point outside the Earth at a distance r from the centre the acceleration f due to gravity is proportional to r^{-2}. Taking the Earth to be a sphere of radius a and given that g is the gravitational acceleration on the surface, find f in terms of a, g and r.

Show that the speed of an artificial satellite in circular orbit at a height h above the surface of the Earth is

$$\sqrt{\left(\frac{ga^2}{a+h}\right)}$$

Find, in terms of a, g and h, the time taken to complete one orbit. [J: 7.9]

4   One end of a light inextensible string of length a is attached to a fixed point A which is at a height h above a smooth horizontal table, where $h < a$. A particle P of mass m is attached to the other end of the string and lies on the table with the string taut. The particle is projected so that it moves on the table in a circle at constant angular speed ω. Show that the tension in the string is $ma\omega^2$.

Find, in terms of m, g, h and ω, the reaction exerted on P by the table. [J: 7.3]

5   A particle is attached to one end of a light string, the other end of which is fixed. When the particle moves in a horizontal circle with speed $2\,\mathrm{m/s}$, the string makes an angle $\tan^{-1}\left(\frac{5}{12}\right)$ with the vertical. Show that the length of the string is approximately $2.5\,\mathrm{m}$. [L: 7.3]

6   The ends of a light inextensible string ABC of length $3l$ are attached to fixed points A and C, C being vertically below A at a distance $\sqrt{3}l$ from A. At a distance $2l$ along the string from A a particle B of mass m is attached. When both portions of the string are taut, B is given a horizontal velocity u, and then continues to move in a circle with constant speed. Find the tensions in the two portions of the string and show that the motion is possible only if

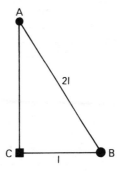

$$u^2 \geqslant \tfrac{1}{3}gl\sqrt{3}$$

[J: 7.3, 7.4]

7   A particle P is attached to one end of a light inextensible string, the other end of which is attached to a fixed point A. The particle describes a horizontal circle, with constant speed (and with the string taut), about the point O vertically below A where $AO = h$. Find the angular speed of P. A second identical string is now attached to P and its other end fastened to the point B at a distance $2h$ vertically below A. The particle P describes a horizontal circle centre O with constant angular speed $3(g/h)^{\frac{1}{2}}$ with both strings taut. Find the ratio of the tension in AP to that in BP.

[A: 7.3, 7.4]

8   A thin straight rod AOB consists of a smooth part AO and a rough part OB. Beads P of mass m and Q of mass M are threaded on to the rod, P on the smooth part AO, with $OP = p$, and Q on the rough part OB, with $OQ = q$. The beads are connected by a light taut inextensible string PQ. The rod is rotating in a horizontal plane with constant angular speed ω about a fixed vertical axis through O, and the beads remain at rest relative to the rod.
Find:

(a)   the tension in the string
(b)   the force of friction between Q and the rod.

Show that, if the coefficient of friction at Q is μ, then

$$\omega^2 |mp - Mq| \leqslant \mu M g$$ [J: 7.2, 3.1]

9   Two small beads P and Q of masses 0.003 kg and 0.006 kg respectively are threaded on a smooth circular wire of radius 0.5 m which is maintained in a horizontal plane. Initially the beads P and Q are at rest at points A and B respectively where A and B are at opposite ends of a diameter of the wire. The coefficient of restitution between the beads is $\frac{1}{4}$. The bead P is then projected towards Q with a speed of u m/s. Find the velocities of the beads immediately after they first collide. Given that $u = 6$ find, as a multiple of π, the time that elapses between the first and second collision. When the wire is rough and P is projected from A with a speed of 6 m/s so as to collide with Q which is at rest at B, the speed of Q, immediately after the collision, is 1.25 m/s. Find the work done by friction as P travels from A to collide with Q and the magnitude of the impulse acting on P during its collision with Q.

[A: 7.3, 5.6, 5.2, 5.4, 3.3]

10   One end of a light inextensible string of length a is fastened to a fixed point and a particle is attached to the other end. The particle is held so that the string is taut and horizontal and is then projected vertically upwards with speed u. Given that the string slackens when it is inclined at an angle of $30°$ to the horizontal, find u in terms of a and g.

[J: 7.6]

11

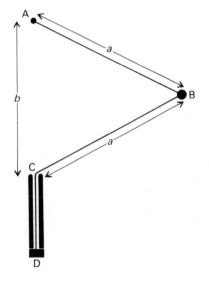

A smooth solid cone of semi-vertical angle α, where $\pi/4 < \alpha < \pi/2$, is fixed with its axis vertical and vertex A upwards, as shown in the diagram. A particle P of mass m is attached by a light inextensible string of length l to a fixed point O vertically above A. The particle moves in a horizontal circle on the surface of the cone with constant angular velocity ω. Given that the string is inclined to the horizontal at a constant angle α, find the tension in the string.

Show that the condition for P to remain on the surface of the cone is

$$\omega^2 \leqslant \frac{g}{l \sin \alpha} \qquad \text{[J: 7.3, 7.4]}$$

*12

One end of a light inextensible string ABCD, in which $AB = BC = a$, is attached to a fixed point A. A smooth narrow tube CD is fixed below A so that ACD is a vertical line and $AC = b$. The end D of the string is threaded through the tube and attached to a body of mass km which cannot pass through the tube. A particle of mass m is fastened to the string at B, and rotates about the line AC with constant angular velocity ω in a horizontal circle, with the string taut and the body in contact with the tube at D. Find the tensions in the two parts of the string and the vertical force exerted at D by the tube on the body, and show that

$$\omega^2 ab \geqslant 2g(a + kb)$$

Given that the greatest tension the string can sustain without breaking is λmg, show that the motion is possible only if

$$(\lambda - k)b \geqslant 2a \qquad \text{[J: 7.3, 7.4]}$$

13 Two particles P and Q of masses m and $2m$, respectively, are connected to a fixed point O by light inextensible strings OP and OQ, each of length l. The particle Q hangs in equilibrium vertically below O and P is held so that the string OP is taut and horizontal and is then released. The coefficient of restitution between P and Q is e. Given that P is brought to rest by its impact with Q, show that $e = \frac{1}{2}$. Given also that Q rises until OQ makes an angle α with the downward vertical, find $\cos \alpha$.

Subsequently Q strikes P. If P rises until OP makes an angle β with the downward vertical, show that $\beta = \alpha$.   [J: 7.9]

*14 A small bead P of mass m is threaded on a smooth thin wire, in the form of a circle of radius a and centre O, which is fixed in a vertical plane. The bead is initially at the lowest point A of the circle and is projected along the wire with a velocity which is just sufficient to carry it to the highest point. Denoting the angle POA by θ, find, in terms of m, g and θ, the magnitude of the reaction of the wire on the bead.

Express $\dfrac{d\theta}{dt}$, where t denotes time, in the form $C \cos(\theta/2)$, where C is a constant.   [J: 7.6, 7.8]

15 Two small smooth pegs O and C are fixed at the same level and $OC = a$. A light inextensible string of length $3a$ has one end attached to O and a particle P of mass m hanging at its other end vertically below O. The particle is projected with speed u parallel to OC. Find the speed of P when OP first makes an acute angle θ with the downward vertical and show that the tension in the string is then

$$mg(3 \cos \theta - 2) + \frac{mu^2}{3a}$$

Show that the string will reach a horizontal position if $u^2 \geqslant 6ga$.
Given that $u^2 > 6ga$, find the tension when, after the string has struck the peg C, the moving portion CP makes an angle ϕ above the horizontal, and show that P will complete a semicircle with centre C if $u^2 \geqslant 12ga$.   [J: 7.6, 7.8]

16 A light rigid rod of length a hangs verically with its upper end smoothly pivoted to a fixed point O and a particle of mass m is attached to its lower end P. The rod is set in motion so that P initially moves horizontally with speed $\frac{5}{2}\sqrt{(ga)}$. Show that, when OP makes an angle θ with the downward vertical,

$$4a\left(\frac{d\theta}{dt}\right)^2 = g(17 + 8 \cos \theta),$$

and find the tension in the rod at this instant.
When P is vertically above O it is brought to rest by striking a fixed inelastic stop. Find the impulse exerted by the stop.   [J: 7.7, 5.4]

*17 A light rod OP of length a is free to rotate in a vertical plane about a fixed point O, and a mass m is attached to it at P. Denote by θ the angle between OP and the downward vertical. When $\theta = 0$, P is moving horizontally with speed u. Find the tension in the rod as a function of θ, and show that the value of the function is positive throughout the motion if $u^2 > 5ag$.
Find the least value of the tension when $u^2 = 6ag$. Show that, if $u^2 < 2ag$, P remains below the level of O and deduce that, in this case also, the value obtained for the tension is positive throughout the motion.   [J: 7.7, 7.8]

*18   A circular cone of semi-vertical angle α is fixed with its axis vertical and its vertex, A, lowest, as shown in the diagram. A particle P of mass m moves on the inner surface of the cone, which is smooth. The particle is joined to A by a light inextensible string AP of length l. The particle moves in a horizontal circle with constant speed v and with the string taut. Find the reaction exerted on P by the cone. Find the tension in the string and show that the motion is possible only if $v^2 > gl\cos\alpha$.

[J: 7.2, 7.3]

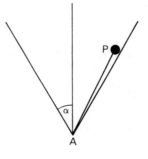

UNIT 8 PROJECTILES

8.1 Horizontal Projection

Projectiles

To investigate the motion of a projectile, its horizontal and vertical motions are considered separately.

Horizontally: There is no horizontal force.
⇒ There is no horizontal acceleration
⇒ Horizontal velocity is constant
⇒ Only equation of motion is $s = ut$

Vertically: The only vertical force is gravity
⇒ Vertical acceleration is $g\downarrow$, which is constant
⇒ The constant acceleration equations of motion, $v = u + at$,

$$s = ut + \tfrac{1}{2}at^2, v^2 = u^2 + 2as, s = \frac{(u + v)t}{2} \text{ can be used.}$$

Horizontal Projection

Example 1   A small boy stands on a rock and throws a stick, pretending it is a spear. He throws it horizontally with a speed of 10 m/s and it sticks in the level ground a distance of 8 m in front of him. Find the height of projection above the level ground. Also calculate the speed of impact and the angle it makes with the ground on impact.

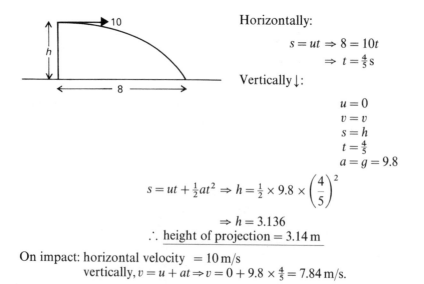

Horizontally:
$$s = ut \Rightarrow 8 = 10t$$
$$\Rightarrow t = \tfrac{4}{5}\,\text{s}$$

Vertically ↓:
$$u = 0$$
$$v = v$$
$$s = h$$
$$t = \tfrac{4}{5}$$
$$a = g = 9.8$$

$$s = ut + \tfrac{1}{2}at^2 \Rightarrow h = \tfrac{1}{2} \times 9.8 \times \left(\frac{4}{5}\right)^2$$

$$\Rightarrow h = 3.136$$
∴ height of projection = 3.14 m

On impact: horizontal velocity = 10 m/s
vertically, $v = u + at \Rightarrow v = 0 + 9.8 \times \tfrac{4}{5} = 7.84$ m/s.

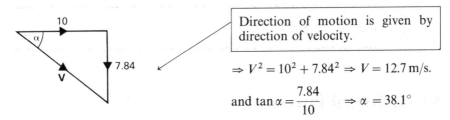

Direction of motion is given by direction of velocity.

$$\Rightarrow V^2 = 10^2 + 7.84^2 \Rightarrow V = 12.7 \, \text{m/s}.$$

$$\text{and } \tan \alpha = \frac{7.84}{10} \Rightarrow \alpha = 38.1°$$

∴ speed of impact = 12.7 m/s and the stick makes an angle of 38.1° with the ground.

Example 2   The ground on top of a vertical cliff is horizontal. The cliff is 90 m high. A car is driven off the top of the cliff and its wreckage is found 60 m from the foot of the cliff. At what speed was it driven off the cliff?

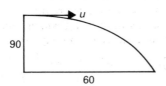

Horizontally: $s = ut \Rightarrow 60 = ut$ ⠀⠀⠀⠀⠀⠀⠀⠀⠀⠀⠀⠀⠀⠀(1)

Vertically ↓:

$u = 0$

$v = -\!-$

$s = 90$ ⠀⠀⠀⠀⠀⠀⠀⠀⠀⠀⠀⠀$s = ut + \frac{1}{2}at^2$

$t = t$ ⠀⠀⠀⠀⠀⠀⠀⠀⠀⠀⠀⠀$\Rightarrow 90 = \frac{1}{2} \times 9.8 \times t^2$

$a = g = 9.8$

$$\Rightarrow t = \sqrt{\frac{90}{4.9}} \approx 4.29 \, \text{s}$$

$$\therefore (1) \Rightarrow u = \frac{60}{t} = \frac{60}{4.29} = 14 \, \text{m/s}$$

∴ the car was driven off the cliff at 14 m/s.

Example 3   A thrush is flying horizontally at 6 m/s, 20 m above the ground, carrying a snail. If it drops the snail, how far forward would the snail travel before hitting the ground?

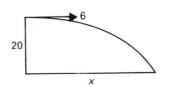

When the snail starts to fall, its initial velocity will be that of the bird: $6\,\text{m/s} \rightarrow$

Horizontally: $s = ut \Rightarrow x = 6t$ (1)

Vertically $\downarrow$:

$u = 0$

$v = \text{—}$

$s = 20$ $s = ut + \frac{1}{2}at^2$

$t = t$ $\Rightarrow 20 = \frac{1}{2} \times 9.8t^2$

$a = g = 9.8$ $\Rightarrow t \approx 2.02\,\text{s}$

and (1) $\Rightarrow x = 6 \times 2.02 = 12.1\,\text{m}.$

$\therefore$ snail travels 12.1 m forwards

- Exercise 8.1   See page 244

8.2 Projection at an Angle I

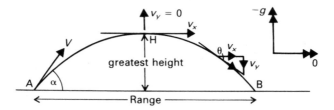

Points to Remember:

(i)   Motion is considered horizontally $\rightarrow$ and vertically $\uparrow$.

(ii)   Horizontally, the only equation of motion is $s = ut$, where $u = V \cos \alpha$.

(iii) The trajectory is symmetrical.

(iv)   The time of flight (from A to B) is twice that taken to reach the highest point.

(v)   To find the greatest height, find s when $v_y = 0$.

(vi)   To find the time to greatest height, find t when $v_y = 0$.

(vii) To find the time of flight, find t when vertical displacement $= 0$.

(viii) To find the range, use $s = uT$ where $T = $ time of flight.

(ix)   Direction of motion is given by the direction of the velocity.

Example 4   A hockey ball is given a speed of 25 m/s at an angle of elevation of 20°. Find:

(a)   the speed and direction of motion of the ball after one second

(b)   the range of the ball

(c)   the maximum height reached in its path.

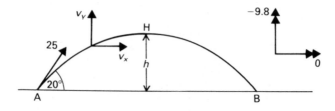

(a) After 1 second.

Horizontally: velocity $=$ constant $\Rightarrow v_x = 25\cos 20° \approx 23.49$

Vertically $\uparrow$:

$$u = 25\sin 20°\uparrow \qquad v = u + at \Rightarrow v_y = 25\sin 20° - 9.8 \times 1$$
$$v = v_y\uparrow \qquad\qquad\qquad\qquad \Rightarrow v_y = -1.25\uparrow$$
$$s = — \qquad\qquad\qquad\qquad\qquad \Rightarrow v_y = 1.25\downarrow$$
$$t = 1$$
$$a = 9.8\downarrow = -9.8\uparrow$$

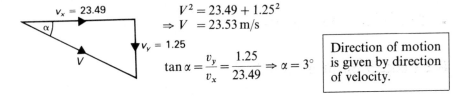

$$V^2 = 23.49 + 1.25^2$$
$$\Rightarrow V = 23.53\,\text{m/s}$$

$$\tan\alpha = \frac{v_y}{v_x} = \frac{1.25}{23.49} \Rightarrow \alpha = 3°$$

Direction of motion is given by direction of velocity.

After 1 second, the speed of the ball is 23.53 m/s and its direction of motion is 3° below the horizontal.

(b)

AB$\uparrow$	AB$\rightarrow$
$u = 25\sin 20°\uparrow$	$25\cos 20°$
$v = —$	—
$s = 0$	R
$t = T$	T
$a = -9.8\uparrow$	0

For AB$\uparrow$:
$$s = ut + \tfrac{1}{2}at^2\uparrow$$
$$\Rightarrow 0 = 25\sin 20°\,T - \tfrac{1}{2}\cdot 9.8T^2$$
$$\Rightarrow T = \frac{25\sin 20°}{4.9}$$
$$\approx 1.745\,\text{s} \leftarrow$$

Before the range can be found, the time of flight must first be cal-culated.

For AB$\rightarrow$:
$$s = at \Rightarrow R = 25\cos 20° \times 1.745$$
$$\Rightarrow R = 41\,\text{m}$$

$\therefore$ range $= 41$ m

(c)

AH$\uparrow$
$u = 25\sin 20°\uparrow$
$v = 0$
$s = h$
$t = —$

For AH$\uparrow$:
$$v^2 = u^2 + 2as\uparrow$$
$$\Rightarrow 0 = (25\sin 20°)^2 - 2 \times 9.8 \times h$$
$$\Rightarrow h = \frac{(25\sin 20°)^2}{2 \times 9.8}$$
$$\Rightarrow h = 3.73\,\text{m}.$$

$\therefore$ maximum height is 3.73 m

Example 5   A golf ball is hit across level ground, with an initial speed of 40 m/s at 10° to the horizontal. Will it clear a shrub 90 cm tall which is growing at a distance of 50 m from the point of projection?

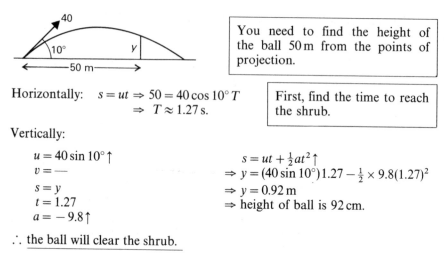

You need to find the height of the ball 50 m from the points of projection.

Horizontally: $s = ut \Rightarrow 50 = 40 \cos 10° \, T$
$\Rightarrow T \approx 1.27 \, \text{s}.$

First, find the time to reach the shrub.

Vertically:

$u = 40 \sin 10° \uparrow$
$v = \text{—}$
$s = y$
$t = 1.27$
$a = -9.8 \uparrow$

$s = ut + \tfrac{1}{2}at^2 \uparrow$
$\Rightarrow y = (40 \sin 10°)1.27 - \tfrac{1}{2} \times 9.8(1.27)^2$
$\Rightarrow y = 0.92 \, \text{m}$
$\Rightarrow$ height of ball is 92 cm.

$\therefore$ the ball will clear the shrub.

- Exercise 8.2   See page 246

8.3 Projection at an Angle II

Example 6   A stone is thrown from the top of a cliff which is 50 m above sea level. If it is projected at 30 m/s at an angle of 30° above the horizontal, find how far it is away from the base of the cliff when it hits the water. (Take $g = 10 \, \text{m/s}^2$).

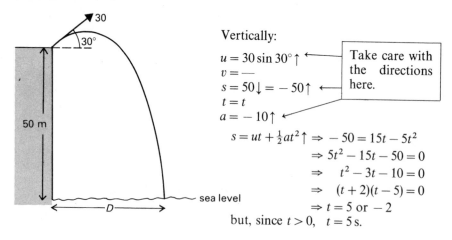

Vertically:

$u = 30 \sin 30° \uparrow$
$v = \text{—}$
$s = 50 \downarrow = -50 \uparrow$
$t = t$
$a = -10 \uparrow$

Take care with the directions here.

$s = ut + \tfrac{1}{2}at^2 \uparrow \Rightarrow -50 = 15t - 5t^2$
$\Rightarrow 5t^2 - 15t - 50 = 0$
$\Rightarrow t^2 - 3t - 10 = 0$
$\Rightarrow (t + 2)(t - 5) = 0$
$\Rightarrow t = 5 \text{ or } -2$
but, since $t > 0$, $t = 5 \, \text{s}.$

Horizontally:

$s = ut \Rightarrow D = 30 \cos 30° t$

$\Rightarrow D = 30 \times \dfrac{\sqrt{3}}{2} \times 5 \approx 130 \, \text{m}.$

$\therefore$ distance from base of cliff $\approx 130 \, \text{m}.$

To Find Speed and Angle of Projection

The initial velocity,

can also be considered in component form,

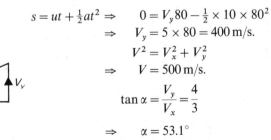

If you are asked to find the speed and angle of projection then either of these two representations of the initial velocity can be used. However, using the component form sometimes makes the algebra used in the question simpler.

Example 7   A gun fires a shell which lands a distance of 24 km away after it has been in the air for 80 seconds. Find the magnitude and direction of the velocity of projection. (Take $g = 10 \, \text{m/s}^2$).

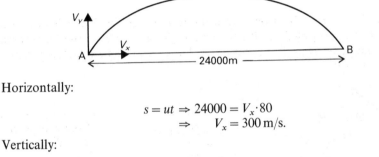

Horizontally:

$$s = ut \Rightarrow 24000 = V_x \cdot 80$$
$$\Rightarrow \quad V_x = 300 \, \text{m/s}.$$

Vertically:

$$s = ut + \tfrac{1}{2}at^2 \Rightarrow \quad 0 = V_y 80 - \tfrac{1}{2} \times 10 \times 80^2$$
$$\Rightarrow \quad V_y = 5 \times 80 = 400 \, \text{m/s}.$$
$$V^2 = V_x^2 + V_y^2$$
$$\Rightarrow \quad V = 500 \, \text{m/s}.$$
$$\tan \alpha = \frac{V_y}{V_x} = \frac{4}{3}$$
$$\Rightarrow \quad \alpha = 53.1°$$

∴ velocity of projection is 500 m/s at 53.1° above the horizontal

---

Note that this question could also have been done using $V \cos \theta$ and $V \sin \theta$ instead of the components V_x and V_y.

●   Exercise 8.3   See page 247

8.4 General Algebraic Results

Time of Flight and Range

Example 8   A body is projected at time $t = 0$, from an origin O on a fixed horizontal plane, with speed U at an angle α to the horizontal. Find the time of flight T and the range R in terms of U, g and α.

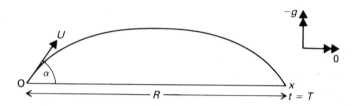

Horizontally:

$s = ut \Rightarrow R = U \cos \alpha T$ (1) $\qquad s = ut + \frac{1}{2}at^2 \uparrow \Rightarrow 0 = U \sin \alpha T - \frac{1}{2}gT^2$

Vertically $\uparrow$:

$u = U \sin \alpha \uparrow$ $\qquad\qquad\qquad\qquad\qquad\qquad\qquad \Rightarrow T = 0$ and $T = \dfrac{2U \sin \alpha}{g}$

$v = —$

$s = 0 \uparrow$

$t = T$

$a = -g \uparrow$

Since $T = 0$ corresponds to the point of projection O, the time of flight is given by

$$T = \frac{2U \sin \alpha}{g}$$

$(1) \Rightarrow R = U \cos \alpha \left(\frac{2U \sin \alpha}{g} \right)$

$$\Rightarrow R = \frac{U^2 \sin 2\alpha}{g}$$

Points to Note:

(i) Maximum range $= \dfrac{U^2}{g}$.

(ii) Angle of projection that gives maximum range is $45°$.

(iii) For each non-maximum value of the range, there are two possible values of the angle of projection, α and $(90° - \alpha)$.

● Exercise 8.4   See page 248

8.5 Path of a Projectile

Consider a general point (x, y) on the path of the projectile.

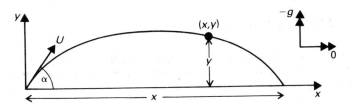

Horizontally:

$s = ut \Rightarrow x = U \cos \alpha t$ (1)

Vertically ↑:
$u = U \sin \alpha \uparrow$
$v = -$
$s = y$
$t = t$
$a = -g \uparrow$

$s = ut + \frac{1}{2}at^2 \uparrow \Rightarrow y = U \sin \alpha t - \frac{1}{2}gt^2$ (2)

$(1) \Rightarrow t = \dfrac{x}{U \cos \alpha}$

and putting this value of t into (2)

$$\Rightarrow y = U \sin \alpha \left(\frac{x}{U \cos \alpha} \right) - \frac{1}{2}g \left(\frac{x}{U \cos \alpha} \right)^2$$

∴ the path of the projectile is given by the equation

$$y = x \tan \alpha - \frac{gx^2}{2U^2} \sec^2 \alpha$$

Example 9   A shell is projected from a gun with a muzzle velocity of $20\sqrt{10}$ m/s at an angle α to the horizontal. When it is at a horizontal distance 200 m from the gun, the shell is 100 m above the ground. If the path of the shell is given by the equation

$$y = x \tan \alpha - \frac{gx^2}{2U^2} \sec^2 \alpha$$

show that the only possible values of $\tan \alpha$ are 1 and 3. In the case when $\tan \alpha = 3$ find the range of the shell. [Take $g = 10$ m/s]

Given that $U = 20\sqrt{10}$, $x = 200$ and $y = 100$ then the equation gives

$$100 = 200 \tan \alpha - \frac{10(200)^2}{2(400)} \sec^2 \alpha$$

$\Rightarrow 100 = 200 \tan \alpha - 50 \sec^2 \alpha$

$\Rightarrow \quad \sec^2 \alpha - 4 \tan \alpha + 2 = 0$

$\Rightarrow \tan^2 \alpha + 1 - 4 \tan \alpha + 2 = 0$ $\boxed{\sec^2 \alpha = \tan^2 \alpha + 1}$

$\Rightarrow \quad \tan^2 \alpha - 4 \tan \alpha + 3 = 0$

$\Rightarrow \qquad (\tan \alpha - 1)(\tan \alpha - 3) = 0$

$\Rightarrow \qquad \tan \alpha = 1$ and 3 [QED]

For $\tan \alpha = 3$, the equation of the path is

$$y = 3x - \frac{10x^2}{2(4000)}(9+1)$$

$$\Rightarrow y = 3x - \frac{x^2}{80}$$

When $y = 0$, $x = $ range

$$\therefore \ 0 = 3x - \frac{x^2}{80}$$

$$\Rightarrow x = 0 \text{ and } x = 240$$

$$\therefore \ \text{range} = 240 \,\text{m}.$$

• **Exercise 8.5** See page 250

8.6 Range on an Inclined Plane

The motion is usually considered either:
(i) perpendicular and parallel to the plane

or

(ii) perpendicular to the plane and horizontally.

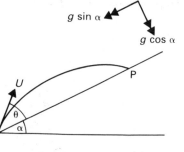

OP is the range on the inclined plane.

Example 10   A body is projected up a plane inclined at 45° to the horizontal with an initial velocity of 40 m/s at 20° to the plane. Find the range of the body.

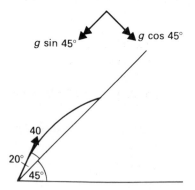

perpendicular to the plane↑:

$u = 40 \sin 20°$

$v = -$

$s = 0$

$t = t_1$

$a = -g \cos 45°$

$s = ut + \frac{1}{2}at^2 \uparrow \Rightarrow 0 = 40 \sin 20° t_1 - \frac{1}{2}\frac{g}{\sqrt{2}}t_1^2$

$$\Rightarrow t_1 = \frac{80\sqrt{2}}{g}\sin 20° \approx 3.95\text{s}$$

parallel to the plane ↗:

$$s = ut + \tfrac{1}{2}at^2 \uparrow$$

$u = 40\cos 20°$
$v = \text{—}$
$s = R$
$t = t_1$
$a = -g\sin 45°$

$$\Rightarrow R = 40\cos 20\,(3.95) - \frac{1}{2} \times \frac{g}{\sqrt{2}} \times (3.95)^2$$

$$\Rightarrow R = 94.4$$

$$\therefore \text{range} = 94.4\,\text{m}$$

Example 11   A body is projected up a plane inclined at 45° to the horizontal with an initial velocity of 40 m/s at an angle θ to the slope. Find the two possible values of θ that give a range up the slope of 60 m.

perpendicular to the plane ↖:

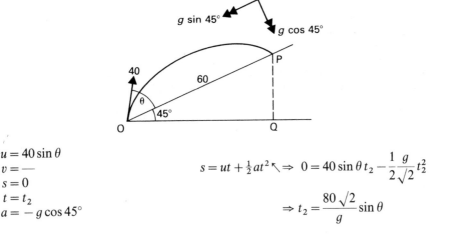

$u = 40\sin\theta$
$v = \text{—}$
$s = 0$
$t = t_2$
$a = -g\cos 45°$

$$s = ut + \tfrac{1}{2}at^2 \nwarrow \Rightarrow 0 = 40\sin\theta\, t_2 - \frac{1}{2}\frac{g}{\sqrt{2}}t_2^2$$

$$\Rightarrow t_2 = \frac{80\sqrt{2}}{g}\sin\theta$$

Horizontally →:

$$u = 40\cos(\theta + 45°)$$
$$s = OQ$$
$$t = t_2$$

$\therefore s = ut \Rightarrow$

$$OQ = 40\cos(\theta + 45°)t_2$$

But,

$$OQ = 60\cos 45°$$

$$\Rightarrow \frac{60}{\sqrt{2}} = 40\cos(\theta + 45°)\left(\frac{80\sqrt{2}}{g}\sin\theta\right)$$

$$\Rightarrow 2\cos(\theta + 45°)\sin\theta = \frac{3g}{160}$$

Applying the factor formulae in reverse.

$$\Rightarrow \sin(2\theta + 45°) - \sin 45° = \frac{3g}{160}$$

$$\Rightarrow 2\theta + 45° = 63° \text{ and } 117°$$

$$\Rightarrow 2\theta \qquad = 18° \text{ and } 72°$$

$$\Rightarrow \theta = 9° \text{ and } 36°$$

- Exercise 8.6   See page 251

8.7 More Complex Algebraic Problems

Example 12   The horizontal and vertical accelerations of a projectile moving freely under gravity are given by

$$\frac{d^2x}{dt^2} = 0 \text{ and } \frac{d^2y}{dt^2} = -g, \text{ respectively.}$$

By integrating these equations, show that

$$\frac{dx}{dt} = u\cos\alpha \text{ and } \frac{dy}{dt} = u\sin\alpha - gt$$

where u is the initial velocity of the projectile and α its angle of projection. Hence find equations for x and y and deduce that

$$y = x\tan\alpha - \frac{gx^2}{2u^2}(\tan^2\alpha + 1)$$

$$\frac{d^2x}{dt^2} = 0$$

Integrating $\Rightarrow$

$$\frac{dx}{dt} = A \quad \text{and when } t = 0, \frac{dx}{dt} = u\cos\alpha = A$$

$$\therefore \frac{dx}{dt} = u\cos\alpha \qquad\qquad\qquad \text{QED}$$

Similarly

$$\frac{d^2y}{dt^2} = -g$$

Integrating $\Rightarrow$

$$\frac{dy}{dt} = -gt + B \quad \text{and when } t = 0, \frac{dy}{dt} = u\sin\alpha = B$$

$$\therefore \frac{dy}{dt} = u\sin\alpha - gt \qquad\qquad\qquad \text{QED}$$

Hence, integrating $\quad \dfrac{dx}{dt} = u\cos\alpha$

gives $x = u\cos\alpha t + C$

and when $r = 0$, $x = 0 \Rightarrow C = 0$

$$\therefore x = u\cos\alpha t \qquad\qquad\qquad\qquad (1)$$

Similarly, integrating $\dfrac{dy}{dt} = u \sin \alpha - gt$

$$y = u \sin \alpha t - \tfrac{1}{2}gt^2 + D$$

and when $r = 0$, $y = 0 \Rightarrow D = 0$

$$\therefore \ y = u \sin \alpha t - \tfrac{1}{2}gt^2 \qquad (2)$$

$(1) \Rightarrow t = \dfrac{x}{u \cos \alpha}$ and putting this into (2)

$$\Rightarrow y = u \sin \alpha \left(\dfrac{x}{u \cos \alpha} \right) - \tfrac{1}{2}g \dfrac{x^2}{u^2 \cos^2 \alpha}$$

$$\Rightarrow y = x \tan \alpha - \dfrac{gx^2}{2u^2}(\tan^2 \alpha + 1)$$

$$\dfrac{1}{\cos^2 \alpha} = \sec^2 \alpha$$
$$= \tan^2 \alpha + 1$$

QED

Example 13   At the same instant, two particles, A and B, are projected from a point O with the same speed u. A is projected at an angle α above the horizontal and B at $(90° - \alpha)$ above the horizontal. Find the coordinates of each of the two particles at a time t after projection.

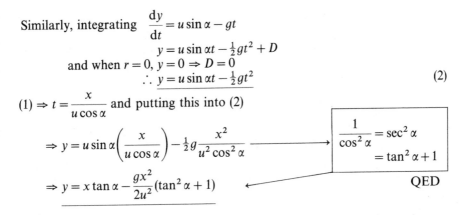

Hence find the coordinates of A when it is at the highest point on its path and find the distance between A and B at this instant. Show that this distance is greatest when $\alpha = 22\tfrac{1}{2}°$.

$s = ut \rightarrow$:

$$x_A = u \cos \alpha t \quad \text{and} \quad x_B = u \sin \alpha t$$

$s = ut + \tfrac{1}{2}at^2 \uparrow$:

$$y_A = u \sin \alpha t - \tfrac{1}{2}gt^2 \quad \text{and} \quad y_B = u \cos \alpha t - \tfrac{1}{2}gt^2$$

$\therefore$ coordinates of A are $(u \cos \alpha t, u \sin \alpha t - \tfrac{1}{2}gt^2)$ (1)

and coordinates of B are $(u \sin \alpha t, u \cos \alpha t - \tfrac{1}{2}gt^2)$ (2)

Note that
$\cos(90 - \alpha) = \sin \alpha$
$\sin(90 - \alpha) = \cos \alpha$

When A is at the highest point, $\dot{y}_A = 0$

$$\Rightarrow u \sin \alpha - gt = 0$$

or use $v = u + at \uparrow$
with $v = 0$

$$\Rightarrow t = \dfrac{u \sin \alpha}{g} \qquad (3)$$

(1) and (3) $\Rightarrow$ coordinates of A are $\left(\dfrac{u^2 \sin \alpha \cos \alpha}{g}, \dfrac{u^2 \sin^2 \alpha}{2g} \right)$

Also coordinates of B are, at this instant, $\left(\dfrac{u^2 \sin^2 \alpha}{g}, \dfrac{u^2 \sin \alpha \cos \alpha}{g} - \dfrac{u^2 \sin^2 \alpha}{g} \right)$

And, $(\text{distance})^2 = (y_B - y_A)^2 + (x_B - x_A)^2$

$$= \left[\frac{u^2}{g} \sin \alpha \cos \alpha - \frac{u^2}{2g} \sin^2 \alpha - \frac{u^2}{2g} \sin^2 \alpha \right]^2$$

$$+ \left[\frac{u^2}{g} \sin^2 \alpha - \frac{u^2}{g} \sin \alpha \cos \alpha \right]^2$$

$$= \left[\frac{u^2 \sin \alpha}{g} \right]^2 (\cos \alpha - \sin \alpha)^2 + \left[\frac{u^2 \sin \alpha}{g} \right]^2 (\sin \alpha - \cos \alpha)^2$$

$$= 2 \left[\frac{u^2 \sin \alpha}{g} (\cos \alpha - \sin \alpha) \right]^2$$

$$\therefore \text{distance} = \frac{\sqrt{2} u^2 \sin \alpha}{g} (\cos \alpha - \sin \alpha)$$

Maximum distance $\Rightarrow \dfrac{d}{d\alpha} (\text{distance}) = 0$

$$\Rightarrow \frac{u^2 \sqrt{2}}{g} [\sin \alpha(-\sin \alpha - \cos \alpha) + (\cos \alpha - \sin \alpha) \cos \alpha] = 0$$

$$\Rightarrow \quad -\sin^2 \alpha - \sin \alpha \cos \alpha + \cos^2 \alpha - \sin \alpha \cos \alpha = 0$$

$$\Rightarrow \quad \cos^2 \alpha - \sin^2 \alpha - 2 \sin \alpha \cos \alpha = 0$$

$$\Rightarrow \quad \cos 2\alpha = \sin 2\alpha$$

$$\Rightarrow \quad \tan 2\alpha = 1$$

$$\Rightarrow \quad 2\alpha = 45°$$

$$\Rightarrow \quad \underline{\alpha = 22\tfrac{1}{2}°} \qquad \text{QED}$$

- Exercise 8.7   See page 252

8.8 + Revision of Unit: A Level Questions

- Miscellaneous Exercise 8B   See page 254

Unit 8 Exercises

(Throughout this unit use $g = 9.8 \, \text{m/s}^2$ unless otherwise instructed.)

Exercise 8.1

1 A stone is thrown horizontally with a speed of 20 m/s from the top of a vertical cliff which is 120 m above the sea. How long will it take to reach the sea and how far from the base of the cliff will it strike the water?

2 A bullet is fired horizontally with a speed of 500 m/s from a gun which is 1.8 m above level ground. What is the range of the bullet?

3 An object is thrown horizontally from a high window with a speed of 24 m/s. Find its horizontal and vertical displacements from its point of projection after 3 seconds. Hence calculate its distance from the point of projection after 3 seconds.

4 A car is driving over level ground at 72 km/h when it is driven straight over the edge of a cliff. The wreckage is found 45 m from the base of the cliff. How high is the cliff?

5 A button is resting on a shelf one metre above the floor. It is knocked horizontally off the shelf and lands on the floor a horizontal distance of 1.5 m away. How long did it take to fall and with what speed did it leave the shelf?

6 A fielder throws a cricket ball horizontally with a speed of 25 m/s to the wicket keeper who is 15 m away. If it leaves the fielder's hands at a height of 2 m above the ground, what is its height when it reaches the wicket keeper?

7 An object is projected horizontally at 100 m/s. Find the horizontal and vertical components of its velocity after 5 seconds.

8 A squash ball is hit horizontally with a speed of 20 m/s directly towards a wall. It strikes the wall 0.4 s later. With what speed and at what angle does it hit the wall?

9 An aircraft flies at 100 m above the ground at 200 m/s in order to bomb a gun emplacement. At what distance before the target must the bomb be released?

10 A boy throws a ball horizontally from the window of a house 4.5 m above the ground. A wall is 10 m from the house and 2.5 m high. If the ball just clears the wall, with what speed was it thrown and how far beyond the wall did it land?

11 A dart is thrown horizontally at 14 m/s from a point 2.4 m in front of the board. At what angle does it strike the board and with what speed?

*12

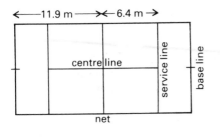

When serving, a certain tennis player stands just to the right of the centre mark on the base line in order to serve into the opposite court. She hits the ball horizontally at a height of 2.4 m above the ground and intends that the ball should land just on the other side of the centre line of the service court. What is the minimum speed at which the ball must be hit in order to clear the net which is 11.9 m away and 90 cm high?

If the ball lands on the correct side of the centre line when hit at 25 m/s, will it land in the service court, given that the service line (the far line of the service court) is 6.4 m from the net?

*13   A particle of mass m is attached by a string of length l to a fixed point O. The particle is held on the same horizontal level as O with the string taut and allowed to fall. When the string is vertical, it breaks and the particle falls under gravity to level ground which is a distance $2l$ below O. Find the horizontal distance of the particle from O when it reaches the ground.

Exercise 8.2

1   A golf ball is hit with a velocity of 45 m/s at an elevation of 30°, along a level fairway. Find:
    (a) the greatest height reached
    (b) the time of flight
    (c) the distance travelled along the fairway.

2   Find the horizontal range, to the nearest metre, of a bullet, fired just above ground level, at an elevation of 5° with a muzzle velocity of 500 m/s.

3   A boy kicks a ball at 15 m/s at 55° to the horizontal. At what height does it hit a wall which is 8 m away?

4   A rugby player takes a place kick at goal from a position 30 m from the goal posts. To score a goal he must kick it between the posts and over a bar which is 3.6 m from the ground. He kicks the ball with a speed of 25 m/s at an elevation of 20°, and it goes between the posts. Show that he does not score a goal.

5   A cricket ball is hit, almost at ground level, with a speed of 30 m/s at an angle of 40° to the horizontal. Find the magnitude and direction of its velocity after 3 seconds.
    In the direction in which the ball is hit, the boundary is 86 m from the batsman. If the ball passes the boundary without bouncing, he scores a six; if it bounces before it passes the boundary he scores a four. If the ball is not intercepted by a fielder, what does the batsman score?

6   Show that a golf ball projected with a velocity of 30 m/s at an angle of elevation of $\tan^{-1}\frac{3}{4}$ will just clear a tree 9.9 m high at a horizontal distance of 72 m from the point of projection. How far past the tree will it first hit the ground?

7   A ball is kicked into the air with an initial velocity of 25 m/s at 55° to the horizontal. Find the magnitude and direction of the velocity, at the two positions when it is 10 m above the ground. Find the length of time that it is more than 10 m above the ground.

*8   A stone is thrown from a point on level ground at a bird sitting on the top of a vertical pole. The velocity of projection is 13 m/s at angle of elevation $\tan^{-1}\frac{12}{5}$. The stone just misses the bird and the greatest height that it reaches in the subsequent motion is twice the height of the pole. Find the height of the pole and the distance of its base from the point of projection. (Let $g = 10\,\text{m/s}^2$)

*9 Firemen wish to aim the jet of water from their hose into an open window which is a horizontal distance of 25 m from the hose and 15 m above it. The speed of the water leaving the hose is 30 m/s. Find the two angles of elevation of the hose that would place the jet of water in through the window. (Let $g = 10\,\text{m/s}^2$)

*10 A gun is positioned such that the horizontal range of its shells is three times the greatest height reached. Find the angle of elevation at which the shell is fired. For a range of 400 m find the muzzle velocity and the time of flight.

Exercise 8.3

1 Geoff Capes, a world class shotputter, releases the shot from a height of about 2 metres. He is reputed to obtain his best distances when he projects the shot at an angle of elevation of 55°. What distance will he put the shot if the speed of projection is 15 m/s?

2 A cricketer hits a ball just above ground level. The highest point of the path of the ball is 20 m above the ground and it hits the ground again at a distance of 75 m from the batsman. Find the horizontal and vertical components of the initial velocity of the ball.
(Let $g = 10\,\text{m/s}^2$)

3 A cricketer hits the ball just above ground level and it is in the air for 2.1 s before it bounces on the ground a distance of 55 m away. Find the initial speed and direction of the ball and its maximum height above the ground.

4 A car is travelling along an undulating road which then levels out into a long straight horizontal stretch. As the car is climbing the last hill before the straight, the driver puts his foot down and the car leaves the road just before the crest of the hill, at 108 km/h and at an angle of elevation of 10°. He lands on the straight stretch which is a distance of 9 m below his point of take-off. How high did the car go above the point of take-off and what horizontal distance had it travelled before it landed?

5 Find the velocity and angle of take-off of a stunt motorcycle, which passes in a horizontal direction just over the top of a bar, which is a horizontal distance of 20 m from the point of take-off and 7 m above it.

6 A stone is thrown from a point C on top of a cliff 52 m above sea level and it strikes the sea at a point S. The velocity of projection is $(24\mathbf{i} + 7\mathbf{j})$ m/s where $\mathbf{i}$ and $\mathbf{j}$ are horizontal and vertically upward unit vectors. Calculate:
(a) the magnitude and angle of elevation of the velocity of projection
(b) the time of flight

(c)   the distance of S from a point at sea level vertically below C
(d)   the magnitude and direction of the velocity of the stone when it enters the water.

(Let $g = 10\,\text{m/s}^2$)

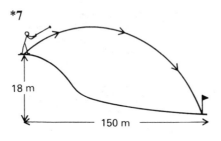

*7   A golfer is standing on a piece of high ground on a golf course. The next hole is a horizontal distance of 150 m away and a vertical distance of 18 m below him. If he gives the ball a speed of 45 m/s, at what angle of elevation should he try to hit the ball in order to get as close to the hole as possible? (Let $g = 10\,\text{m/s}^2$)

*8   During a military exercise, planes are dive-bombing a target buoy. One plane, flying at 210 m/s, and diving at an angle of 20° below the horizontal, releases its bomb at a height of 100 m, when it is a horizontal distance of 260 m from the target. Show that it misses the target by approximately 7 m.

*9   A golfer wishes to place her ball as close as possible to the hole which is 20 m away from her on the same horizontal level. Unfortunately, to do this she has to play the ball over the top of a tree which is between her and the hole. The tree is 10 m tall and 13 m away from the golfer. What angle of elevation and what speed should she be trying to give to the ball?

Exercise 8.4

1   What is the maximum range of a gun that fires a bullet at 250 m/s from just above ground level? (Let $g = 10\,\text{m/s}^2$)

2   In the First World War, the German 21-centimetre gun, known as "Big Bertha" was used to shell Paris. Its maximum range was 122 km. What was the initial velocity of the shell? (Let $g = 10\,\text{m/s}^2$)

3   A long-jumper leaves the take-off board with a speed of 8 m/s and he rises to a maximum height of 50 cm above the point of take-off. Show that his initial angle of elevation is approximately 23° and estimate how far he jumps.

4   World class long-jumpers achieve jumps of 8 m or more. Bob Beamon made a new world record in 1968 when he jumped 8.90 m in Mexico city. This was an incredible jump because he raised the previous world record by a stagggering 55 cm and his record is still unbeaten 20 years later. Show that he must have left the take off board with a speed of at least 9.34 m/s.

5   A gun fires a shell with a muzzle velocity of 600 m/s. Show that the maximum range is achieved when the shell is fired at 45° to the horizontal and find the maximum range. Also calculate the two angles of elevation for which the shell has a range of 20 km.
(Let $g = 10\,\text{m/s}^2$)

6 A body is projected from a point on a horizontal plane with speed U at an angle α to the horizontal. Show that the maximum height reached is $U^2 \sin^2 \alpha/2g$ and the range is $U^2 \sin 2\alpha/g$.
Find the range, in terms of U and g, if the body must not exceed a height of (a) $U^2/8g$ (b) $U^2/3g$.

*7 Find the greatest possible range of a projectile inside a tunnel which is 3 m high, if the initial speed of the projectile is 100 m/s.

8 Show that when a particle is projected with speed V at an inclination α the range on a horizontal plane through the point of projection is

$$\frac{V^2 \sin 2\alpha}{g}$$

Given that the maximum range is 90 m, find V.
Find also the speed of projection U for which the maximum range is $60\sqrt{3}$ m.
In a cricket match a man must throw a ball to reach a wicketkeeper, who is 90 m away. Assuming that the ball arrives in the wicketkeeper's hands at the same height as it left the man's hand, find the time taken for the ball to reach the wicketkeeper when the man throws it with speed V.
Show that when the man throws the ball with speed U there are two possible angles of inclination and find the two corresponding times of flight. (Take g as 10 m/s².)   [L]

9 A particle P is projected from a point O on a horizontal plane with speed u at an angle α to the horizontal and it next strikes the plane at a distance R from O.

(a) In the special case when $u = 40$ m/s and $R = 100$ m, find, correct to the nearest degree, the possible values of α.

(b) The angle between the direction of motion of P and the horizontal at time t after projection is denoted by β. In the general case when u and R are not given, express $\tan \beta$ in terms of u, α, t and g.
Given that $\tan \beta$ is of magnitude $\frac{1}{2}\tan^2 \alpha$ at two points whose horizontal distance apart is $R/5$, find $\tan \alpha$.
(Take g to be 10 m/s².)   [A]

10 A particle is projected with speed u at an angle of elevation α to the horizontal. Given that R is the range attained on a horizontal plane through the point of projection, and h is the maximum height of the trajectory, prove that

$$R = 2c \sin \alpha \cos \alpha$$
$$2h = c \sin^2 \alpha$$

where $c = u^2/g$.

Hence prove that $R^2 = c^2 - (c - 4h)^2$.

If u is held fixed while α varies, so that R and h vary, deduce from the last equation that R is an increasing function of h when $h < c/4$. Hence, prove that, if a particle is projected with speed 30 m/s from the floor of a horizontal

tunnel of height 20 m, the greatest range which can be attained in the tunnel is about 89.4 m.
(Take g as $10 \,\text{m/s}^2$.)   [L]

Exercise 8.5

1 A particle is projected from a point O and, some time later, passes through a point with coordinates (x, y), where Ox and Oy are cartesian axes with Ox horizontal and Oy vertically upwards. Given that the velocity of projection has a horizontal component u and a vertical component v, show that

$$2yu^2 - 2uvx + gx^2 = 0$$

Given that the particle passes through the points with coordinates $(36a, 6a)$ and $(48a, 4a)$, show that the velocity of projection is $13\sqrt{(ag/2)}$ at an elevation arctan $\frac{5}{12}$.   [L]

2 At time $t = 0$, a particle is projected from a point O with speed u at an angle of elevation α. At time t, the horizontal and vertical distances of the particle from O are x and y respectively. Express x and y in terms of u, α, t and g. Hence show that

$$y = x \tan \alpha - \frac{gx^2}{2u^2}(1 + \tan^2 \alpha)$$

A golf ball is struck from a point A, leaving A with speed 30 m/s at an angle of elevation θ, and lands, without bouncing, in a bunker at a point B, which is at the same horizontal level as A. Before landing in the bunker, the ball just clears the top of a tree which is at a horizontal distance of 72 m from A, the top of the tree being 9 m above the level of AB. Show that one of the possible values of θ is $\tan^{-1}\frac{3}{4}$ and find the other value. Given that θ was in fact $\tan^{-1}\frac{3}{4}$, find the distance AB.
(Take g as $10 \,\text{m/s}^2$.)   [L]

3 The components of the initial velocity of a particle, projected under gravity from the origin O, are (u, v) referred to horizontal and upward vertical axes Ox, Oy, respectively. Show that the equation of its path referred to these axes is

$$2u^2 y = 2uvx - gx^2$$

A particle is projected from a point on a horizontal plane so that it just clears a vertical wall of height $a/2$ at a horizontal distance a from the point of projection and strikes the plane at a horizontal distance $3a$ beyond the wall. Show that the angle of projection with the horizontal is $\tan^{-1}\frac{2}{3}$.
Find, in terms of a and g, the speed at which the particle is projected.   [J]

4 A point O lies on a horizontal plane, and the point A is at a height h vertically above O. A particle is projected from A with speed V at an angle α above the horizontal. Taking O as the origin and Oy vertically upwards, show that the equation of the path of the particle can be written in the form

$$y = h - \frac{gx^2}{2V^2} + x \tan \alpha - \frac{gx^2}{2V^2} \tan^2 \alpha$$

The particle hits the plane at the point B(r, 0). In the case when $V^2 = gh$ derive a quadratic equation for $\tan \alpha$ in terms of h and r, and show that $r \leqslant \sqrt{3h}$. For the same value of V show that $r = \sqrt{3h}$ when $\alpha = 30°$.

[J]

5 A small particle P is projected from a point O on a horizontal plane so that it first lands again at a point on the plane at a distance $2a$ from O. The maximum height reached by the particle is b and in this position the particle is directly above a point B on the plane. Find, in terms of a, b and g, the tangent of the angle of projection and the square of the speed of projection. At time t the horizontal and vertical displacements of P from O are x and y respectively. Show, that

$$y = \frac{2bx}{a} - \frac{bx^2}{a^2}$$

Hence determine, in terms of a and b, constants p, q and r such that

$$y = -p(x - q)^2 + r$$

and so show that

$$BP^2 = \frac{b^2}{a^4}(x - a)^4 - \frac{2b^2}{a^2}(x - a)^2 + (x - a)^2 + b^2$$

Hence find the greatest value of b/a such that BP is never less than b throughout the motion.

[A]

Exercise 8.6

In this exercise, you may assume that all motion takes place up or down a line of greatest slope.

1 A body is projected from the foot of a plane inclined at 30° to the horizontal. The body is given an initial velocity of 60 m/s at an angle of 30° to the plane. Find the time of flight and the range up the plane. (Let $g = 10$ m/s²)

2 A body is projected *down* a plane inclined at 45° to the horizontal. It is given an initial velocity of 20 m/s at an angle of 60° to the plane. Find the range down the plane.

3 A ball is thrown with a velocity of 20 m/s at an angle of 45° to the horizontal. How far does it go along a plane inclined at 30° to the horizontal if it is thrown (a) up (b) down the plane.

4 A body is projected up a plane inclined at 30° to the horizontal. It is given a velocity of 50 m/s at 30° to the plane. Show that it strikes the plane at an angle of 60° to the plane.

5 A jet of water is projected down a slope inclined at 30° to the horizontal. It has an initial velocity of 25 m/s at an angle θ to the slope. If it hits the slope 120 m below the point of projection, measured along the slope, find the two possible values of θ.

*6 A body is projected up a plane inclined at an angle α to the horizontal. It is projected with velocity U at an angle θ to the horizontal and it strikes the inclined plane horizontally.
Show that:
(a) its time of flight is $2U \sin(\theta - \alpha)/g \cos \alpha$
(b) $\tan \theta = 2 \tan \alpha$
(c) its range is $2U^2 \sin(\theta - \alpha) \cos \theta / g \cos^2 \alpha$.

*7 A body is projected up a plane inclined at an angle α to the horizontal, with a speed U at an angle β to the plane. Show that, when it returns to the plane, it makes an angle ϕ with the plane where

$$\tan \phi = \frac{\tan \beta}{1 - 2 \tan \beta \tan \alpha}$$

*8 Show that the greatest range of a projectile up an inclined plane at angle α is

$$\frac{U^2}{g(1 + \sin \alpha)}$$

where U is the initial speed of the projectile.

Exercise 8.7

1 A particle P moves freely under gravity in the plane of a fixed horizontal axis Ox and a fixed upward vertical axis Oy. At time t the coordinates of P are (x, y). Write down the values

$$\frac{d^2 x}{dt^2} \text{ and } \frac{d^2 y}{dt^2}$$

The particle is projected from O, at time $t = 0$, with speed V at an angle α to Ox. Find by integration the values of x and y at time t, and deduce that

$$y = x \tan \alpha - \frac{gx^2}{2V^2}(1 + \tan^2 \alpha).$$

The point O is on horizontal ground and AB is a vertical post of height $2h$ whose foot A is at the point $(3h, 0)$. Given that

$$V^2 = 9gh$$

show that the particle will pass over the post provided that

$$1 < \tan \alpha < 5$$

Show that in this case the particle will hit the ground at a distance greater than $6h/13$ from A.   [J]

2

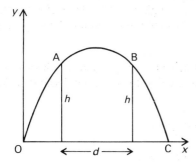

A particle P moves freely under gravity in the plane of a fixed horizontal axis Ox, which lies on flat ground, and a fixed upward vertical axis Oy. The particle is projected from O with a velocity whose components along Ox and Oy are u and v, respectively. Prove that the particle returns to the ground at a point C, at a distance $2uv/g$ from O. The particle passes, in succession, through two points A and B, at a height h above Ox and at a distance d apart, as shown in the diagram.

Write down the horizontal and vertical components of the velocity of the particle at A. Hence show that

$$d = 2u(v^2 - 2gh)^{\frac{1}{2}}/g$$

Given that the direction of motion of the particle as it passes through A is inclined to the horizontal at an angle of $\tan^{-1}\frac{1}{2}$, show that

$$u = \sqrt{gd}$$

and obtain an expression for v in terms of g, d and h. [J]

3 A particle is projected from a point O with speed V at an angle of elevation α. Prove that the equation of its path referred to horizontal and vertical axes Ox, Oy is

$$y = nx - \frac{gx^2}{2V^2}(1 + n^2)$$

where $n = \tan\alpha$.
There is a vertical wall perpendicular to Ox at a distance b from O and the particle hits it at a height h above the level of O. Show that, for fixed value of V, h is maximum when α is chosen so that $ngb = V^2$, and that in this case the particle is travelling in a direction making an angle α with the wall when it hits it. Find the speed of the particle at the moment of impact in terms of V, b and g. [J]

4 Two particles, P and Q are simultaneously projected from a point O with the same speed but at different angles of elevation, and they both pass through a point C which is at a horizontal distance $2h$ from O and at a height h above the level of O. The particle P is projected at an angle $\tan^{-1}2$ above the horizontal. Find the speed of projection in terms of g and h, and show that Q is projected at an angle $\tan^{-1}\frac{4}{3}$ to the horizontal.
Show that the time interval between the arrivals of the two particles at C is

$$(3 - \sqrt{5})\sqrt{\left(\frac{2h}{3g}\right)}$$

[J]

5   A shot putter releases the shot at a height h metres above the ground. It is given a velocity of V m/s and an angle of elevation α. If the horizontal range is R, show that R satisfies the equation

$$\frac{R^2 g}{2V^2} \sec^2 \alpha - R \tan \alpha - h = 0$$

Hence show that the maximum range is

$$\frac{V^2}{g}\left(1 + \frac{2gh}{V^2}\right)^{1/2}$$

and that, when the range is a maximum,

$$\tan \alpha = \left(1 + \frac{2gh}{V^2}\right)^{-1/2}$$

Miscellaneous Exercise 8B

1   A ball is kicked from a point A on level ground and hits a goalpost at a point 4 m above the ground. The goalpost is at a distance of 32 m from A. Initially the velocity of the ball makes an angle $\tan^{-1}\frac{3}{4}$ with the ground. Show that the initial speed of the ball is 20 m/s. Find the speed of the ball when it hits the goalpost.
(Take g as $10 \,\text{m/s}^2$.)   [L: 8.2]

2   A particle is projected with speed v at an angle α above the horizontal from a point O on a horizontal plane. Find expressions for:
(a)   the maximum height h reached by the particle
(b)   the distance r from O of the point at which the particle returns to the plane.
Show that $4h = r \tan \alpha$.   [J: 8.2, 8.4]

3   A particle P projected from a point O on level ground strikes the ground again at a distance of 120 m from O after a time 6 s. Find the horizontal and vertical components of the initial velocity of P. The particle passes through a point Q whose horizontal displacement from O is 30 m. Find:
(a)   the height of Q above the ground
(b)   the tangent of the angle between the horizontal and the direction of motion of P when it is at Q and also the speed of P at this instant
(c)   the horizontal displacement from O of the point at which P is next at the level of Q.
(Take g to be $10 \,\text{m/s}^2$.)   [A: 8.2, 8.3]

4   A particle projected from a point O on level ground strikes the ground again at a distance $4a$ from O after a time T. Find the horizontal and vertical components of its initial velocity.
The particle just clears a vertical post, of height h, at a horizontal distance a from O. Show that $h = 3gT^2/32$ and find, in terms of a, g and T, the speed of the particle as it passes over the post.   [J: 8.3, 8.5]

5   A particle of mass 2 kg is attached to one end of a light inextensible cord, the other end of which is fixed at a point O. The particle moves with speed

5 m/s in a horizontal circle, which is of radius 6 m and whose centre is vertically below O. Calculate:

(a)   the magnitude of the tension in the cord
(b)   the length of the cord.

The path of the particle is at a height of 20 m above level ground when the cord breaks. Find the tangent of the angle which the velocity of the particle makes with the horizontal when the particle strikes the ground. (Take g as 10 m/s^2.)   [A: 7.3, 8.1]

6   A particle is projected from a point O on horizontal ground with speed v at an angle of elevation α. The point O is 20 m away from the foot, F, of a vertical post. The particle subsequently strikes the post at a point 15 m above the ground and at the moment of impact, the particle is travelling horizontally. Find α and v.
The coefficient of restitution between the particle and the post is $\frac{1}{2}$. The particle next meets the ground at a point B where O, B and F are in a straight line. Calculate OB and the angle which the direction of motion of the particle makes at B with the horizontal.
(Take g to be 9.8 m/s^2.)   [A: 8.2, 5.8]

7   A particle P, of mass m, is projected from a point O with speed u at an angle of inclination α to the horizontal. Find the time taken for P to reach the highest point A of its path.
When P is at A it collides directly, and coalesces, with a particle Q, of mass $3m$. Just before impact Q was moving horizontally with the same speed as P but in the opposite direction. Find the speed of the composite particle immediately after impact and show that the loss of kinetic energy is $\frac{3}{2}mu^2\cos^2\alpha$.
Given that the composite particle meets the horizontal plane through O at a point B, express OB in terms of u, α and g.
    [L: 5.1, 5.5, 8.2]

8   A particle P of mass m moves freely under gravity in the plane of a horizontal axis Ox and an upward vertical axis Oy. The particle is projected from O, at time $t = 0$, with speed u and at an angle α above Ox. Write down expressions for the coordinates (x, y) of P at time t. Given that u is fixed and that α may vary between 0 and $\frac{1}{2}\pi$ radians, derive, in terms of u and g, the maximum range R along Ox.
In the case when α is such that the range along Ox is a maximum, show that:

(i)   the greatest height reached is $\frac{1}{4}R$
(ii)   the kinetic energy of the particle at any instant is proportional to the difference between $\frac{1}{2}R$ and y.
    [J: 8.2, 8.4, 8.5]

9   At time $t = 0$ s a small smooth sphere P is projected in a vertical plane from a point O on horizontal ground. At time $t = 2$ s P is moving at an acute angle β to the horizontal and at time $t = 3$ s P is moving at an acute angle γ to the horizontal where $\tan\beta = 2$ and $\tan\gamma = \frac{3}{2}$.
Given that it takes more than three seconds for P to reach its maximum height find the horizontal and vertical components of the velocity of P at O. At the point where P attains its maximum height it collides with a small

sphere Q of equal radius and moving directly towards it with the same speed as itself. The mass of Q is twice that of P and the coefficient of restitution for collisions between the spheres is $\frac{1}{4}$. Find the speed of P immediately after collision and determine the distance from O at which it hits the ground.

(Take the acceleration due to gravity to be $10\,\text{m/s}^2$.)   [A: 8.2, 8.3, 5.6]

10 A point A is at a height h vertically above a point B, which is on a smooth horizontal plane. A ball is projected from A with speed u in a direction parallel to a line BC on the plane. The length of BC is a, and at C there is a vertical post CD of height $h/2$. Find the time taken for the ball, moving under gravity, to reach the level of D, and show that the ball cannot pass over the post before hitting the plane if

$$u < a\sqrt{(g/h)}$$

The ball hits the plane at a point E between B and C. Find the time taken for the ball to reach E and the vertical component of its velocity when it reaches E. Given that the coefficient of restitution between the ball and the plane is e, show that the ball cannot clear the post if

$$e^2 < \tfrac{1}{2}$$

Given that $e^2 = \tfrac{1}{2}$ and that the ball hits the post at D, show that

$$u = \frac{a}{\sqrt{2}+1}\sqrt{\frac{g}{h}}$$

[J: 8.1, 8.2, 5.8]

*11 Two equal particles are projected at the same instant from points A and B on horizontal ground, the first from A with speed u at an angle of elevation α and the second from B with speed v at an angle of elevation β. They collide directly when they are moving horizontally in opposite directions. Find v in terms of u, α and β, and show that

$$AB = \frac{u^2 \sin \alpha \sin (\alpha + \beta)}{g \sin \beta}$$

The particles coalesce on impact. Find, in terms of u, α and β, the speed of the combined particle immediately after the collision. In the case when $\alpha = 30°$ and $\beta = 60°$ find AC, where C is the point where the combined particle returns to the ground.   [J: 5.5, 8.7]

*12 Water is ejected in a narrow jet from a nozzle N and hits horizontal ground, 15 m below the level of N, at a point G. The water emerges from N at the rate of 5 kg/s and is moving with speed 20 m/s at an angle of 30° above the horizontal. Find:

(a) the maximum height above N reached by the water
(b) the horizontal displacement of G from N
(c) the vertical component of the velocity of the water at G
(d) the minimum value of the force exerted on the ground by the water.

Given also that the water is pumped from rest at a depth of 25 m below N find the effective rate at which the pump is working.
(Take g to be $10\,\text{m/s}^2$.)   [A: 8.2, 8.3, 5.1]

*13 A particle P is projected horizontally, with speed V, from a point O on a plane which is inclined at an angle β to the horizontal. The particle hits the plane at a point A which is on the line of greatest slope through O. Show that the time of flight is

$$\frac{2V}{g}\tan\beta$$

Find the tangent of the acute angle between the horizontal and the direction of motion of P when P reaches A.

A second particle Q is projected from O, with speed V, in a direction perpendicular to the plane. Find the time taken for Q to return to the plane and show that Q hits the plane at A. [J: 8.7, 8.6]

*14 A particle P is projected with speed V at an angle of elevation α from a point O on a horizontal plane and moves under gravity in a plane Oxy, where the axis Ox is horizontal and the axis Oy is vertical. Starting from the horizontal and vertical components of the acceleration of P obtain, by integration, expressions for the coordinates x, y of P at a time t after the instant of projection. Hence prove that P will return to the plane at a point A, where $OA = 2(V^2/g)\sin\alpha\cos\alpha$. Find the height above the plane of the highest point H of the trajectory of P.

Another particle Q is projected from O with speed V in the direction OH. Find the distance from O of the point B where Q returns to the plane.

Find the value of $\cos\alpha$ for which B coincides with A and show that, in this case,

$$OA = (2\sqrt{2/3})r,$$

where r is the maximum range, on the plane, for any particle projected from O with speed V. [J: 8.7, 8.5]

*15 A particle is projected with speed V and angle of elevation α from a point A on a horizontal plane and it returns to the plane at B. The direction of motion of the particle makes an angle β with the horizontal, first at a point C and then at a point D. Show that the time taken to travel from C to D is

$$\frac{2V}{g}\cos\alpha\tan\beta$$

Hence, or otherwise, show that

$$\frac{CD}{AB} = \frac{\tan\beta}{\tan\alpha}$$

Find, in terms of α, β, g and V, the height above A of C and D. [J: 8.7, 8.2]

**16 A particle is projected from a point O with speed V at an angle α above the horizontal. Prove that the equation of the path of the particle referred to horizontal and vertically upward axes Ox, Oy, is

$$y = x\tan\alpha - \frac{gx^2}{2V^2}\sec^2\alpha$$

A particle is projected at an angle α above the horizontal from a point on a table of height h standing on a horizontal floor. The particle reaches the floor at a point whose horizontal distance from the point of projection is $2h$. Show that the inclination β below the horizontal of its direction of motion when it strikes the floor is given by

$$\tan \beta = \tan \alpha + 1$$

Find the speed of projection and the time of flight of the particle in terms of g, α and h.

In the case when the speed of projection is $\sqrt{(\frac{4}{3}gh)}$, find the two possible values of $\tan \alpha$. 			[J: 8.7, 8.3]

**17 A particle is projected from a point O with a velocity u at an angle α above the horizontal. Show that the equation of its path can be written in the form

$$y = -\frac{1}{2}\frac{g \sec^2 \alpha}{u^2}\left(x - \frac{u^2}{2g}\sin 2\alpha\right)^2 + \frac{u^2}{2g}\sin^2 \alpha$$

where O is the origin and the y-axis is vertically upwards. Hence, or otherwise, determine, in terms of u, α and g, the coordinates (b, h) of the highest point of the path, and show that $\tan \alpha = 2h/b$. In the case when $\alpha \neq \frac{1}{4}\pi$ find a second angle of projection such that a particle projected with the same initial speed from O attains its maximum height when $x = b$, and express this maximum height in terms of h and b. 			[J: 8.7, 85]

UNIT 9 VARIABLE ACCELERATION AND FORCES: SIMPLE HARMONIC MOTION

9.1 Variable Acceleration I

Acceleration as a Function of Time

If the displacement, x is given by

then the velocity, v, is given by

and the acceleration, a, is given by

$$x = f(t)$$

$$v = \frac{dx}{dt} = f'(t)$$

$$a = \frac{dv}{dt} = f''(t)$$

Example 1   The distance of a particle from a fixed point O, at time t, is given by $x = 3t + \sin t$. Find expressions for its velocity and acceleration in terms of t.

$$x = 3t + \sin t$$

Differentiating: $\underline{\text{velocity} = \dfrac{dx}{dt} = 3 + \cos t}$

Differentiating: $\underline{\text{acceleration} = \dfrac{dv}{dt} = \dfrac{d^2x}{dt^2} = -\sin t}$

Example 2   A particle moving in a straight line has an acceleration $2t\,\text{m/s}^2$ where t is the time in seconds. When $t = 0$, the particle is passing through a fixed point A with speed $2\,\text{m/s}$. Find the velocity and displacement from A (a) at time t and   (b) after 3 seconds.

Method 1

(a)
$$\frac{dv}{dt} = 2t$$

Integrating;
$$v = t^2 + A$$

when $t = 0$, $v = 2 \Rightarrow A = 2$

$$\therefore v = t^2 + 2 \tag{1}$$

Hence
$$\frac{dx}{dt} = t + 2$$

Integrating:
$$x = \frac{t^3}{3} + 2t + B$$

when $t = 0$, $x = 0 \Rightarrow B = 0$

$$\therefore x = \tfrac{1}{3}t^3 + 2t \tag{2}$$

Method 2

(a) $$\frac{dv}{dt} = 2t$$

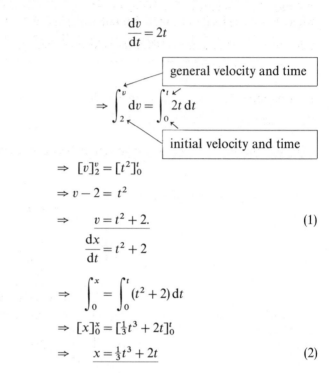

$$\Rightarrow \int_{2}^{v} dv = \int_{0}^{t} 2t \, dt$$

general velocity and time

initial velocity and time

$$\Rightarrow [v]_{2}^{v} = [t^2]_{0}^{t}$$

$$\Rightarrow v - 2 = t^2$$

$$\Rightarrow \quad v = t^2 + 2. \tag{1}$$

Also $$\frac{dx}{dt} = t^2 + 2$$

$$\Rightarrow \int_{0}^{x} = \int_{0}^{t} (t^2 + 2) \, dt$$

$$\Rightarrow [x]_{0}^{x} = [\tfrac{1}{3}t^3 + 2t]_{0}^{t}$$

$$\Rightarrow \quad x = \tfrac{1}{3}t^3 + 2t \tag{2}$$

(b) where $t = 3$, (1) $\Rightarrow \underline{v = 11 \, \text{m/s}}$ and (2) $\Rightarrow \underline{x = 15 \, \text{m/s}}$

Both methods ought to be known, but the second method will be used in examples from now on.

Acceleration as a Function of Velocity or Displacement

Acceleration $= \dfrac{dv}{dt}$

but, since $v = \dfrac{dx}{dt}$, $\dfrac{dv}{dt}$ can be written as

$$\frac{dv}{dt} = \frac{dx}{dt} \cdot \frac{dv}{dx} = v \frac{dv}{dx}.$$

Acceleration $= \dfrac{dv}{dt}$

is used when a relationship between velocity and time is required.

Acceleration $= v \dfrac{dv}{dx}$

is used when a relationship between velocity and displacement is required.

Example 3   A particle is moving in a straight line with a retardation $2v^2$ m/s^2, where v is the velocity at time t. Initially the particle is at a fixed point O, with velocity u. Find the time taken for the velocity to become $\dfrac{u}{2}$ and the displacement from O at that time.

Retardation $= 2v^2 \Rightarrow$ Acceleration $= -2v^2$

To find the time, we need a connection between v and t and so use $\dfrac{dv}{dt}$.

$$\frac{dv}{dt} = -2v^2$$

$$\Rightarrow \int_u^{\frac{u}{2}} -\frac{dv}{v^2} = \int_0^t 2\,dt$$

initial values of velocity and time

$$\Rightarrow \left[\frac{1}{v}\right]_u^{\frac{u}{2}} = [2t]_0^t$$

$$\Rightarrow \frac{2}{u} - \frac{1}{u} = 2t$$

$$\Rightarrow \quad t = \frac{1}{2u} \text{ seconds}$$

To find the displacement, we need a connection between v and x, and so use $v\dfrac{dv}{dx}$.

$$v\frac{dv}{dx} = -2v^2$$

$$\Rightarrow \int_u^{\frac{u}{2}} -\frac{dv}{v} = \int_0^x 2\,dx$$

$$\Rightarrow [-\ln v]_u^{u/2} = [2x]_0^x$$

$$\Rightarrow -\ln \tfrac{u}{2} + \ln u = 2x$$

$$\Rightarrow \quad x = \tfrac{1}{2}\ln 2 \text{ m}$$

Example 4   A particle moves in a straight line which passes through a point O. The acceleration of the particle is proportional to the distance of the particle from O, and directed towards O. When $t = 0$, the particle is at rest and its displacement from O is a. Show that the particle has maximum velocity when at O, and show that the velocity, when the particle is a distance x from O, is given by

$$v^2 = k(a^2 - x^2) \quad \text{where } k \text{ is a constant.}$$

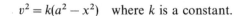

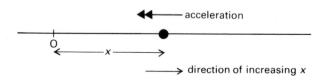

Acceleration $= -kx$
Maximum velocity occurs when acceleration is zero: i.e. $x = 0$
$\quad\quad\quad\quad$ ∴ maximum velocity occurs when particle is at O $\quad\quad\quad$ QED

$$v\frac{dv}{dx} = -kx$$

| Since we want a connec-tion between v and x. |

$$\Rightarrow \int_0^v v\,dv = \int_a^x -kx\,dx$$

$$\Rightarrow \left[\frac{v^2}{2}\right]_0^v = \left[-\frac{kx^2}{2}\right]_a^x$$

$$\Rightarrow \quad v^2 = -kx^2 + ka^2$$

$$\Rightarrow \quad v^2 = k(a^2 - x^2) \quad\quad\quad\quad\quad\quad\quad\quad\quad\quad \text{QED}$$

● **Exercise   9.1**   See page 272

9.2 Variable Acceleration II: More Complex Problems

Example 5   A particle moves in a straight line with acceleration $3(1 + v)\,\text{m/s}^2$ where v is the speed of the particle at time t. Initially the particle passes through a fixed point O with speed $4\,\text{m/s}$. Find its displacement from O when its speed is $9\,\text{m/s}$.

$$v\frac{dv}{dx} = 3(1+v)$$

$$\Rightarrow \qquad \int_4^9 \frac{v\,dv}{1+v} = \int_0^x 3\,dx$$

$$\frac{v}{1+v} \text{ is top heavy.}$$
It cannot be integrated in this form. By division,

$$\Rightarrow \qquad \int_4^9 \left(1 - \frac{1}{1+v}\right) dv = \int_0^x 3\,dx$$

$$\frac{v}{1+v} = 1 - \frac{1}{1+v}$$

$$\Rightarrow \qquad [v - \ln(1+v)]_4^9 = [3x]_0^x$$

$$\Rightarrow 9 - \ln 10 - (4 - \ln 5) = 3x$$

$$\Rightarrow x = \tfrac{1}{3}(5 - \ln 2) \quad \text{(or } 1.44\,\text{m)}$$

Example 6   A particle moves in a straight line with acceleration ae^{-bv}, where v is the velocity of the particle and a, b are constants. Initially its speed is w. How long does it take to double its initial speed.

$$\frac{dv}{dt} = ae^{-bv}$$

$$\Rightarrow \int_w^{2w} e^{bv}\,dv = \int_0^t a\,dt$$

$$\Rightarrow \left[\frac{1}{b}e^{bv}\right]_w^{2w} = [at]^t$$

$$\Rightarrow \qquad at = \frac{1}{b}(e^{2bw} - e^{bw})$$

$$\Rightarrow \qquad \text{time} = \frac{1}{ab}(e^{2bw} - e^{bw})$$

Example 7   A particle is moving in a straight line. Initially its speed is $\sqrt{6}\,\text{m/s}$ and it is subject to a retardation of $4\,\text{m/s}^2$. The retardation increases in proportion to the square of the speed of the particle and is equal to $12\,\text{m/s}^2$ when the speed is $2\,\text{m/s}$. Find an expression for the retardation when the speed of the particle is $v\,\text{m/s}$. Hence show that the time taken for the speed to fall to $\sqrt{2}\,\text{m/s}$ is $\pi\sqrt{2}/48$ s.

Given:

$$\text{initial retardation} = 4\,\text{m/s}^2$$
$$\text{retardation increases in proportion to } v^2$$
$$\text{when } v = 2, \text{ retardation} = 12 \tag{1}$$

Let

$$\text{retardation} = 4 + kv^2$$
$$(1) \Rightarrow \qquad 12 = 4 + k\cdot 4$$
$$\Rightarrow \qquad k = 2$$
$$\therefore \text{retardation} = 4 + 2v^2$$

Hence

$$\frac{dv}{dt} = -(4+2v^2)$$

$$\Rightarrow \int_{\sqrt{6}}^{\sqrt{2}} \frac{dv}{2+v^2} = \int_0^t -2\,dt$$

$$\Rightarrow \left[\frac{1}{\sqrt{2}}\tan^{-1}\frac{v}{\sqrt{2}}\right]_{\sqrt{6}}^{\sqrt{2}} = [-2t]_0^t$$

$$\Rightarrow t = \frac{1}{2\sqrt{2}}[\tan^{-1}\sqrt{3} - \tan^{-1}1] = \frac{1}{2\sqrt{2}}\left[\frac{\pi}{3} - \frac{\pi}{4}\right]$$

$$\Rightarrow t = \frac{\pi\sqrt{2}}{48}\,s \qquad\qquad \text{QED}$$

- **Exercise 9.2** See page 272

9.3 Motion Under A Variable Force

Example 8   A particle of mass m is projected vertically downwards with a speed v_0. Its subsequent motion is subject to a resistance which is mk times its speed. Show that the speed of the particle, at time t after projection, is given by

$$v = \frac{g}{k}[1 - e^{-kt}] + v_0 e^{-kt}$$

Hence find the distance fallen after time t and its terminal (or limiting) speed.

$F = ma$:

$$mg - mkv = m\frac{dv}{dt} \qquad (1)$$

$$\Rightarrow \int_0^t dt = \int_{v_0}^v \frac{dv}{g - kv}$$

$$\Rightarrow [t]_0^t = \left[-\frac{1}{k}\ln(g - kv)\right]_{v_0}^v$$

$$\Rightarrow t = -\frac{1}{k}[\ln(g - kv) - \ln(g - kv_0)]$$

$$\Rightarrow t = -\frac{1}{k}\left[\frac{g - kv}{g - kv_0}\right]$$

$$\Rightarrow \frac{g - kv}{g - kv_0} = e^{-kt}$$

$$\Rightarrow \quad g - kv = (g - kv_0)e^{-kt}$$

$$\Rightarrow \quad kv = g - (g - kv_0)e^{-kt}$$

$$\Rightarrow \quad v = \frac{g}{k}[1 - e^{-kt}] + v_0 e^{-kt} \qquad\qquad \text{QED}$$

To find the distance fallen from this equation, let $v = \dfrac{dx}{dt}$.

Hence
$$\frac{dx}{dt} = \frac{g}{k}[1 - e^{-kt}] + v_0 e^{-kt}$$

Integrating:
$$x = \frac{g}{k}\left[t + \frac{e^{-kt}}{k}\right] - v_0 \frac{e^{-kt}}{k} + A$$

When $t = 0$, $x = 0$ $\therefore$ $A = 0$

$$\Rightarrow \text{distance fallen} = x = \frac{g}{k}\left[t + \frac{e^{-kt}}{k}\right] - \frac{v_0 e^{-kt}}{k}.$$

Its terminal (or limiting) speed occurs when acceleration is zero
and (1) $\Rightarrow g = kv$.

$$\therefore \text{ terminal speed} = \frac{g}{k}$$

Example 9   The resistance to a speed boat, of mass m, travelling across a
lake, is proportional to the square of the speed. When travelling at a constant
speed u, the engine exerts constant power P watts. If the boat is travelling under
constant power P, find the distance travelled in accelerating from speed $u/4$ to
$u/2$. When the boat reaches the speed $u/2$, its engine is switched off. Find the
time taken for the speed to drop to $u/4$ again.

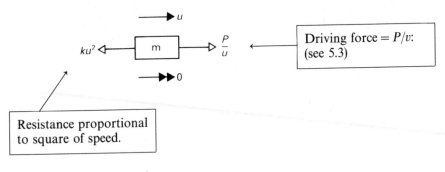

Resistance proportional
to square of speed.

$F = ma$ at constant speed u:

$$\frac{P}{u} = ku^2 \Rightarrow P = ku^3 \qquad\qquad (1)$$

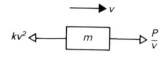

$F = ma$:

$$\frac{P}{v} - kv^2 = mv\frac{dv}{dx}$$

$$(1) \Rightarrow \quad \frac{ku^3}{v} - kv^2 = mv\frac{dv}{dx}$$

$$\Rightarrow \frac{k}{v}[u^3 - v^3] = mv\frac{dv}{dx}$$

$$\Rightarrow \quad \int_0^x k\,dx = \int_{u/4}^{u/2} \frac{mv^2\,dx}{u^3 - v^3} = -\frac{m}{3}\int_{u/4}^{u/2} \frac{-3v^2\,dv}{u^3 - v^3}$$

$$\Rightarrow \quad [kx]_0^x = \left[-\frac{m}{3}\ln(u^3 - v^3) \right]_{u/4}^{u/2}$$

$$\Rightarrow \quad kx = -\frac{m}{3}\ln\left(\frac{u^3 - u^3/8}{u^3 - u^3/64} \right) = -\frac{m}{3}\ln\frac{8}{9}$$

$$\Rightarrow \quad x = \frac{m}{3k}\ln\frac{9}{8} \leftarrow \boxed{-\ln\tfrac{8}{9} = +\ln\tfrac{9}{8}}$$

> **But,** k is not in the original question. So we need to use (1) again.

$$k = P/u^3$$

$$\therefore\ x = \frac{mu^3}{3P}\ln\frac{9}{8}$$

Engine switched off:

$$kv^2 \leftarrow \boxed{m} \qquad \longrightarrow v$$

$F = ma$:

$$-kv^2 = m\frac{dv}{dt}$$

$$\Rightarrow \quad \int_0^t -k\,dt = \int_{u/2}^{u/4} \frac{m\,dv}{v^2}$$

$$\Rightarrow \qquad [-kt]_0^t = \left[-\frac{m}{v} \right]_{u/2}^{u/4}$$

$$\Rightarrow \qquad kt = m\left[\frac{4}{u} - \frac{2}{u} \right] = \frac{2m}{u}$$

$$\Rightarrow \qquad t = \frac{2m}{uk}$$

and (1) $\Rightarrow$ $\qquad \underline{\text{time} = \frac{2m}{u} \times \frac{1}{P/u^3} = \frac{2mu^2}{P}}$

- **Exercise 9.3**   See page 274

9.4 Simple Harmonic Motion (SHM)

Definition of Simple Harmonic Motion

If a particle moves so that its acceleration is directed towards a fixed point in its path, and varies directly as the distance from this fixed point, then the motion is said to be **simple harmonic**.

$$\ddot{x} = -\omega^2 x$$

Terminology and Notation

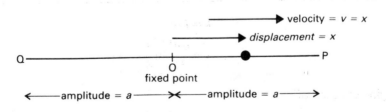

velocity $= v = x$

displacement $= x$

O
fixed point

amplitude $= a$ amplitude $= a$

The particle moves from O to P and back through O to Q, then back to O again. This kind of motion is called a **periodic oscillation**. The time interval after which the motion is repeated is called the **period**.

The fixed point, O, is the **centre of the oscillation** or **mean position**.

The greatest distance that the particle reaches from O is called the **amplitude**.

The velocity is positive when the particle is moving away from O and negative when moving towards O.

Equations of Simple Harmonic Motion (may be quoted)

1    $\ddot{x} = -\omega^2 x$

2   Displacement and velocity:
   (a) $x = a\sin(\omega t + \varepsilon)$ and $v = \dot{x} = a\omega\cos(\omega t + \varepsilon)$ (general equation: ε = phase angle)

   (b) $x = a\sin\omega t$ and $v = \dot{x} = a\omega\cos\omega t$ (motion starts from centre of oscillation)

   (c) $x = a\cos\omega t$ and $v = \dot{x} = -a\omega\sin\omega t$ (motion starts from edge of oscillation)

3    $v_{max} = a\omega$

4    $v^2 = \omega^2(a^2 - x^2)$

5   Period $= T = \dfrac{2\pi}{\omega}$

Note: 1   Differentiating any of the above equations for x, twice, shows that they satisfy the equation $\ddot{x} = -\omega^2 x$.

2    $v^2 = \omega^2(a^2 - x^2)$ is obtained by eliminating t from the equations for x and v.

3   Values of x start to repeat when $\omega t = 2\pi \Rightarrow$ period $= 2\pi/\omega$.

Example 10   Find the period of SHM defined by $\ddot{x} = -64x$.

Compare with
$$\ddot{x} = -\omega^2 x. \quad \Rightarrow \quad \omega = 8.$$

$$\text{Period} = T = \frac{2\pi}{\omega} = \frac{2\pi}{8}$$

$$\therefore \text{ Period} = \pi/4 \text{ s.}$$

Example 11   A particle moves with SHM about a centre of oscillation O. The motion has period 8 seconds and amplitude 2 m. Find the maximum velocity and the velocity 1 m from O.

$$\text{Period} = 8\,\text{s} \Rightarrow \frac{2\pi}{\omega} = 8 \Rightarrow \omega = \frac{\pi}{4}$$

$$v_{max} = a\omega \Rightarrow v_{max} = 2 \times \frac{\pi}{4}$$

$$\Rightarrow \text{ max. velocity} = \pi/2 \text{ m/s.}$$

$$v^2 = \omega^2(a^2 - x^2) \Rightarrow v^2 = \left(\frac{\pi}{4}\right)^2 (2^2 - 1^2)$$

$$\Rightarrow v^2 = \left(\frac{\pi}{4}\right)^2 3$$

$$\Rightarrow \text{ velocity} = \frac{\pi\sqrt{3}}{4} \text{ m/s.}$$

Example 12 A particle performs simple harmonic oscillations about its mean position A. Initially it is projected from A with speed π m/s. The distance between its extreme positions is 2.4 m. Find:

(a) the period of the motion
(b) how far the particle is from A after $\frac{1}{2}$ second
(c) how long it takes to first reach a point 60 cm from A.

(a) Distance between extreme positions is 2.4 m
$$\Rightarrow \text{amplitude} = a = 1.2 \text{ m}.$$
Velocity at $A = \pi \text{ m/s} \Rightarrow \text{max. vel.} = \pi \text{ m/s}$

$$\Rightarrow a\omega = \pi \text{ m/s}$$

$$\Rightarrow \omega = \frac{\pi}{1.2}.$$

Period $= \dfrac{2\pi}{\omega} = \dfrac{2\pi}{\pi/1.2} = \underline{2.4 \text{ s}}$

(b) Since it starts from the centre of oscillation, x is given by

$$x = a \sin \omega t \qquad \longleftarrow \boxed{\text{Note that this angle is in \textbf{radians}.}}$$

$$\Rightarrow x = 1.2 \sin \frac{\pi}{1.2} t$$

After $\frac{1}{2}$ second, $x = 1.2 \sin\left(\dfrac{\pi}{1.2} \times \dfrac{1}{2}\right) = 1.16 \text{ m}.$

$\therefore$ particle is $\underline{116 \text{ cm from A}}$

(c) When $x = 0.6 \text{ m}$

$$0.6 = 1.2 \sin \frac{\pi}{1.2} t$$

$$\Rightarrow \sin \frac{\pi t}{1.2} = 0.5 \Rightarrow \frac{\pi t}{1.2} = \frac{\pi}{6} \Rightarrow \underline{t = 0.2 \text{ s}}$$

● **Exercise 9.4** See page 275

9.5 More Difficult Problems on SHM

Example 13 A particle performs simple harmonic oscillations of period 2π seconds and amplitude 1.5 m. P and Q are two points on the same side of the centre of oscillation A, such that $PQ = QA = 50 \text{ cm}$. Find the time taken to travel from P to Q.

$$\left.\begin{array}{l}\text{Period} = 2\pi\,\text{s} \quad \Rightarrow \omega = 1 \\ \qquad\qquad\quad \text{Also } a = 1.5\end{array}\right\}\qquad(1)$$

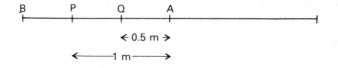

Method 1

Let motion start at A

$$\Rightarrow x = a \sin \omega t$$
$$\text{and (1)} \quad \Rightarrow x = 1.5 \sin t$$

At Q, $x = 0.5 \Rightarrow \quad 0.5 = 1.5 \sin t_{AQ}$

$$\Rightarrow \sin t_{AQ} = \tfrac{1}{3}$$

$$\Rightarrow \quad t_{AQ} = 0.34\,\text{s}$$

At P, $x = 1 \Rightarrow \quad\quad 1 = 1.5 \sin t_{AP}$

$$\Rightarrow \sin t_{AP} = \tfrac{2}{3}$$

$$\Rightarrow \quad t_{AP} = 0.73\,\text{s}.$$

> Calculator must be in radian mode.

$$t_{PQ} = t_{QP} = t_{AP} - t_{AQ} = 0.73 - 0.34 = 0.39$$

$$\therefore \text{ time to travel from P to Q} = 0.39\,\text{s}$$

Method 2

Let motion start at edge of oscillation, B.

$$\Rightarrow x = a \cos \omega t$$
$$\text{and (1)} \quad \Rightarrow x = 1.5 \cos t.$$

At P, $x = 1 \Rightarrow \quad\quad 1 = 1.5 \cos t_{BP}$

$$\Rightarrow \cos t_{BP} = \tfrac{2}{3}$$

$$\Rightarrow \quad t_{BP} = 0.84\,\text{s}$$

At Q, $x = 0.5 \Rightarrow \quad 0.5 = 1.5 \cos t_{BQ}$

$$\Rightarrow \cos t_{BQ} = \tfrac{1}{3}$$

$$\Rightarrow \quad t_{BQ} = 1.23\,\text{s}$$

> Note that x is always measured from the centre.

$$t_{PQ} = t_{BQ} - t_{BP} = 1.23 - 0.84 = 0.39$$

$$\therefore \text{ time to travel from P to Q} = 0.39\,\text{s}$$

Note that, if you are finding a time interval within the motion, it does not matter where you consider the motion to start.

● **Exercise 9.5**   See page 277

9.6+ Properties of Forces both Constant & Variable: Revision of Unit: A Level Questions.

Constant forces	Variable forces
$F = ma$	$F = ma$
Impulse $= Ft$	Impulse $= \displaystyle\int_0^t F\,dt$
Impulse $= mv - mu$	Impulse $= mv - mu$
Work done $= Fx$	Work done $= \displaystyle\int_0^x F\,dx$
Work done $=$ increase in KE	Work done $=$ increase in KE

Example 14 A particle of mass 2 kg moves in a straight line under the action of a force F, where $F = 3t^2$ at time t. Given that the particle has a velocity 1.5 m/s when $t = 0$, find its velocity after 2 seconds.

$$\int_0^2 F\,dt = mv - mu$$

$$\Rightarrow \int_0^2 3t^2\,dt = 2v - 2 \times 1.5$$

$$\Rightarrow \quad [t^3]_0^2 = 2v - 3$$

$$\Rightarrow \quad 2v = 8 + 3 = 11$$

$$\Rightarrow \text{ velocity} = 5.5 \text{ m/s}$$

Example 15 A particle of mass m, falling through a resisting medium, undergoes a resistance of ms, where s is the distance travelled through the medium. If the particle is at rest when it first enters the medium, find its velocity when $s = 1$. (Take $g = 10$ m/s^2).

Work done $=$ increase in KE

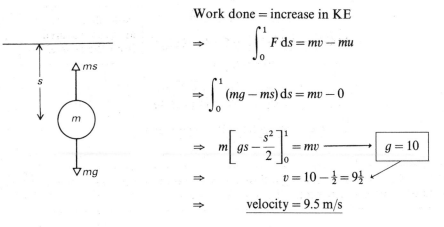

$$\Rightarrow \quad \int_0^1 F\,ds = mv - mu$$

$$\Rightarrow \int_0^1 (mg - ms)\,ds = mv - 0$$

$$\Rightarrow \quad m\left[gs - \frac{s^2}{2} \right]_0^1 = mv \longrightarrow \boxed{g = 10}$$

$$\Rightarrow \quad v = 10 - \tfrac{1}{2} = 9\tfrac{1}{2}$$

$$\Rightarrow \quad \text{velocity} = 9.5 \text{ m/s}$$

● Miscellaneous Exercise 9B See page 278

Unit 9 Exercises

Exercise 9.1

1   The distance of a particle from a point O at time t is given by
$$x = 3 \sin 2t + 4 \cos 2t.$$
Find expressions for the velocity and acceleration in terms of t. Show that the particles's acceleration, $\dfrac{d^2x}{dt^2}$, and its displacement from O, x satisfies $\dfrac{d^2x}{dt^2} = -k^2 x$ and find the value of k.

2   The acceleration of a particle at time t seconds is given by $\frac{6}{5}t + 2$. When $t = 0$, the particle is passing through a fixed point A with a velocity of 3. Find, in terms of t, the velocity of the particle and its displacement from A.

3   Initially a particle is 1 m from a fixed point O and travelling with a speed of 4 m/s. Throughout its motion, its acceleration, at time t seconds, is given by $(3t^2 - 1)$ m/s^2. Find its velocity at time t and after 2 seconds.

4   A particle is moving with an acceleration of $(6t - 10)$ m/s^2, where t is the time in seconds. When $t = 0$, its velocity is 3 m/s and it passes through a fixed point O. Find the times when its velocity is zero and its displacement from O after 5 seconds.

5   A particle moving in a straight line has acceleration $5/v$ m/s^2 where v = velocity at time t. Initially the particle is at a fixed point O, travelling with speed 2 m/s. Find the velocity of the particle at time t and find its velocity when its displacement from O is 3 m.

6   A particle moves in a straight line with acceleration $5x^2$ m/s^2, where x metres is the displacement of the particle from a fixed point O in the line, at time t seconds. When its displacement is -3 m its velocity is zero. Find the velocity of the particle when it passes through O.

7   A particle moves in a straight line with acceleration $3v$ m/s^2 where v is the velocity at time t. Initially the particle is at a fixed point O with velocity 2 m/s. Show that its velocity after 5 seconds is $2e^{15}$ m/s and find its displacement from O when its velocity is 8 m/s.

*8   A particle moves in a straight line with an acceleration $2v^3$ where v is its velocity at time t. Initially the particle is at a fixed point A, with a velocity of u m/s. Find the velocity of the particle 5 seconds later, in terms of u, find its displacement from O when the speed has doubled.

Exercise 9.2

1   A particle moves in a straight line with an acceleration $25 + v^2$ where v is the velocity at time t. If the particle is initially at rest, find its velocity when $t = \pi/4$.

2 A particle moves in a straight line with an acceleration $v + 2$ where v is its velocity at time t. When $t = 2$, $v = 2$. Show that its velocity after 3 s is $4e - 2$ and that the time to reach a speed of 4 m/s is $2 + \ln 3/2$ s.

3 A particle moves in a straight line with an acceleration $1 - 3/v$ where v is its velocity at time t. Initially it passes through a fixed point O with velocity 4 m/s. Show that when its velocity is 5 m/s its displacement from O is $9 \ln 2 + 15/2$.

4 A particle moves in a straight line with a retardation $(g/k) - v$ where v is its velocity at time t, and k, g are constants. When $t = 3$, $v = u$. Show that, when $t = 5$,

$$v = \frac{g}{k}(1 - e^2) + ue^2$$

5 A particle P starts from rest the point O and moves along Ox. Initially P has an acceleration of magnitude 6 m/s². After t seconds, when P is at a distance x m from O and has speed v m/s, the acceleration of P has magnitude $\lambda(2 + v)$ m/s², where λ is a positive constant. Find the value of λ.
When $t = T$, the speed of the particle is 10 m/s and $x = d$. Show that:

(a) $T = \frac{1}{3}\ln 6$ (b) $d = \frac{1}{3}(10 - 2\ln 6)$. [L]

6 A particle starts with speed 20 m/s and moves in a straight line. The particle is subjected to a resistance which produces a retardation which is initially 5 m/s² and which increases uniformly with the distance moved, having a value of 11 m/s² when the particle has moved a distance of 12 m. Given that the particle has speed v m/s when it has moved a distance of x m, show that, while the particle is in motion,

$$v\frac{dv}{dx} = -(5 + \tfrac{1}{2}x)$$

Hence, or otherwise, calculate the distance moved by the particle in coming to rest. [L]

7 A particle moving on a straight line with speed v experiences a retardation of magnitude $be^{v/u}$, where b and u are constants. Given that the particle is travelling with speed u at time $t = 0$, show that the time t_1 for the speed to decrease to $\frac{1}{2}u$ is given by

$$bt_1 = u(e^{-\frac{1}{2}}, -e^{-1})$$

Find the further time t_2 for the particle to come to rest. Deduce that $t_2/t_1 = e^{\frac{1}{2}}$. Find, in terms of b and u, an expression for the distance travelled in decelerating from speed u to rest. [L]

8 A particle moves on the positive x-axis. The particle is moving towards the origin O when it passes through the point A, where $x = 2a$, with speed $\sqrt{(k/a)}$, where k is constant. Given that the particle experiences an acceleration $k/(2x^2) + k(4a^2)$ in a direction away from O, show that it comes instantaneously to rest at a point B, where $x = a$.

Immediately the particle reaches B the acceleration changes to $k/(2x^2) - k/(4a^2)$ in a direction away from O. Show that the particle next comes instantaneously to rest at A. [L]

Exercise 9.3

1   A car of mass 900 kg moves along a straight level road. It may be assumed that the total resistance to its motion has the constant value of 1500 N and that the car engine is working at a constant rate of 60 kW. Denoting the velocity of the car at time t seconds by v ms^{-1}, show that

$$\frac{3}{5} v \frac{dv}{dt} = 40 - v$$

By using the substitution $40 - v = y$, or otherwise, find the time, to the nearest tenth of a second, for the speed of the car to increase from 10 m/s to 20 m/s. [J]

2   A particle of mass m is moving along a straight groove in a horizontal plane. It passes through a point O of the groove at speed u, and when it has travelled a distance x from O its speed is v and the resistance to its motion is kx^2v^3, where k is a constant. Write down the equation of motion, and show that

$$\frac{1}{v} = \frac{k}{3m} x^3 + \frac{1}{u}$$

Determine the time taken for the particle to travel a distance b from O. [J]

3   The engine of a car of mass m works at a constant rate S. The car moves on a level road and is subject to a constant resistance R. Show that the time taken for its speed to increase from U to V, where $RU < RV < S$, is

$$\frac{mS}{R^2} \log_e \frac{S - RU}{S - RV} - \frac{m}{R}(V - U)$$ [J]

4   A car of mass m moves along a straight level road with its engine switched off. The resistance to its motion at speed v is proportional to $v^2 + U^2$, where U is a constant. It comes to rest from speed U in time T. Show that the resistance at zero speed is $\frac{1}{4}m\pi U/T$.
The car now starts from rest and moves, under the same law of resistance, with its engine exerting a constant force of magnitude $m\pi U/T$. Show that it reaches speed U after travelling a distance $(2UT/\pi)\log_e(\frac{3}{2})$. State the limiting speed under this force. [J]

5   The resistance to the motion of a lorry of mass m is kv at speed v, where k is a constant.
(i)   Find the acceleration of the lorry when it is climbing a hill of inclination α to the horizontal at speed v under full power S. Given that it can climb the same hill under full power at a steady speed u, show that

$$ku^2 = S - mgu \sin \alpha$$

(ii)   The lorry can travel at steady speed w on a level road under full power and the same law of resistance. Show that the time taken for it to accelerate under full power from speed v_1 to speed v_2 (where $v_1 < v_2 < w$) on the level road is

$$\frac{mw^2}{2S}\log_e\frac{w^2 - v_1^2}{w^2 - v_2^2} \qquad \text{[J]}$$

6 A particle of mass m is released from rest and moves under gravity in a medium which exerts a resistance k times the speed, where k is constant. Show that after time t its downward velocity v is given by

$$kv = mg(1 - e^{-\frac{kt}{m}})$$

Find the distance x that the particle has travelled in terms of m, g, k and t. Show that $kx = mgt - mv$.

Obtain, in terms of m, g, k and t, an expression (which need not be simplified) for the energy absorbed by the resistance during the time t. [J]

Exercise 9.4

1 Find the period of SHM defined by

    (a) $\ddot{x} = -36x$ (b) $\ddot{x} = 4x$ (c) $\ddot{x} = -4\pi^2 x$

2 If a particle moving with SHM has acceleration $4.8\,\text{m/s}^2$ when it is $2.4\,\text{m}$ from the centre of oscillation, find the period of the motion.

3 A particle performs SHM with period $\frac{\pi}{4}$ s. Find the magnitude of the acceleration when it is $25\,\text{cm}$ from the mean position.

4 If the acceleration of a particle moving under SHM has magnitude $4\,\text{m/s}^2$ when at a distance of $20\,\text{cm}$ from the mean position, find the magnitude of the acceleration when it is $25\,\text{cm}$ from the mean position.

5 At what position(s) in a simple harmonic oscillation does the body have:
    (a) maximum velocity   (b) maximum acceleration?

6 If the amplitude of oscillations of a body moving with SHM is $80\,\text{cm}$, and the magnitude of the acceleration is $2.4\,\text{m/s}^2$ when it is $60\,\text{cm}$ from the mean position, find:

    (a) the maximum velocity
    (b) the maximum acceleration
    (c) the velocity when it is $50\,\text{cm}$ from the mean position.

7 A particle moves with SHM about a mean position O. The amplitude is $3\,\text{m}$ and the period is 4π seconds. Find:

    (a) the maximum velocity
    (b) the velocity when it is $2\,\text{m}$ from O
    (c) the distance from O when the velocity is $1\,\text{m/s}$.

8 A particle moves with SHM about a centre of oscillation O. When it is $50\,\text{cm}$ from O its velocity is $2.4\,\text{m/s}$, and when it is $1.2\,\text{m}$ from O its velocity is $1\,\text{m/s}$.

Find:

(a) the maximum displacement of the particle from O
(b) the period of the motion
(c) the magnitude of the maximum acceleration.

9 A particle of mass 2 kg moves with SHM about a mean position A. Its velocity is zero when it is 60 cm from A and 3 m/s at A. Find:

(a) the maximum velocity
(b) the amplitude
(c) the period of the motion
(d) the maximum kinetic energy.

10 A particle moves with SHM about a mean position O. If its maximum velocity is 10 m/s and the time it takes to move between its two extreme positions is $\frac{\pi}{2}$ s, find:

(a) the amplitude of the motion
(b) the velocity when it is 3 m from O.

11 The distance, x, of a particle moving under SHM, from the centre of oscillations is given by

$$x = a \cos \omega t$$

where t is the time in seconds.
Where is the particle at the point when timing commences? Show that the velocity at time t is given by $v = -a\omega \sin \omega t$.
Hence show that v and x both satisfy

$$v^2 = \omega^2(a^2 - x^2)$$

12 A particle is projected from a point O with a speed $\frac{\pi}{3}$ m/s and subsequently moves with simple harmonic oscillations about O. It just reaches a point P which is 2 m from O. Find:

(a) the period of the motion
(b) how far the particle is from O after 1 second
(c) how far the particle is from O after 3 seconds.

13 A particle performing simple harmonic oscillations about O is initially released from rest at a point 2 m from O. The period of the oscillations is 8 s. Find the time when it is 1 m from O for the:
(a) first time   (b) second time   (c) third time   (d) fourth time.

14 A particle is initially projected from a point Q and performs SHM with Q as the centre of oscillations. It performs 20 oscillations per minute whose amplitude is 40 cm. Find:

(a) the initial speed
(b) the speed after one second
(c) the distance that the particle is from Q after 1.5 seconds.

15 If the amplitude of a certain simple harmonic motion is $2\sqrt{3}$ m find the distance of the body from the mean position when its velocity is half the maximum velocity.

Exercise 9.5

1 A particle moves with SHM of amplitude 1.2 m. How far will it be from the centre of oscillation when its velocity is $\frac{3}{5}$ of its maximum velocity?

2 A particle is projected from a point P with speed 4 m/s and it performs SHM about P as the mean position. A point Q is 50 cm from P. If the amplitude of the oscillations is 1.6 m, how long does the particle take between the first and second times that it passes through Q?

3 A particle is released from rest 1.2 m from a point O and performs simple harmonic oscillations about O with period 6 seconds. Find its velocity when it is 1 m from O. At this point it is given a blow which doubles its velocity. Find the new amplitude of the motion if the period remains the same.

4 A body moves with SHM about O. The displacement of the body, x, at time t is given by

$$x = a \sin(\omega t + \varepsilon)$$

If the amplitude of the motion is 2 m, its period is 10 s and, initially, it is 50 cm from O, find the smallest positive value of ε to 3 decimal places. Hence find the time when it is first instantaneously at rest.

5 Show that if a particle moves so that its displacement from a point O at time t is given by

$$x = 2 \cos \omega t + 3 \sin \omega t$$

then it is moving with SHM.

6 A particle is moving with SHM and performing 40 oscillations per minute. At a point 5 cm from the centre of the oscillations, O, the velocity is 50 cm/s. Show that the furthest displacement from O is approximately 13 cm.
If P and Q are two points 2 cm and 12 cm respectively away from O, find the time taken in going from P to Q.

*7 A particle moves along the x-axis under SHM about a centre of oscillation at $x = 2$. The time to complete one oscillation is π seconds and the particle has a speed of 3 m/s when $x = 4$. Find the maximum displacement from $x = 2$.
Initially the particle is moving through the centre of oscillation in the positive x-direction. Find an equation that gives the position of the particle at time t.

8 A particle of mass 5 kilograms executes simple harmonic motion with amplitude 2 metres and period 12 seconds. Find the maximum kinetic energy of the particle, leaving your answer in terms of π.

Initially the particle is moving with its maximum kinetic energy. Find the time that elapses until the kinetic energy is reduced to one quarter of the maximum value, and show that the distance moved in this time is $\sqrt{3}$ metres.   [J]

9 A particle describes simple harmonic motion with period $\frac{1}{2}\pi$ s about a point O as centre and the maximum speed in the motion is $20\,\text{m s}^{-1}$. Find:

(a) the amplitude of the motion
(b) the speed of the particle when at a point Q distance $2\,\text{m}$ from O
(c) the time taken to travel directly from O to Q.

Given that the particle is of mass 0.5 kg find the rate at which the force acting on the particle is working at a time $(\pi/12)$ s after it has passed through O. Find also the maximum rate of working of this force.   [A]

Miscellaneous Exercise 9B

1 The tension in a rubber band varies according to the extension, the amount it is stretched beyond its natural length. If the tension in a particular rubber band is given by $T = 12\,x$, where x metres is the extension, show that, in stretching it from its natural length to an extension of 20 cm, the work done against the tension is 0.24J.

2 A particle P moves in a straight line and experiences a retardation of $0.01\,v^3\,\text{m/s}^2$, where $v\,\text{m/s}$ is the speed of P. Given that P passes through a point O with speed 10 m/s show that, when it is a distance 10 m from O, its speed is 5 m/s.
Find the time taken for the speed of P to be reduced from 10 m/s to 5 m/s.
[L:9.1,9.2]

3 A particle P, of mass 0.01 kg, moves in a horizontal straight line under the action of a resistive force whose magnitude is proportional to the speed of P. The force is of magnitude 2N when P is moving with speed 8 m/s. Find the distance travelled by P during the interval in which its speed reduces from 8 m/s to 3 m/s.   [A:9.3]

4 A particle moves along a straight line with acceleration k/v, where k is a constant and v is the velocity of the particle. When the particle is at a point O its velocity is 1 m/s, and when its displacement from O is 13 metres its velocity is 3 m/s. Show that $k = \frac{2}{3}$. Find the velocity of the particle when it is at the point A whose displacement from O is 62 metres.
Find also the time taken for the particle to travel from O to A.
[J:9.1, 9.2]

5 A particle P of mass 0.04 kg describes simple harmonic motion about a point O as centre. When P is at a distance of 3 m from O its acceleration is of magnitude $48\,\text{m/s}^2$ and its kinetic energy is 5.12 J. Find the amplitude of the motion.   [A:9.4]

6 A particle moves in a straight line with variable velocity. When its displacement from a fixed point O on the line is x its velocity is v. Show that the acceleration is then

$$v\frac{dv}{dx}$$

The particle is projected from O with velocity u (where $u > 0$) and moves under the action of a force directed towards O and proportional to $v^{3/2}$. When $x = a$ the velocity of the particle is $u/4$. Find its velocity when $x = 4a/3$. [J: 9.3]

7 A particle moves in a straight line with variable acceleration $\dfrac{k}{1+v}$ m/s^2, where k is a constant and v m/s is the speed of the particle when it has travelled a distance x m. Find the distance moved by the particle as its speed increases from 0 to u m/s. [L: 9.1, 9.2]

8 A particle of mass 4 kilograms executes simple harmonic motion with amplitude 2 metres and period 10 seconds. The particle starts from rest at time $t = 0$. Find its maximum speed and the time at which half the maximum speed is first attained. Find also the maximum value of the magnitude of the force required to maintain the motion. (You may leave your answers in terms of π.) [J: 9.5]

9 A man pushes a loaded cart of total mass m along a straight horizontal path against a constant resisting force λm. The cart starts from rest, and the man exerts a force, in the direction of the motion, of magnitude $Ke^{-\alpha x}$, where x is the distance moved, α is a positive constant and K is a constant greater than λm. Write down a differential equation, involving x and the speed v, for the motion of the cart (while it continues to move), and hence obtain v in terms of x and the given constants. [J: 9.3]

10 A small smooth sphere S, of mass 0.2 kg, moves in a straight line with variable acceleration $(2 + v)$ m/s^2, where v m/s is the speed of the sphere. Find the distance, to the nearest 0.1 m, moved by S as its speed increases from 1 m/s to 5 m/s.
When S is moving with speed 5 m/s it overtakes and collides with a second small smooth sphere T, of mass 0.3 kg, which is moving with speed 2 m/s in the same direction as S. The coefficient of restitution between the spheres is 2/3. Calculate:
(a) the speed of each sphere immediately after the impact
(b) the magnitude of the impulse received by each sphere on impact. [L: 5.4, 9.1, 9.2]

11 A particle of mass m is projected vertically upwards, with speed V, in a medium which exerts a resisting force of magnitude $mg\,v^2/c^2$, where v is the speed of the particle and c is a positive constant. Show that the greatest height attained above the point of projection is

$$\frac{c^2}{2g}\ln\left(1 + \frac{V^2}{c^2}\right)$$

Find an expression, in terms of V, c and g, for the time taken to reach this height. [L: 9.3]

12 A particle moves in a straight line. At time t its displacement from a fixed point O on the line is x, its velocity is v, and its acceleration is $k \sin \omega t$, where k and ω are constant. The displacement, velocity and acceleration are all measured in the same direction. Initially the particle is at O and is moving with velocity u. Show that

$$v = \frac{k}{\omega}(1 - \cos \omega t) + u$$

and hence find x in terms of t, k, ω and u.

(i) Show that, if $u = 0$, the particle first comes to instantaneous rest after travelling a distance $2\pi k/\omega^2$.
(ii) If $u \neq 0$ find the relation which holds between u, k and ω if the motion is simple harmonic.

[J: 9.1, 9.2, 9.4]

13 A man of mass m stands on a horizontal platform which begins to move at time $t = 0$ and thereafter performs vertical oscillations such that its depth x below a fixed level is given by

$$x = a \cos \omega t,$$

where a and ω are positive constants. Find the force at time t which he exerts on the platform while he remains on it. Deduce that he never loses contact with the platform if

$$a\omega^2 < g$$

In the case when $a\omega^2 = 2g$ find, in terms of ω, the time at which the man first loses contact with the platform.   [J: 9.3]

14 A particle of mass m moves in a straight line on a rough horizontal table, the coefficient of friction being μ. In addition to the frictional force, the motion of the particle is subject to an air resistance mkv, where v is the velocity of the particle and k is a constant. At time $t = 0$ the particle is projected along the table with velocity V. Show that, at any time t before the particle comes to rest,

$$v = Ve^{-kt} - \frac{\mu g}{k}(1 - e^{-kt})$$

Hence find the distance travelled at time t.   [J: 9.3]

15 A particle which falls vertically through a resistive medium has speed V m/s at time t seconds. The table gives measured values of V for various values of t.

V	0	4.297	4.901	4.986	4.998
t	0	1	2	3	4

Use Simpson's rule to estimate the distance h m fallen in the first four seconds of motion, giving your answer to 2 decimal places.

The resistance to motion of the medium is of magnitude $kV\,\mathrm{m/s^2}$ per unit mass, where k is a positive constant. Given that the particle is at rest initially and has a termimal speed of $5\,\mathrm{m/s}$, (i.e., $V \to 5$ as $t \to \infty$) show that its speed at any time t is given by

$$V = 5(1 - e^{-gt/5}),$$

where $g\,\mathrm{m/s^2}$ is the acceleration due to gravity.

Integrate the expression for V to show that the Simpson's rule estimate for h is in error by an amount $0.14\,\mathrm{m}$, when g is taken as 9.8.

[A: 9.3]

16 A vehicle travelling on a straight track joining two points A and B can accelerate at a constant rate $0.2\,\mathrm{m/s^2}$ and can decelerate at a constant rate of $1\,\mathrm{m/s^2}$. It covers a distance of $1.8\,\mathrm{km}$ from rest at A to rest at B by accelerating uniformly to a fixed speed and then travelling at that speed until it starts to decelerate. Given that the journey from rest to rest takes $2\frac{1}{2}$ minutes find the steady speed, in $\mathrm{m/s}$, at which the vehicle travels. Explain clearly the reason for your choice of the value for the speed.

The braking system of the vehicle is then changed so that, until the speed drops to $5\,\mathrm{m/s}$, the deceleration at any instant is proportional to the cube of the speed at that instant. When braking, the speed drops from $15\,\mathrm{m/s}$ to $5\,\mathrm{m/s}$ in a distance of $500\,\mathrm{m}$. Find, in $\mathrm{m/s^2}$, the retardation of the vehicle when the speed is $10\,\mathrm{m/s}$.   [A: 1.6, 9.1, 9.2]

*17 A vehicle P moves along a straight track which passes through two points A and B which are at a distance of $80\,\mathrm{m}$ apart. A vehicle Q moves along a straight parallel track whose perpendicular distance from the first track is to be neglected. The vehicle P has an acceleration of $3\,\mathrm{m/s^2}$ in the sense from A to B whilst Q has an acceleration of $1\,\mathrm{m/s^2}$ in the sense from B to A. At time $t = 0\,\mathrm{s}$ P passes through A moving towards B with a speed of $12\,\mathrm{m/s}$. Find the distance of P from A six seconds later and its speed at this time.

Also at time $t = 0\,\mathrm{s}$ Q passes B moving towards A with a speed of $8\,\mathrm{m/s}$. Find an expression for the distance PQ at time t and determine when P and Q are at a distance of $32\,\mathrm{m}$ apart.

When P is moving at a speed of $30\,\mathrm{m/s}$ the acceleration ceases, the brakes are applied and continue to be applied until the speed has dropped to $10\,\mathrm{m/s}$. During this period the retardation produced when the speed is $v\,\mathrm{m/s}$ is $v^2/150\,\mathrm{m/s^2}$. Find the time taken for the speed to drop to $10\,\mathrm{m/s}$ from $30\,\mathrm{m/s}$.   [A: 1.5, 9.1, 9.2]

18 A load P of mass m is being raised vertically by an engine working at a constant rate kmg. Denoting by v the speed of the load when it has been raised a distance x, show that

$$v^2 \frac{\mathrm{d}v}{\mathrm{d}x} = (k - v)g$$

Initially P is at rest. Show that

$$gx = k^2 \log_e \frac{k}{k-v} - kv - \tfrac{1}{2}v^2$$

Write down the work done by the engine in time t and hence, or otherwise, show that the time taken for the load to reach the speed v is

$$\frac{k}{g}\log_e\frac{k}{k-v} - \frac{v}{g}$$

[J: 5.2, 9.3]

19 A particle P is projected vertically upwards at speed b from a point A in a medium which offers a resistance proportional to the square of the speed. Show that, while the particle is ascending, its speed v and the distance AP are connected by a relation of the form

$$\frac{dv}{dx} = -\frac{g}{v}\left(1 + \frac{v^2}{c^2}\right)$$

where x denotes AP and c is constant.

Derive an expression for x in terms of g, c, b and v, and show that the particle is instantaneously at rest at a point H, where

$$AH = \frac{c^2}{2g}\log_e\left(1 + \frac{b^2}{c^2}\right)$$

Write down a differential equation connecting the speed w and the distance $HP = y$ for the motion of the particle when it is descending from H, and find y in terms of g, c and w.

Given that the particle returns to A with speed $\frac{1}{2}b$, show that $3c^2 = b^2$.

[J: 9.3]

20 A particle which falls vertically through a resistive medium has speed v m/s at time t s. The table below gives measured values of v for various values of t.

t	0	0.05	0.10	0.15	0.20	0.25	0.30
v	0	0.35	0.51	0.59	0.62	0.64	0.65

Use the trapezium rule to estimate the distance travelled between $t = 0$ and $t = 0.3$.

The medium actually produces a resistance to motion of magnitude kv m/s² per unit mass where k is a positive constant. Show that, in this case, the speed v is given by the equation

$$\frac{dv}{dt} = g - kv$$

where g m/s² is the acceleration due to gravity.

Hence find v in terms of g, k and t given that the particle is at rest initially. Given also that $k = 15$ and $g = 9.8$, integrate your expression for v to calculate the distance travelled between $t = 0$ and $t = 0.3$.   [A: 9.3]

*21 A particle of mass m moving vertically upwards with speed u enters a fixed horizontal layer of material of thickness a which resists its motion with a constant force of magnitude R. Show that it will pass through the layer if $u > u_0$, where

$$u_0^2 = 2a\left(\frac{R}{m} + g\right)$$

The particle emerges from the layer and moves freely under gravity until it re-enters the layer, which resists its motion as before. Show that the particle will pass completely through the layer in its downward motion only if $u^2 \geqslant 4aR/m$. [J: 9.3]

*22

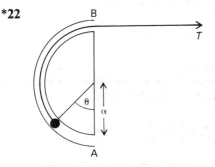

The diagram shows a narrow smooth tube in the form of a semicircle of radius a which is fixed in a vertical plane with its diameter vertical. A particle of mass m is initially at rest inside the tube at the lowest point A. The particle is attached to one end of a light inextensible ring which passes through the tube and out at the highest point B. The string is taut and its other end is pulled horizontally by a constant force T. Show that if, at a time t after the particle has left A, the radius of the semicircle through the particle makes an angle θ with BA, then

$$ma\left(\frac{d\theta}{dt}\right)^2 = 2T\theta - 2mg(1 - \cos\theta)$$

Find the reaction of the tube on the particle in terms of a, m, g, T and θ. Find also the value of $d^2\theta/dt^2$.

Show that, if $T = \tfrac{1}{2}mg$, then $\dfrac{d^2\theta}{dt^2} = 0$ when $\theta = \dfrac{\pi}{6}$.

By writing $\theta = \varphi + \dfrac{\pi}{6}$ show that

$$\frac{d^2\varphi}{dt^2} = \frac{g}{2a}(1 - \cos\varphi - \sqrt{3}\sin\varphi).$$

Deduce that, if φ is so small that φ^2 may be neglected in comparison with φ, then the motion is approximately simple harmonic, and find the period of the motion. [J: 7.8, 9.4]

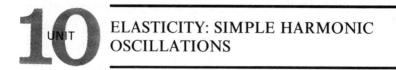

10 ELASTICITY: SIMPLE HARMONIC OSCILLATIONS

UNIT

10.1 Elasticity

Hooke's Law (Experimental)

For an elastic string:

Tension is proportional to extension

For an elastic spring:

Tension is proportional to compression/extension

$$T = \lambda \frac{x}{l}$$

where x = extension or compression
l = natural length
λ = modulus of elasticity or elastic constant (measured in N).

Example 1   Find the tension in an elastic string of modulus 10 N and natural length 2 m when it is stretched to a length of 2.5 m.

$\lambda = 10\,\text{N}$ Hooke's Law $\Rightarrow T = \lambda \dfrac{x}{l}$
$x = 0.5$
$l = 2\,\text{m}$
$T = ?$ $\Rightarrow T = 10 \times \dfrac{0.5}{2}$

$\Rightarrow T = 2.5\,\text{N}$

Example 2   A spring of natural length 40 cm is compressed 10 cm. Find the modulus of elasticity if the thrust in the spring is 5 N.

$\lambda = ?$ Hooke's Law $\Rightarrow T = \lambda \dfrac{x}{l}$
$x = 0.1$
$l = 0.4$
$T = 5\,\text{N}$ $\Rightarrow 5 = \lambda \dfrac{0.1}{0.4}$

$\Rightarrow \lambda = 20\,\text{N}$

Equilibrium of a Hanging Body

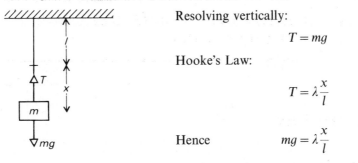

Resolving vertically:

$$T = mg$$

Hooke's Law:

$$T = \lambda \frac{x}{l}$$

Hence

$$mg = \lambda \frac{x}{l}$$

Example 3   A light elastic string has one end fastened to a fixed point and hangs freely with a body of mass 500 grams attached to the other end. If the natural length of the string is 1.2 m and its modulus is $2g$ N, find its extension when in equilibrium. If the hanging mass is pulled down below the point of equilibrium and released, find the acceleration when the string is 1.6 m long.

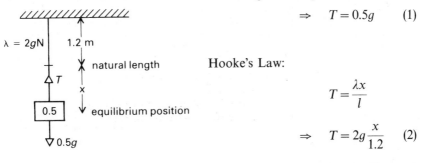

equilibrium position

Resolving vertically:

$$\Rightarrow \quad T = 0.5g \qquad (1)$$

Hooke's Law:

$$T = \frac{\lambda x}{l}$$

$$\Rightarrow \quad T = 2g \frac{x}{1.2} \qquad (2)$$

Hence (1) and (2) $\Rightarrow 0.5g = 2g \dfrac{x}{1.2}$

$$\Rightarrow \quad x = 0.3$$
$$\Rightarrow \text{extension} = 30 \, \text{cm}$$

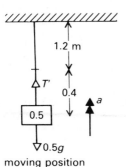

moving position

Hooke's Law:

$$T = \frac{\lambda x}{l}$$

$$\Rightarrow T' = 2g \frac{0.4}{1.2} = \frac{2}{3} g \qquad (3)$$

$F = ma \uparrow$:

$$T' - 0.5g = 0.5a \quad (4)$$

Hence (3) and (4) $\Rightarrow \frac{2}{3}g - \frac{1}{2}g = 0.5a$
$$\Rightarrow \text{acc}^n = a = \frac{9}{3} \text{m/s}^2.$$

• Exercise 10.1   See page 295

10.2 Work, Energy and Elastic Strings

Workdone in Stretching an Elastic String or Spring

Method 1

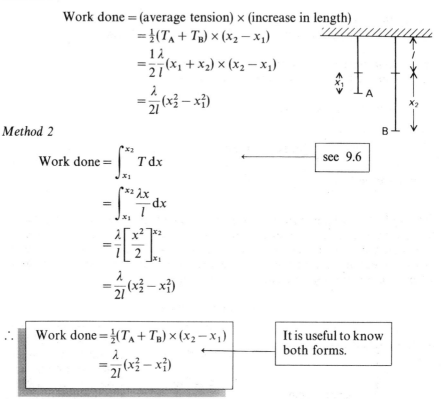

Work done = (average tension) × (increase in length)

$$= \tfrac{1}{2}(T_A + T_B) \times (x_2 - x_1)$$

$$= \frac{1}{2}\frac{\lambda}{l}(x_1 + x_2) \times (x_2 - x_1)$$

$$= \frac{\lambda}{2l}(x_2^2 - x_1^2)$$

Method 2

$$\text{Work done} = \int_{x_1}^{x_2} T\,dx$$

see 9.6

$$= \int_{x_1}^{x_2} \frac{\lambda x}{l}\,dx$$

$$= \frac{\lambda}{l}\left[\frac{x^2}{2}\right]_{x_1}^{x_2}$$

$$= \frac{\lambda}{2l}(x_2^2 - x_1^2)$$

$\therefore$
$$\text{Work done} = \tfrac{1}{2}(T_A + T_B) \times (x_2 - x_1)$$
$$= \frac{\lambda}{2l}(x_2^2 - x_1^2)$$

It is useful to know both forms.

Elastic Potential Energy (EPE)

Elastic Potential Energy = energy stored in the string or spring
= work done in stretching the string or spring from its natural length.

$\therefore$ $\text{EPE} = \dfrac{\lambda x^2}{2l}$

Note that the work done (above) is equal to the difference between the initial and final energies.

Example 4   Find the energy stored in an elastic string of length 1.5 m if its unstretched length is 1.2 m and its modulus of elasticity is 20 N.

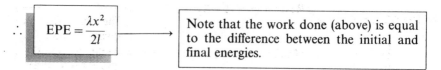

Extension   = 1.5 m − 1.2 m = 0.3 m

EPE $= \dfrac{\lambda x^2}{2l}$

$$= \frac{20(0.3)^2}{2.4}$$

$\Rightarrow$ Elastic potential energy $= 0.75$ J

Example 5 An elastic spring has natural length 40 cm and modulus 40 N. Find the work done in compressing it from 35 cm to 30 cm.

Initial compression $= x_1 = 40$ cm $- 35$ cm $= 0.05$ m
Final compression $= x_2 = 40$ cm $- 30$ cm $= 0.1$ m

$$\text{Work done} = \frac{\lambda}{2l}(x_2^2 - x_1^2)$$

$$= \frac{40}{2(0.4)}[(0.1)^2 - (0.05)^2]$$

$\therefore$ Work done $= 0.375$ J

Example 6 A light spring is extended 10 cm when a weight of mass 2 kg is hung from it. The spring is suspended from a fixed point and a weight of mass 3 kg attached to the lower end. The weight is pulled down a distance of 10 cm below the equilibrium position and then released. Find the speed of the mass when it passes through the equilibrium position.

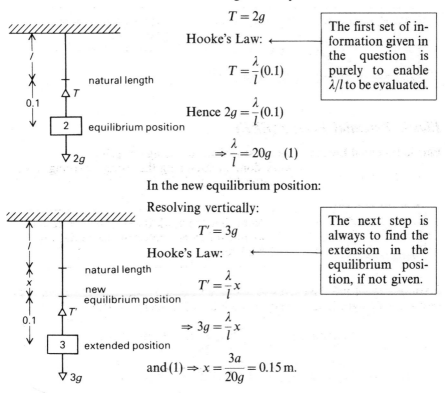

Resolving vertically:

$$T = 2g$$

Hooke's Law: ←

$$T = \frac{\lambda}{l}(0.1)$$

> The first set of information given in the question is purely to enable λ/l to be evaluated.

Hence $2g = \frac{\lambda}{l}(0.1)$

$$\Rightarrow \frac{\lambda}{l} = 20g \quad (1)$$

In the new equilibrium position:

Resolving vertically:

$$T' = 3g$$

Hooke's Law: ←

$$T' = \frac{\lambda}{l}x$$

> The next step is always to find the extension in the equilibrium position, if not given.

$$\Rightarrow 3g = \frac{\lambda}{l}x$$

and $(1) \Rightarrow x = \frac{3a}{20g} = 0.15$ m.

During the motion from the extended position to the new equilibrium position:

$$\text{Gain in KE of mass} = \tfrac{1}{2} \times 3 \times v^2$$
$$\text{Gain in PE of mass} = 3g(0.1)$$
$$\text{Loss in EPE of spring} = \frac{1}{2}\frac{\lambda}{l}[(0.25)^2 - (0.15)^2]$$
$$= 10g(0.04) = 0.4g \qquad \text{using (1)}$$

Conservation of energy:

$$\text{Loss in EPE} = \text{gain in KE} + \text{gain in PE}$$
$$\Rightarrow \qquad 0.4g = \tfrac{3}{2}v^2 + 0.3g$$
$$\Rightarrow \qquad \tfrac{3}{2}v^2 = 0.1g$$
$$\Rightarrow \qquad \text{velocity} = v \simeq 0.8 \text{ m/s}$$

• **Exercise 10.2**   See page 296

10.3 More Difficult Problems on Elastic Strings

Example 7   A particle of mass 400 grams is attached to one end of a light elastic string of natural length 1.5 m and modulus 8.4 N. The other end of the string is fixed and the particle moves as a conical pendulum, performing horizontal circles with an angular speed of 2 rad/s. Find the extension of the string.

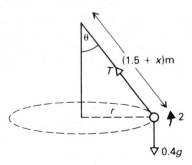

$F = ma$ radially:

$$T \sin\theta = 0.4r(2)^2$$

but $r = (1.5 + x)\sin\theta$

$$\Rightarrow T\sin\theta = 0.4(1.5 + x)\sin\theta \times 4$$
$$\Rightarrow T \qquad = 1.6(1.5 + x) \qquad (1)$$

Hooke's Law:

$$T = \frac{\lambda x}{l} \Rightarrow T = \frac{8.4x}{1.5} \qquad (2)$$

(1) and (2) $\Rightarrow$

$$\frac{8.4x}{1.5} = 1.6(1.5 + x)$$
$$\Rightarrow \qquad 8.4x = 2.4(1.5 + x)$$
$$\Rightarrow 8.4x - 2.4x = 3.6$$
$$\Rightarrow \qquad 6x = 3.6$$
$$\Rightarrow \qquad x = 0.6$$
$$\therefore \text{ extension} = 60 \text{ cm}$$

• **Exercise 10.3**   See page 297

10.4 Horizontal Simple Harmonic Oscillations

Oscillations: Points to Remember

1 If it is stated in a question that the motion is simple harmonic then the equations of SHM (9.4) may be quoted and used.

2 If, however, it is not stated that the motion is simple harmonic then this fact will first have to be proved.

3 To show whether motion is simple harmonic:

(i) Find the equilibrium position

(ii) Consider a general displacement, x, from the equilibrium position

(iii) Apply $F = m\ddot{x}$ in this position, **where $\ddot{x}$ is in the direction of x increasing.**

$$\text{Reminder:} \quad \ddot{x} = \frac{d^2x}{dt^2}$$

(iv) If an equation of the form $\ddot{x} = -\omega^2 x$ is obtained then motion is simple harmonic.

4 To find the amplitude, calculate the distance of the point where the velocity is zero from the equilibrium position.

5 An elastic spring is under tension both when extended and when compressed but, if an elastic string goes slack, then there is no tension.

Example 8 A light spring of natural length 50 cm and modulus 20 N has one end attached to a fixed point X on a smooth horizontal surface. A body of means 400 grams is attached to the other end and released from rest at a distance 65 cm from X. Show that subsequently the body moves under simple harmonic motion and find the speed of the body when the spring is 40 cm long.

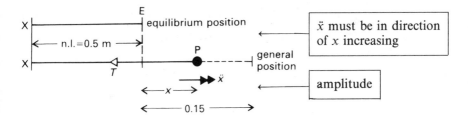

In the general position:
Hooke's Law:

$$T = \frac{\lambda x}{l} = \frac{20x}{0.5} = 40x$$

$F = m\ddot{x}$:

$$-T = 0.4\ddot{x}$$
$$\Rightarrow \quad -40x = 0.4\ddot{x}$$
$$\Rightarrow \quad \ddot{x} = -100x \text{ which is SHM} \qquad \text{[QED]}$$

When the spring is 40 cm long, the compression is 10 cm.

$$\Rightarrow x = 0.1 \, m$$

Also, $\omega = 10$ and $a = 0.15$

$\therefore$, using $v^2 = \omega^2(a^2 - x^2)$, $v^2 = 100 \, (0.15^2 - 0.1^2)$

$$\Rightarrow \text{speed} \approx 1.12 \, m/s.$$

Example 9   A light elastic string 1.6 m long and with modulus 20 N is stretched between two points X and Y, 2 m apart on a smooth horizontal surface. A body of mass 3 kg is attached to a point one quarter way along the string from X. The body is pulled 9 cm towards X and then released. Show that the ensuing motion is simple harmonic and state the magnitude of the maximum acceleration.

Since the body is fastened to the string, we consider the body as if it were attached to two strings, of natural lengths 0.4 m and 1.2 m, both with modulus 20 N.

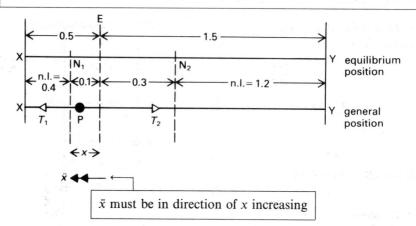

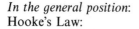

$\ddot{x}$ must be in direction of x increasing

In the general position:
Hooke's Law:

$$T_1 = \frac{\lambda}{l}(\text{extension}) = \frac{20}{0.4}(0.1 - x)$$

$$T_2 = \frac{20}{1.2}(0.3 + x)$$

$F = m\ddot{x}$:

$$T_1 - T_2 = 3\ddot{x}$$

$$\Rightarrow \frac{20}{0.4}(0.1 - x) - \frac{20}{1.2}(0.3 + x) = 3\ddot{x}$$

$$\Rightarrow \frac{-20}{0.4}x - \frac{20}{1.2}x = 3\ddot{x}$$

$$\Rightarrow \qquad \ddot{x} = -\frac{500}{9}x \text{ which is SHM}$$

Maximum acceleration occurs when x is a maximum, that is $x = 9\,\text{cm} = 0.09\,\text{m}$.

$$\therefore \text{ magnitude of maximum acceleration} = \left|\frac{-500}{9} \times 0.09\right| = 5\,\text{m/s}^2$$

- Exercise 10.4   See page 299

10.5 Vertical Simple Harmonic Oscillations

Example 10   AB is a light spring of natural length 50 cm and modulus $20g$ N. The end A is attached to a fixed point and the spring hangs vertically with a weight of mass 5 kg fastened at B. The weight is pulled down until it is 70 cm below A and released. Prove that the motion is simple harmonic and find the period of the motion and the maximum speed.

In the equilibrium position
Hooke's Law:

$$T_E = \frac{\lambda d}{l} = \frac{20gd}{0.5}$$

$$\Rightarrow T_E = 40gd \qquad (1)$$

$F = ma$:

$$T_E = 5g \qquad (2)$$

(1) and (2) $\Rightarrow 5g = 40gd$

$$\Rightarrow d = \tfrac{1}{8}\text{m} \qquad (3)$$

In the general position
Hook's Law:

$$T = \frac{\lambda}{l}(d + x)$$

$$\Rightarrow \qquad T = \frac{20g}{0.5}(d + x) = 40g(d + x) \qquad (4)$$

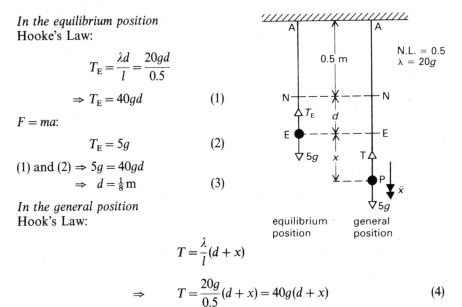

N.L. = 0.5
λ = 20g

equilibrium position   general position

$F = ma\downarrow$:

$$5g - T = 5\ddot{x}$$

(4) $\Rightarrow 5g - 40gd - 40gx = 5\ddot{x}$
(3) $\Rightarrow 5g - 5g - 40gx = 5\ddot{x}$
$\Rightarrow \ddot{x} = -8gx$ which is SHM

In equilibrium, the extension is $\tfrac{1}{8}\text{m} = 12\tfrac{1}{2}\,\text{cm}$ so the particle is pulled down $70 - (50 + 12\tfrac{1}{2})\,\text{cm} = 7\tfrac{1}{2}\,\text{cm}$ below the equilibrium position, before being released from rest. Thus the amplitude is $7\tfrac{1}{2}\,\text{cm}$.

Also, $\omega = \sqrt{8g}$ $\therefore$ period $= \dfrac{2\pi}{\sqrt{8g}} = \pi\sqrt{\dfrac{2}{g}}\,\text{s} \approx 1.42\,\text{s}$

Maximum speed $= a\omega = 0.075 \times \sqrt{8g}$
$\Rightarrow$ Maximum speed $\approx 0.664\,\text{m/s}$ or $66.4\,\text{cm/s}$.

- **Exercise 10.5**   See page 300

10.6 Further Oscillations

Example 11   A light elastic string of natural length 80 cm hangs vertically with its upper end fixed. A weight of mass 2 kg is fastened to its lower end. It rests in equilibrium with the string extended to 1 m long. The mass is pulled down a further 25 cm and released. Show that initially the motion is simple harmonic. After how long does the motion cease to be simple harmonic?

In the equilibrium position
Hook's Law:

$$T_E = \frac{\lambda 0.2}{0.8} \qquad (1)$$

$F = ma$:

$$T_E = 2g \qquad (2)$$

$$(1)\ \text{and}\ (2) \Rightarrow 2g = \frac{\lambda}{4}$$

$$\Rightarrow \lambda = 8g \qquad (3)$$

In the general position
$F = ma\downarrow$:

$$2g - T = 2\ddot{x} \qquad (4)$$

Hooke's Law:

$$T = \frac{\lambda}{0.8}(0.2 + x) \qquad (5)$$

$$(3)\ \text{and}\ (5) \Rightarrow T = 2g + 10gx \qquad (6)$$
$$(4)\ \text{and}\ (6) \Rightarrow 2g - 2g - 10gx = 2\ddot{x}$$
$$\Rightarrow \ddot{x} = -5gx \text{ which is SHM}$$

Thus the motion is simple harmonic unless, and until, the string goes slack. The weight moves with simple harmonic motion until it reaches N. We need to calculate the time from X to N.

$$\text{Period} = \frac{2\pi}{\omega} = \frac{2\pi}{\sqrt{5g}} = \frac{2\pi}{7}$$

Time for $X \to E = \frac{1}{4}$ period $= \dfrac{\pi}{14}\,\text{s}$.

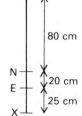

To calculate time for E→N, let $x = a \sin \omega t$
with $x = 0.2$, $a = 0.25$, $\omega = 7$

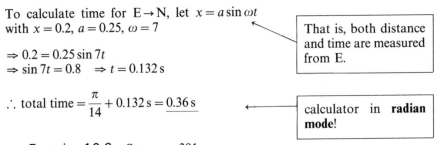

That is, both distance and time are measured from E.

$\Rightarrow 0.2 = 0.25 \sin 7t$
$\Rightarrow \sin 7t = 0.8 \quad \Rightarrow t = 0.132\,\text{s}$

$\therefore$ total time $= \dfrac{\pi}{14} + 0.132\,\text{s} = \underline{0.36\,\text{s}}$

calculator in **radian** mode!

- Exercise 10.6   See page 301

10.7 + The Simple Pendulum: Revision of Unit: A Level Questions

The Simple Pendulum

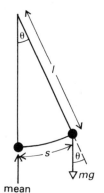

mean position

A simple pendulum consists of a heavy bob attached to one end of a light string whose other end is fixed. The particle and string make small oscillations in a vertical plane.

The accelerating force towards the mean position is $mg \sin\theta$.

The displacement along the path of the bob is s where $s = l\theta$

$$\Rightarrow \frac{d^2 s}{dt^2} = l\frac{d^2\theta}{dt^2}$$

$F = ma$:

$$-mg\sin\theta = ml\frac{d^2\theta}{dt^2}$$

$-$ sign because force and acceleration are in opposite directions.

For small oscillations, $\sin\theta \approx \theta$

$$\therefore l\frac{d^2\theta}{dt^2} = -g\theta$$

$$\Rightarrow \frac{d^2\theta}{dt^2} = -\frac{g}{l}\theta \quad \text{which is SHM of period } 2\pi\sqrt{\frac{l}{g}}.$$

Seconds Pendulum

A seconds pendulum takes exactly one second to perform one swing. Thus its period is 2 seconds.

Example 12   A simple pendulum which is meant to beat seconds loses $\frac{1}{2}$ minute per day. By what fraction of its length should it be shortened to make it beat accurately?

If accurate, the pendulum should make $(24 \times 60 \times 60)$ beats (half period) per day. This pendulum makes $[(24 \times 60 \times 60) - 30]$ beats per day.

$$\text{Time for one beat (half period)} = \frac{24 \times 60 \times 60}{(24 \times 60 \times 60) - 30} \text{ s}$$

$$= \frac{24 \times 60 \times 2}{(24 \times 60 \times 2) - 1} \text{ s} = \frac{2880}{2879} \text{ s.}$$

$$\therefore \pi\sqrt{\frac{l}{g}} = \frac{2880}{2879} \tag{1}$$

let the new length $= l - kl = l(1 - k)$

For this length, the beat $= \pi\sqrt{\dfrac{l(1-k)}{g}} = 1$

Thus
$$\pi\sqrt{\frac{l}{g}}\sqrt{1-k} = 1$$

$$(1) \Rightarrow \quad \frac{2880}{2879}\sqrt{1-k} = 1$$

$$\Rightarrow \quad \sqrt{1-k} = \frac{2879}{2880}$$

$$\Rightarrow \quad k = 1 - \left(\frac{2879}{2880}\right)^2 = 0.0007$$

$\therefore$ pendulum should be decreased by $\dfrac{7}{10000}$ of its length

Example 13   At a place where $g = 9.80\,\text{m/s}^2$, a simple pendulum beats exact seconds. If it is removed to a place where $g = 9.81$ m/s, by how many seconds per hour will it be inaccurate?

$$\text{Period} = 2\pi\sqrt{\frac{l}{g}} \Rightarrow \pi\sqrt{\frac{l}{9.80}} = 1$$

$$\Rightarrow \quad \pi\sqrt{l} = \sqrt{9.80} \tag{1}$$

When $g = 9.81$, the time of one beat $= \pi\sqrt{\dfrac{l}{9.81}} = \pi\sqrt{l} \times \dfrac{1}{\sqrt{9.81}}$

$$= \sqrt{9.80/9.81} \qquad \text{using (1)}$$

Number of beats in 1 hour $= (60 \times 60) \div \sqrt{9.80/9.81} \approx 3601.8$

Number of seconds in 1 hour $= 60 \times 60 = 3600$
$\therefore$ it gains 1.8 seconds each hour

• Miscellaneous Exercise 10B See page 302

Unit 10 Exercises

Exercise 10.1

1 An elastic string of natural length 2 m has modulus 12 N.
Find: (a) the tension in the string when the extension is 20 cm
      (b) the length of the string when the tension is 1.5 N.

2 A spring of natural length 120 cm has modulus 18 N. Find the thrust in the spring when it is compressed to a length of 1 m.

3 A spring of natural length 50 cm is compressed 10 cm by a weight of 5 N. Find its modulus.

4 The tension in an elastic string is 10 N when the string is extended by 20 cm. If the modulus is 9 N, find the natural length of the string.

5 If the modulus of an elastic string is 10 N and its tension is 2.5 N when the string is 2 m long, find its natural length.

6 If the length of a light elastic string is doubled when a mass of 8 kg is hung from one end, find the modulus of the string.

7 A light elastic string of natural length 80 cm and modulus $2g$ N has one end fixed and a mass of 600 grams is hung from the other end. Find the length of the string when the mass hangs in equilibrium.

8 A light elastic spring hangs vertically with its upper end fixed and a body of mass 800 grams attached to its other end. If the natural length of the spring is 1.6 m and the body hangs in equilibrium 2 m below the fixed end, show that the modulus is $3.2g$ N. If the body is then pulled down a distance of 20 cm below the equilibrium position and then released, find its initial acceleration.

9 A body of mass 3 kg lies on a smooth horizontal surface and is fastened to a fixed point O by a light elastic string of natural length 50 cm and modulus 20 N. The body moves in a circle of radius 80 cm with constant speed v. Find v correct to 2 decimal places.

10 A light elastic string AB has A fastened to a fixed point and a body of mass $2\sqrt{3}$ kg hanging freely from B. The string is pulled away from the vertical by a horizontal force P acting at B, and is in equilibrium with the string making 30° with the vertical and extended to one and a half times its natural length. Find the modulus of the string and the magnitude of P.

11 The ends of a light elastic string of natural length $2a$ are fastened to two points A and B, on the same horizontal level and a distance $2a$ apart. When a particle of mass 2 kg is fastened to the midpoint of the string it is found to hang in equilibrium at a distance a below AB. Show that the modulus of the string is

$$\frac{\sqrt{2g}}{\sqrt{2}-1}$$

12 An elastic string has natural length 30 cm and modulus 12 N. One end of the string is fastened to a fixed point, O, on a horizontal surface and the other end is attached to a body of mass 3 kg. The body is held at rest 45 cm from O and then released. Find the initial acceleration of the body in the cases where:
(a)   the surface is smooth
(b)   the coefficient of friction between the body and the surface is $\frac{1}{6}$.

*13 A light elastic string, of natural length $2a$, is hung, at one end, from a fixed point. When a particle of weight W is hung from the other end, the string extends a distance d. Show that the modulus is $2aW/d$.

The particle is removed and attached to the midpoint of the string. The ends of the string are now tied to two points A and B, where B is vertically below A and $AB > 2a$. If, in the equilibrium position, the lower part of the string remains taut, show that the displacement of the weight W from the midpoint of AB is $d/4$.

Exercise 10.2

1 Find the energy stored in a string of natural length 1.6 m when it is stretched to a length of 2 m, if the modulus of the string is 20 N.

2 Find the work done in compressing a spring of natural length 20 cm to a length of 15 cm, if its modulus is 30 N.

3 An elastic string has natural length 50 cm and modulus 40 N. Find the work done in stretching it from a length of 60 cm to 80 cm.

4 When an elastic string is 80 cm long its tension in 6 N and, when stretched to 1 m, its tension is 10 N. Find the work done in stretching if from 80 cm long to 1 m long, if its unstretched length is 60 cm.

5 If it requires a force of 16 N to compress a certain spring 1 cm, how much work is done in compressing it a further centimetre?

6 A light elastic string is of natural length 2 m and modulus $10g$ N. A weight of mass 2 kg is hung from one end of the string, whose other end is fixed. The weight is pulled down 20 cm below the equilibrium position and then released. Find its velocity when it is 10 cm above the equilibrium position.

7 A toy gun fires a 5 gram pellet by means of a spring mechanism. When loaded and ready to fire, the spring is compressed 3 cm and the thrust in it is 80 N. If a third of the energy stored in the spring is given to the pellet, with what speed does the pellet start to move?

8 A spring buffer on a model railway compresses 2 cm under a thrust of 30 N. If a model train of mass 500 grams runs into the buffer at 60 m/s, show that the spring will be compressed approximately 11 mm before the train is halted.

9 A mass of 1 kg hung in equilibrium on the end of a light elastic string stretches it 5 cm. If the mass is pulled down to 10 cm below the equilibrium position and then released, show that its speed when the string goes slack is $\sqrt{0.15g}$. Hence find the height above the equilibrium position to which the mass will rise.

10 A light elastic string is suspended from a fixed point. When a mass M is hung from the other end of the string, in equilibrium, the string is extended by an amount a. In this position a mass m is added to the original mass and both are allowed to fall. Using energy considerations, show that the distance both particles fall before coming instantaneously to rest is

$$\left(\frac{M + 2m}{M}\right)a$$

*11 A light elastic string has one end fastened to a fixed point O and it hangs freely with a small mass m attached to the other end. In equilibrium, the length of the string is $5l/3$ where l is the unstretched length of the string. If the mass is raised to the point O and allowed to fall, show that the greatest length of string subsequently is $3l$. Find also the speed of the mass when it is passing through the equilibrium position.

Exercise 10.3

1 A light elastic string of natural length $3m$ stretches to $4m$ when a weight of mass 8 kg is hung from one end. Find its modulus of elasticity. This same string then has its two ends fastened to two points, A and B, at the same horizontal level and at a distance of $4m$ apart. A mass of 10 kg is fastened to the midpoint of the string and held, level with A and B, by a vertical force. Find the tension in each of the two parts of the string. The vertical force is removed and the mass allowed to fall. Find the tension in the string when the mass has fallen 1.5 m and hence find the velocity of the mass at that point.

2 A light elastic string of natural length $2a$ and modulus λ has its two ends fastened to A and B where is a distance $3a$ vertically below A. A particle of mass m is fastened to the midpoint of the string. If both parts of the string remain taut, show that the particle rests in equilibrium a distance

$$\frac{3a}{2} + \frac{mga}{2\lambda}$$

below A.

3

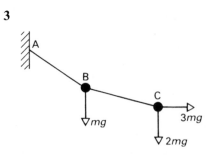

A light elastic string AC has a particle of mass m fastened to its midpoint B. The end A is fixed and a particle of mass $2m$ is fastened at C, to which a horizontal force of magnitude $3mg$ is applied. When the system is in equilibrium find:

(a) the tensions in the two parts of the string
(b) the inclination of each part of the string to the vertical.

If $\lambda = 4mg$, show that the ratio of the lengths of AB:BC is

$$(3\sqrt{2}+4):(\sqrt{13}+4)$$

4 An elastic string of natural length a and modulus λ lies on a smooth horizontal table with one end attached to a fixed point O on the table. To the other end is attached a particle of mass m and the string is stretched so that the particle is at a point C, where $OC = a + b$. The particle is then projected from C along the table in a direction perpendicular to the string at such a speed that its path is a circle with centre O. Calculate the time taken for the string to turn through a right angle.   [J]

5 A particle, of mass m, is suspended from a fixed point O by a light elastic string, of natural length l and modulus λ. The particle moves with constant angular speed ω in a horizontal circular path with the string making a constant angle θ with the downward direction of the vertical. Show that x, the extension of the string, is given by

$$\lambda x = \omega^2 \, lm(l + x).$$

Deduce that the motion described cannot take place unless $\omega^2 < \lambda/(lm)$. Show further that, for a given value of ω, the depth of the horizontal circle below O is independent of λ.   [L]

6 A small rough ring P of weight W is threaded on to a fixed horizontal straight wire which passes through a point A, as shown in the diagram. The ring is joined by a light elastic string of natural length l and modulus λ to a fixed point B which is at a distance $2l$ vertically below A. The ring is in equilibrium with the angle ABP $= \theta$. Find the tension in the string and the vertical and horizontal components of the force exerted by the wire on the ring.
Show that

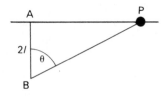

$$W \geqslant \lambda(2 - \cos\theta)\left(\frac{\tan\theta}{\mu} - 1\right),$$

where μ is the coefficient of friction between the ring and the wire.

[J]

7   A light elastic string is held with one end fixed. When a force of magnitude 50 N is applied at the free end, with the line of action of the force being along the string, the length of the string is 0.34 m. When the magnitude of the force is increased to 125 N the string is 0.37 m long. Find the natural length of the string and hence deduce that when a force of magnitude 100 N is applied at its free end the string is of length 0.36 m.

One end of the string is attached to a fixed point O on a smooth horizontal table and a particle P of mass 0.04 kg is attached to the other end. This particle is projected along the table so as to describe, with constant angular speed, a circle centre O and radius 0.36 m. Find, as a multiple of π, the time taken for one complete revolution.

Referred to axes Ox, Oy in the plane of the table P is, at time $t = 0$ s, at the point $(0.36, 0)$ m and moving parallel to the positive sense of Oy. Find the coordinates of P at time t s. The coordinates of a second particle Q at time t s are $(0.5 \cos 60t, 0.5 \sin 60t)$ m. Verify that OPQ are collinear when $t = 0$ and determine the value of t when they are next collinear with P between O and Q.   [A]

Exercise 10.4

1   A light spring of natural length 1 m and modulus 8 N has one end attached to a fixed point A on a smooth horizontal surface. A body of mass 2 kg is attached to the other end. The body is released from rest at a distance of 1.2 m from A. Show that the ensuing motion is simple harmonic and find the maximum speed during the motion.

2   A light elastic string of natural length 1.2 m and modulus 20 N has its ends attached to two points A and B 2 m apart on a smooth horizontal surface. A body of mass 2 kg is attached to the midpoint of the string and released from rest at a distance of 70 cm from B. Show that the subsequent motion is simple harmonic and find the speed of the body when it is 80 cm from A.

3   When a weight of mass 2 kg is hung from vertical spring of natural length 40 cm, its extension is 5 cm. Find the modulus of the spring. The same spring is now placed on a smooth horizontal surface with one end fixed to the surface. The 2 kg mass on the other end is pulled so that it is 50 cm from the fixed point and released from rest. Find the velocity at the point when the tension in the spring is instantaneously zero.

4   A light elastic string of natural length 2 m and modulus 25 N is stretched between two points A and B which are 3 m apart. A body of 4 kg is attached to the midpoint of the string and then pulled 20 cm towards A and released from rest. Show that the subsequent motion is simple harmonic. If the same body is then removed from the centre, refastened to the string at a point one third of the way along AB from A, pulled 20 cm towards A and then released, show that this motion will also be simple harmonic. Show that the ratio of the periods of the two motions is $3\sqrt{2} : 4$.

5 X and Y are two points on a smooth horizontal surface a distance of 2 m apart. One end of an elastic rubber band of natural length 75 cm and modulus 16 N is fastened to X. One end of another elastic rubber band of natural length 60 cm and modulus 8 N is fastened to Y. The other ends of both bands are fastened to a body P of mass 2 kg. When P is at rest, show that it is midway between X and Y.

P is pulled 20 cm towards Y and released from rest. Show that the subsequent motion is simple harmonic and that its period is $2\pi\sqrt{(3/52)}$.

*6 In the last question, if P is pulled 35 cm towards Y and then released from rest, the motion is the same as before until the rubber band attached to X goes slack. Show that whilst it is slack, the motion is given by the relationship

$$\ddot{y} = -\left(4 + \frac{20y}{3}\right)$$

where y is the extension of the stretched rubber band.

Using the substitution $z = 4 + \frac{20}{3}y$

show that the motion reduces to the simple harmonic relationship

$$\ddot{z} = -\frac{20z}{3}$$

Exercise 10.5

1 A light spring of natural length 90 cm and modulus $2mg$ N hangs vertically with its top end fixed and a weight of mass m kg fastened to the lower end. Find the extension of the spring when the system is in equilibrium. The body is then pulled down vertically until the spring is 1.5 m long and released. Show that the motion is SHM and find the period and amplitude of the oscillations.

The spring is replaced by a string with the same natural length and modulus. It is pulled down a distance e from the equilibrium position and released from rest. For what values of e will the motion be simple harmonic for the whole of each oscillation?

2 A body of mass 400 grams is attached to the lower end of a light elastic string of natural length 80 cm, whose upper end is fixed so that the string hangs vertically. The body is pulled down a small distance below the equilibrium position and then released. If the subsequent motion is simple harmonic with period 0.4 s, calculate the modulus of the string.

3 A light spring of natural length 20 cm and modulus $2g$ N hangs vertically with its upper end fixed and a particle of mass m fastened to its other end. When in equilibrium, the extension of the string is 5 cm. Another particle of mass m is hung from a second vertical spring of 25 cm fixed at its top. When this spring is in equilibrium, the particle is pulled down a short distance and released. The ensuing motion is simple harmonic with period 0.25 s. Find the value of m and the modulus of the second spring.

*4 A light elastic string of natural length a and modulus λ has one end fixed at a point A and a particle of mass m attached to its other end. The particle is pulled down to a distance $a + b$ vertically below A and then released. Prove that, if b lies within certain limits, the whole motion is simple harmonic. Find the distance below A of the centre of oscillation. Find the period of the oscillations and show that their amplitude is $b - mga/\lambda$.

Explain why $\dfrac{mga}{\lambda} < b \leqslant \dfrac{2mga}{\lambda}$

*5 A light elastic spring, of natural length a and modulus $8mg$ N, stands vertically with its *lower end* fixed and carries a particle of mass m fastened to its upper end. This particle is resting in equilibrium when a second particle, also of mass m, is dropped onto it from rest at a height of $3a/8$ above it. The particles coalesce on impact. Show that the composite particle oscillates with SHM about a point which is at a height $3a/4$ above the lower end of the spring with period $\pi\sqrt{(a/g)}$ and find the amplitude of the motion.

Exercise 10.6

1 A light elastic string of natural length l and modulus $4mg$ has one end fixed to a point A and a particle of mass m fastened to the other end. If the particle hangs in equilibrium vertically below A, find the extension of the string.

The particle is now held at a point B, which is a distance l vertically below A, and projected vertically downwards with speed $\sqrt{(6gl)}$. If C is the lowest point reached by the particle, prove that the motion from B to C is SHM of amplitude $5l/4$. Prove also that the time taken by the particle to move from B to C is

$$\frac{1}{2}\left(\frac{\pi}{2} + \sin^{-1}\tfrac{1}{5}\right)\sqrt{\frac{l}{g}}$$

2 A particle P, of mass m, is attached to two points X and Y on a smooth horizontal table, where $XY = 5a$, by two light elastic strings XP and PY. XP has modulus $2mg$ and PY has modulus mg and both strings have natural length a. In equilibrium, P rests at O. Find the length of XO.

If, at time t, P is on the straight line between X and Y, both strings are taut and $XP = 2a + x$, show that the equation of motion is

$$\ddot{x} = -\frac{3gx}{a}$$

If L and M are points between X and Y such that $XL = MY = a$, and P starts from rest at M, find its velocity at L. Find also the time taken by P to travel from M to L.

3 A light elastic string of natural length a and modulus $2mg$ is fastened at one end to a fixed point A. It then passes over a smooth peg B, on the same horizontal level as A, where AB is of length a. A ring, P, of mass m is attached to the other end of the string. P is threaded onto a smooth fixed horizontal wire, which is $3a/4$ below AB and parallel to it. Initially the ring is held so that angle ABP is acute, and then released from rest. Prove that the ring will perform SHM of period

$$\pi\sqrt{\frac{2a}{g}}$$

and find the reaction between the ring and the wire at the mean position. If the angle ABP is initially 45°, determine the greatest speed of the ring during the motion.

4   A light elastic string of natural length a and modulus of elasticity $2mg$ is fastened at one end to a fixed point O. A particle P, of mass m, is hung from the other end of the string. Find the extension of the string when P is in its equilibrium position A.
If P moves along the line OA, show that its displacement, x, from A at time t is given by

$$\frac{d^2x}{dt^2} = -\frac{2gx}{a}$$

provided the string is taut.
If P is released from rest at a distance d below A, show that it will come to instantaneous rest at O, provided that $4d^2 = 5a^2$.
Show that the time taken to reach O is

$$\left(2 + \frac{\pi}{2} + \sin^{-1}\frac{1}{\sqrt{5}}\right)\sqrt{\frac{a}{2g}}$$

5   A light elastic string PQ of natural length 60 cm has a weight of mass 1.5 kg attached to Q and P is fixed so that the string hangs vertically. If it hangs in equilibrium with an extension of 15 cm, find the modulus of the string. In the equilibrium position the weight is projected vertically down with a speed of 1.7 m/s. Show that the ensuing motion is initially simple harmonic and find the period and the amplitude of the motion. After how long does the string become slack and what is the closest that the particle gets to P?

Miscellaneous Exercise 10B

1   What is the length of a seconds pendulum that is accurate when $g = 9.81 \text{ m/s}^2$?

2   A simple pendulum that is meant to beat seconds gains 2.5 seconds per hour. To make it beat accurately, by what fraction of its length should it be lengthened?

3   A pendulum in a clock is made of a fine cord with a brass bob on the end. It beats exact seconds in a house at sea level where $g = 9.81 \text{ m/s}^2$. If the owners move into the mountains where $g = 9.80 \text{ m/s}^2$, by how many seconds a day will it be wrong?

4   A particle P of mass m is suspended from a fixed point O by a light inextensible string of length l and is released from test with the string taut and making a small angle α radians with the downward vertical. Show that the period of oscillations of P is approximately $2\pi\sqrt{(l/g)}$. Show also that the average speed of P, with respect to time, during one complete oscillation is approximately

$$\frac{2\alpha}{\pi}\sqrt{(lg)} \qquad\qquad \text{[J: 10.7]}$$

5   The diagram shows a small bead B of mass 0.1 kg free to slide on a smooth fixed vertical rod AV. One end of a light elastic string of modulus 12 N and natural length 1 m is attached to B and the other end is attached to a fixed point O which is at a perpendicular distance of 1.2 m from AV.

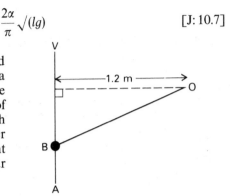

The bead is released from rest from the position where OB is horizontal. Show that when B has dropped a distance of 0.5 m the energy stored in the string has increased by 0.3 J and find the speed of B at this time. (Take g to be $10\,\text{m/s}^2$)   [A: 10.2]

6   A light elastic spring of natural length a is such that a mass m suspended from it produces an extension $\frac14 a$. Finds its modulus of elasticity.
    The spring is placed on a smooth horizontal table with one end fixed to a point O of the table. A particle P of mass m is attached to the other end of the spring. At time $t = 0$ the system is at rest with $OP = a$ and an impulse of magnitude $3m(ga)^{1/2}$ is then applied to P in the direction OP. Obtain an expression for the extension of the spring at any subsequent time and find the maximum extension. Find an expression for the time taken until the extension first becomes $\frac34 a$.   [J: 10.4]

7   An elastic string of natural length $2a$ and modulus λ has its ends attached to two points A, B on a smooth horizontal table. The distance AB is $4a$ and C is the mid-point of AB. A particle of mass m is attached to the mid-point of the string. The particle is released from rest at D, the mid-point of CB.
    Denoting by x the displacement of the particle from C, show that the equation of motion of the particle is

$$\frac{d^2x}{dt^2} + \frac{2\lambda}{ma}x = 0$$

Find the maximum speed of the particle and show that the time taken for the particle to move from D directly to the mid-point of CD is

$$\frac{\pi}{3}\left(\frac{ma}{2\lambda}\right)^{1/2} \qquad\qquad \text{[J: 10.4]}$$

8   A uniform rod AB, of length $2a$ and mass m, rests in equilibrium with its

lower end A on a rough horizontal floor. Equilibrium is maintained by a horizontal elastic string, of natural length a and modulus λ. One end of the string is attached to B and the other end to a point vertically above A. Given that θ, where $\theta < \pi/3$, is the inclination of the rod to the horizontal, show that the magnitude of the tension in the string is $\frac{1}{2}mg \cot \theta$.
Prove also that

$$2\lambda = (mg \cot \theta)/(2 \cos \theta - 1)$$

Given that the system is in limiting equilibrium and that the coefficient of friction between the floor and the rod is $\frac{2}{3}$, find $\tan \theta$. Hence show that $\lambda = 10mg/9$.
[L: 10.3]

9 A particle of mass m is attached to one end of a light elastic string of natural length l and modulus $2mg$. The other end of the string is fixed at a point A. The particle rests on a support B vertically below A, with $AB = \frac{5}{4}l$. Find the tension in the string and the reaction exerted on the particle by the support B.
The support B is suddenly removed. Show that the particle will execute simple harmonic motion and find:

(i) the depth below A of the centre of oscillation
(ii) the period of the motion.   [J: 10.5]

10 A particle P of mass m is joined to a fixed point O by a light inextensible string of length l, and is released from rest with the string taut and making an acute angle α with the downward vertical. At a time t after P is released the string makes an angle θ with the downward vertical and the tension in the string is T. Show that

$$\left(\frac{d\theta}{dt}\right)^2 = \frac{2g}{l}\cos \theta + A$$

and

$$T = 3mg \cos \theta + B,$$

where A and B are constants, and find A and B.
Find $d^2\theta/dt^2$ in terms of g, l and θ.
Given that θ is always small enough for θ^2 to be negligible show that:

(i) $T \approx mg$
(ii) the motion of P is approximately simple harmonic with period $2\pi \sqrt{(l/g)}$

and find θ in terms of α, g, l and t.   [J: 7.6, 9.5, 10.7]

*11 A particle of mass m is attached to one end of a light inextensible string of length l, the other end of which is attached to a fixed point O at a height h above a smooth horizontal table. With the string taut the particle describes a circle on the table, at a constant angular speed ω. The centre of the circle is vertically below O. Find the tension in the string and the magnitude of the reaction of the table. Determine the greatest value of ω^2 for which such a motion is possible.
The inextensible string is now replaced by a light elastic string of modulus

mg and natural length *a* and the table is removed. Find the greatest and least values of ω^2, in terms of *g* and *a*, such that P can still describe horizontal circles with constant angular speed ω.   [A: 7.4, 10.1, 10.3]

*12 A particle of mass *m* is attached to one end of an elastic spring of natural length *l* and modulus $m\omega^2 l$. The other end of the spring is fastened to a point O on a smooth horizontal table. The particle is released from rest at a point on the table at a distance $l + a$ from O, where $a < l$. Show that the subsequent motion of the particle is simple harmonic.
Derive the relation

$$v^2 = \omega^2(a^2 - x^2),$$

where *v* is the velocity of the particle and *x* is its displacement from the equilibrium position A.
The particle, still attached to the spring, is now placed at rest at A and another particle, of mass 3*m*, lying in the line OA produced is projected towards it with speed $\frac{1}{3}\omega l$.
Given that the particles adhere on impact, show that the combined particle moves through a distance $\frac{1}{2}l$ before first coming to rest, and find the time it takes to move through this distance.   [J: 5.5, 9.4, 10.4]

13 A uniform rod AB, of length 2*l* and weight *W*, has its end A in contact with a rough vertical wall and its end B connected by a string BC to a point C vertically above A with $AC = 2l$. The rod is in equilibrium in a vertical plane perpendicular to the wall with the angle $ACB = \alpha$, where $\alpha < 45°$. Show that the tension in the string is $W \cos \alpha$ and that the frictional force on the rod at A must act upwards. Show also that the least possible value of the coefficient of friction at A is $\tan \alpha$.
Given that $\cos \alpha = \frac{3}{4}$ and that the string is an elastic one with modulus of elasticity $3W/2$, find the natural length of the string.   [L: 10.3]

*14 A particle of mass *m* is attached to one end of a light elastic string of natural length 6*a* and modulus of elasticity 3*mg*. The other end of the string is fixed to a point O on a smooth plane inclined at an angle of 30° to the horizontal. The string lies along a line of greatest slope of the plane and the particle rests in equilibrium at a point C on the plane. Calculate the distance OC.
The particle is now pulled a further distance 2*a* down the line of greatest slope through C and released from rest. At time *t* later, the displacement of the particle from C is *x*, where the positive direction of *x* is down the plane. Show that until the string slackens, *x* satisfies the differential equation

$$\ddot{x} + \frac{gx}{2a} = 0$$

Find the time at which the string slackens and determine the speed of the particle at this time.   [J: 9.5, 10.4, 10.6]

15 The end A of a light elastic spring AB of natural length 1 m is held fixed. A particle P is attached at B and where P hangs vertically in equilibrium $AB = 1.4$ m. The spring is placed on a smooth horizontal table with P still

attached to B and the end A held fixed. The particle P is then set in motion in the line AB so that at time t s the distance AB is equal to $(1 + x)$ m. Show, by taking g to be $10\,\text{m/s}^2$, that

$$\frac{d^2x}{dt^2} + 25x = 0,$$

and find the positive constant ω so that

$$x = a\cos(\omega t - \alpha)$$

satisfies this equation for all constants a and α.
Find a and $\tan\alpha$ given that, when $t = 0$, $x = 0.3$ and that P is moving in the direction AB with speed $2\,\text{m/s}$. Determine also the maximum value of x and the maximum speed of P.   [A: 10.4]

16 State Hooke's Law for a stretched elastic string.
Prove, by integration, that the work done in stretching an elastic string, of natural length l and modulus of elasticity λ, to a length $l + x$ is

$$\tfrac{1}{2}\lambda x^2/l$$

A particle of mass m is suspended from a fixed point O by a light elastic string of natural length l. When the mass is hanging freely at rest the length of the string is $13l/12$. The particle is allowed to fall from rest at O. Find the greatest extension of the string in the subsequent motion. Show that the maximum kinetic energy of the particle during this fall occurs when it passes through the equilibrium position.   [J: 10.1, 10.2, 10.3]

*17 One end of a light elastic string, of natural length l, is attached to a fixed point O. To the other end of the string is attached a particle of mass m. When the particle hangs in equilibrium the length of the string is $3l/2$.
(a) The particle is pulled vertically downwards from its equilibrium position and released from rest when the length of the string is $7l/4$. Find the period and the amplitude of the subsequent oscillations.
(b) If instead the particle is released from rest at a point A vertically below O, where OA $= 9l/4$, show that the time taken for the string to first become slack is

$$\left(\frac{l}{2g}\right)^{1/2}\left[\pi - \cos^{-1}\left(\frac{2}{3}\right)\right]$$ [L: 10.5, 10.6]

*18 A light elastic string of modulus mln^2 and natural length l has one end fixed at O and carries at its other end two particles, P and Q, of masses m and $2m$, respectively. Find, in terms of g and n, the extension of the string when the particles hang in equilibrium. When the system is in equilibrium the particle Q is gently removed. In the subsequent motion of P, let x denote the displacement at time t of P below the point at which it would hang in equilibrium.

Show that, whilst the string is taut,

$$\ddot{x} + n^2 x = 0$$

Find the time that elapses after the particle Q is removed before the string slackens.

Given that when P reaches its greatest height it is still below O, show that:

(a) in the upward motion of P the ratio of the time for which the string is taut to the time for which it is slack is $2\pi:3\sqrt{3}$

(b) $l > \dfrac{3g}{2n^2}$. [J: 10.5, 10.6]

*19 Prove that the potential energy of a light elastic string of natural length l and modulus λ when stretched to a length $l + x$ is $\frac{1}{2}\lambda x^2 / l$.

One end of a light elastic string of natural length $4a$ and modulus $4mg$ is attached to a fixed point A, and a particle of mass m is attached to the other end. The particle is allowed to fall from rest at A. When it has fallen through a distance $6a$ it collides with a particle of mass $3m$ which is instantaneously at rest but free to move, and the two particles coalesce. Show that the speed of the composite particle immediately after the collision is $\sqrt{(ga/2)}$ and that the lowest point reached by the particle is at a distance $(8 + \sqrt{6})a$ from A. Prove that the string remains taut throughout the ensuing motion. [J: 10.6, 5.5]

*20 A particle A, of mass m, is moving on a smooth horizontal plane with speed v when it strikes directly a particle B, of mass $3m$, which is at rest on the plane. The coefficient of restitution between A and B is $\frac{1}{2}$ and after impact B moves with speed $3u$. Find, in terms of u, the value of v and the speed of A after impact. Find also, in terms of m and u:

(a) the magnitude of the impulse of the blow received by B
(b) the kinetic energy lost in the impact.

A third particle C, of mass $6m$, is at rest on the plane and is connected by an elastic spring, of natural length l and modulus of elasticity kmg, to a point D on the plane so that BCD is a straight line and $CD = l$. Particle B strikes C and is brought to rest by the impact. Particle C then moves a distance $\frac{1}{4}l$ before coming momentarily to rest. Show that

$$k = \frac{216u^2}{gl}$$ [L: 5.4, 5.6, 10.1, 10.2]

*21 One end of a light elastic string of natural length l and modulus $4mg$ is attached to a fixed point O. The other end is attached to a particle P of mass m. The particle is projected vertically downwards from O with speed $\sqrt{(4gl)}$. By using the principle of conservation of energy, or otherwise find the speed of P when P is at a depth x below O, where $x > l$, and show that the greatest depth below O attained by P is $5l/2$. Find also the maximum value of the speed.

Show that the particle subsequently rises to a maximum height $3l/2$ above O. [J: 10.2, 10.3]

*22

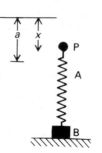

The diagram shows a light coiled spring PB which is subject to Hooke's Law both when extended and when compressed. The lower end is attached to a block B of mass m which rests on a fixed horizontal table. The upper end carries a particle P, also of mass m, which can move in a vertical line. In the equilibrium position, in which P is at O, the length of the spring is less than its natural length by an amount h. Initially P is released from rest at A, which is at a depth a below O. Show that at a subsequent time t the depth x of P below O is

$$a \cos \omega t, \text{ where } \omega = \sqrt{(g/h)}$$

provided that B remains on the table. Find, in terms of m, g, a, h, ω and t:

(i)   the upward force exerted on P by the spring
(ii) the reaction between B and the table.

Show that the base B will remain on the table throughout the motion if

$$a \leqslant 2h$$

Show that if $a = 4h$ then B will leave the table when

$$t = \frac{2\pi}{3} \sqrt{\left(\frac{h}{g}\right)}$$

[J: 10.5, 10.6]

*23 A trolley of mass m runs down a smooth track of constant inclination $\pi/6$ to the horizontal, carrying at its front a light spring of natural length a and modulus mga/c, where c is constant. When the spring is fully compressed it is of length $a/4$, and it obeys Hooke's law up to this point. After the trolley has travelled a distance b from rest the spring meets a fixed stop. Show that, when the spring has been compressed a distance x, where $x < 3a/4$, the speed v of the trolley is given by

$$cv^2/g = c(b + x) - x^2$$

Given that $c = a/10$ and $b = 2a$, find the total distance covered by the trolley before it momentarily comes to rest for the first time.

[L: 10.1, 10.2, 10.3]

**24 A scale pan is suspended from a fixed point A by a light elastic spring and a particle P of mass 0.3 kg is placed in the pan and attached to it with adhesive.

The pan is pulled down from its equilibrium position and set in motion so that P moves in a vertical line through A with the base of the pan remaining horizontal. Given that the motion of P is simple harmonic with period

$\frac{\pi}{5}$ s and that the maximum and minimum

distances of P below A are 1.35 m and 0.85 m respectively, find:

(a) the distance below A of the centre O of the oscillation, the amplitude of the oscillation and the maximum speed attained by P
(b) the time to move directly upwards a distance of 0.125 m from O
(c) the maximum force, normal to the scale pan, that the adhesive has to exert on P
(d) the length of the spring, when, in the absence of adhesive, P would leave the pan.
(Take g to be $10 \, \text{m/s}^2$)   [A: 10.5, 10.6]

**25 A particle P of mass m is attached to one end of a light elastic string of natural length l and modulus $3mg$. The other end of the string is fastened to a fixed point A. The particle is projected from A vertically upwards with speed $\sqrt{(3gl)}$. Find its speed when it reaches the point B which is at a height l above A.

At a time t after passing through B, but before it first returns to B, the particle is at a distance x above B. Find $\dfrac{d^2 x}{dt^2}$ in terms of g, l and x, and verify that the equation you have found is satisfied if

$$x = -\frac{l}{3} + H \cos \omega t + K \sin \omega t,$$

where $\omega^2 = 3g/l$ and H and K are constant.
Find H and K in terms of l.
Find the time that elapses from the projection of the particle from A to its return to A.   [J: 9.1, 10.6]

**26 Two small smooth pegs A and B are fixed at the same horizontal level at a distance $2a$ apart, and a light elastic band of natural length $2a$ is stretched round them. A particle P of mass m is attached to the band and the system is in equilibrium with P at a point O which is at a distance b vertically below AB. Show that the modulus of the string is $\frac{1}{2}mga/b$.
The particle is then given a vertical velocity. Show that its equation of motion is

$$\frac{d^2 x}{dt^2} + \frac{g}{b} x = 0$$

where $x + b$ is the depth of P below AB.
At a certain instant $x = b$ and P has a downward velocity of magnitude $\sqrt{(gb)}$. Find the speed of P as it passes through O and the time taken for P to reach O.   [J: 10.3, 10.4, 10.6]

11 VECTOR APPLICATIONS TO MECHANICS: RELATIVE VELOCITY

11.1 Force, Velocity and Acceleration

Some Vector Reminders

(i) A two-dimensional vector can be expressed as $p\mathbf{i} + q\mathbf{j}$ or $\begin{pmatrix} p \\ q \end{pmatrix}$

(ii) A three-dimensional vector can be expressed as $p\mathbf{i} + q\mathbf{j} + r\mathbf{k}$ or $\begin{pmatrix} p \\ q \\ r \end{pmatrix}$

(iii) The scalar product of two vectors $\mathbf{a}$ and $\mathbf{b}$ is $\mathbf{a} \cdot \mathbf{b}$ and $\mathbf{a} \cdot \mathbf{b} = ab \cos \theta$ where θ is the angle between $\mathbf{a}$ and $\mathbf{b}$

(iv) If $\mathbf{a} = p\mathbf{i} + q\mathbf{j} + r\mathbf{k}$ and $\mathbf{b} = x\mathbf{i} + y\mathbf{j} + z\mathbf{k}$ then $\mathbf{a} \cdot \mathbf{b} = \begin{pmatrix} p \\ q \\ r \end{pmatrix} \cdot \begin{pmatrix} x \\ y \\ z \end{pmatrix} = px + qy + rz$

1 Velocity and Acceleration

If the displacement of a body, at time t, is given by

then the
and the
$$\begin{array}{l} \mathbf{r} = \mathbf{f}(t) \\ \text{velocity} \quad = \mathbf{v} = \dot{\mathbf{r}} = \mathbf{f}'(t) \\ \text{acceleration} = \mathbf{a} = \ddot{\mathbf{r}} = \mathbf{f}''(t) \end{array}$$
$$\dot{\mathbf{r}} \equiv \frac{d\mathbf{r}}{dt}$$

Example 1 If the position vector of a particle of mass m, relative to the origin, is $\mathbf{r}$, where

$$\mathbf{r} = a \sin \omega t \mathbf{i} + b \cos \omega t \mathbf{j} + c\mathbf{k}$$

find expression for the velocity and acceleration of the particle in terms of t.

$$\mathbf{r} = a \sin \omega t \mathbf{i} + b \cos \omega t \mathbf{j} + c\mathbf{k}$$
velocity $\quad = \mathbf{v} = \dot{\mathbf{r}} = a\omega \cos \omega t \mathbf{i} - b\omega \sin \omega t \mathbf{j}$
acceleration $= \mathbf{a} = \ddot{\mathbf{r}} = -a\omega^2 \sin \omega t \mathbf{i} - b\omega^2 \cos \omega t \mathbf{j}$

Example 2 If a particle has an acceleration $2\mathbf{i} + (2t - 1)\mathbf{j}$ find its velocity at time t, given that its initial velocity is $3\mathbf{i} - 4\mathbf{j}$.

$$a = \frac{dv}{dt} = 2i + (2t - 1)j$$

Integrating
when $t = 0$
$\therefore$
$\Rightarrow$

$v = 2ti + (t^2 - t)j + Ai + Bj$ ⟵ $\quad$ Constant of integration must also be a vector.
$v = 3i - 4j$
$v = 2ti + (t^2 - t)j + 3i - 4j$
$v = (2t + 3)i + (t^2 - t - 4)j$

2 Newton's Second Law

$$\boxed{F = ma \quad \text{or} \quad F = m\frac{dv}{dt}}$$

Example 3   A force, which at time t is given by $2mi + 30tmj$, acts upon a particle of mass m. Initially the particle is at rest at the point with position vector $i + 2j$. Find the position vector of the particle at time t.

$F = ma$
Integrating

$\quad\quad\quad F = m(2i + 30tj)$
$\Rightarrow a = 2i + 30tj$
$v = 2ti + 15t^2j + Ai + Bj$

When $t = 0$, the particle is at rest $\Rightarrow A = B = 0$

Integrating
What $t = 0$

$\therefore v = 2ti + 15t^2j$
$r = t^2i + 5t^3j + Ci + Dj$
$r = i + 2j$
$\Rightarrow r = t^2i + 5t^3j + i + 2j$
$\Rightarrow r = (t^2 + 1)i + (5t^3 + 2)j$

3 Components of Forces and Velocities

$$\boxed{\text{Component of } F \text{ in direction of } a = \frac{F \cdot a}{|a|}}$$

Example 4   Find the component of the velocity $3i + 4j + k$ in the direction of $2i - 2j + k$.

$$\text{Component} = \frac{\left[\begin{pmatrix} 3 \\ 4 \\ 1 \end{pmatrix} \cdot \begin{pmatrix} 2 \\ -2 \\ 1 \end{pmatrix}\right]}{\sqrt{2^2 + 2^2 + 1^2}} = \left(\frac{6 - 8 + 1}{3}\right) = -\frac{1}{3}$$

4   *Angle Between Two Forces or Velocities*

$$\text{Angle between } \mathbf{v}_1 \text{ and } \mathbf{v}_2 = \cos^{-1}\left(\frac{\mathbf{v}_1 \cdot \mathbf{v}_2}{|\mathbf{v}_1||\mathbf{v}_2|}\right)$$

Example 5   Find the angle between the directions of the two forces $\mathbf{F}_1 = (2\mathbf{i} + 3\mathbf{j})\,\text{N}$ and $\mathbf{F}_2 = (3\mathbf{i} - 4\mathbf{j})\,\text{N}$

$$\text{Angle} = \cos^{-1}\frac{\left[\binom{2}{3}\cdot\binom{3}{-4}\right]}{\sqrt{2^2+3^2}\,\sqrt{3^2+4^2}} = \cos^{-1}\left[\frac{(6-12)}{\sqrt{13}\times 5}\right] = \underline{109.4^\circ}$$

5   *Perpendicular Vectors*

$$\mathbf{a} \text{ is perpendicular to } \mathbf{b} \Leftrightarrow \mathbf{a}\cdot\mathbf{b} = 0$$

6   *Resultant of Two Forces*

If two forces intersect than the resultant passes through the point of intersection of the two forces.

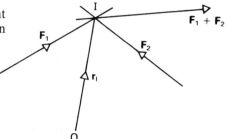

Example 6   A force $\mathbf{F}_1 = 2\mathbf{i} + \mathbf{j} + \mathbf{k}$ acts through the point with position vector $\mathbf{i} - \mathbf{j}$. A second force $\mathbf{F}_2 = -\mathbf{j} + 2\mathbf{k}$ acts through the point with position vector $3\mathbf{i} + 2\mathbf{j} - 3\mathbf{k}$. Find the resultant force. Show that the lines of action of the forces intersect and find the position vector of the point of intersection. Hence find the line of action of the resultant.

$$\text{Resultant force} = \mathbf{F}_1 + \mathbf{F}_2 = \begin{pmatrix} 2 \\ 1 \\ 1 \end{pmatrix} + \begin{pmatrix} 0 \\ -1 \\ 2 \end{pmatrix} = \begin{pmatrix} 2 \\ 0 \\ 3 \end{pmatrix}$$

$\therefore$ resultant force $= 2\mathbf{i} + 3\mathbf{k}$

Let $\mathbf{r}_1$, $\mathbf{r}_2$ be the position vectors of general points on the lines of action of $\mathbf{F}_1$, $\mathbf{F}_2$ respectively.

$$\mathbf{r}_1 = \begin{pmatrix} 1 \\ -1 \\ 0 \end{pmatrix} + \lambda\begin{pmatrix} 2 \\ 1 \\ 1 \end{pmatrix} \qquad \mathbf{r}_2 = \begin{pmatrix} 3 \\ 2 \\ -3 \end{pmatrix} + \mu\begin{pmatrix} 0 \\ -1 \\ 2 \end{pmatrix}$$

Let $\mathbf{r}_1 = \mathbf{r}_2$. Then $\begin{pmatrix} 1 \\ -1 \\ 0 \end{pmatrix} + \lambda\begin{pmatrix} 2 \\ 1 \\ 1 \end{pmatrix} = \begin{pmatrix} 3 \\ 2 \\ -3 \end{pmatrix} + \mu\begin{pmatrix} 0 \\ -1 \\ 2 \end{pmatrix}$

$$\Rightarrow \qquad 2\lambda = 2 \qquad\qquad (1)$$
$$\lambda = 3 - \mu \qquad\qquad (2)$$
$$\lambda = -3 + 2\mu \qquad\qquad (3)$$

$(1) \Rightarrow \lambda = 1$, $(2) \Rightarrow \mu = 3 - \lambda = 2$.

Putting $\lambda = 1$, $\mu = 2$ into $(3) \Rightarrow$ LHS $= 1$, RHS $= -3 + 4 = 1 =$ LHS.

$\therefore$ lines of action do intersect

Putting $\lambda = 1$ into $\mathbf{r}_1 \Rightarrow \mathbf{r}_T = \begin{pmatrix} 1 \\ -1 \\ 0 \end{pmatrix} + 1 \begin{pmatrix} 2 \\ 1 \\ 1 \end{pmatrix} = \begin{pmatrix} 3 \\ 0 \\ 1 \end{pmatrix}$

$\therefore$ position vector of point of intersection is $3\mathbf{i} + \mathbf{k}$

Let P be a general point on the line of action of the resultant. Let $\mathbf{r} = \overrightarrow{OP}$

$$\mathbf{r} = \mathbf{r}_T + \lambda(\mathbf{F}_1 + \mathbf{F}_2)$$

$$\Rightarrow \mathbf{r} = \begin{pmatrix} 3 \\ 0 \\ 1 \end{pmatrix} + \lambda \begin{pmatrix} 2 \\ 0 \\ 3 \end{pmatrix} = \begin{pmatrix} 3 + 2\lambda \\ 0 \\ 1 + 3\lambda \end{pmatrix}$$

$\therefore$ line of action of resultant is given by $\mathbf{r} = (3 + 2\lambda)\mathbf{i} + (1 + 3\lambda)\mathbf{k}$

- Exercise 11.1 See page 331

11.2 Work, Energy, Power, Impulse and Motion Under Constant Forces

7 *Work done by a Constant Forces*

Work done by $\mathbf{F} = F \cos\theta \times d$

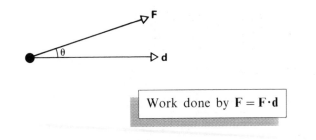

Work done by $\mathbf{F} = \mathbf{F} \cdot \mathbf{d}$

8 *Kinetic Energy*

Kinetic Energy $= \frac{1}{2} m \mathbf{v} \cdot \mathbf{v}$

9   The Work–Energy Principle

Work done = change in KE + change in PE
Provided no sudden change in motion takes place.
If there is no change in PE

$$\sum \mathbf{F} \cdot \mathbf{d} = \tfrac{1}{2}m(\mathbf{v} \cdot \mathbf{v} - \mathbf{u} \cdot \mathbf{u})$$

Example 7   A force $\mathbf{F}$, of magnitude 20 N, acts in the direction of $3\mathbf{i} + 4\mathbf{j}$, where $\mathbf{i}$ and $\mathbf{j}$ are unit vectors in a horizontal plane. It is applied to a particle of mass 4 kg which is initially at rest. If the particle undergoes a displacement of $(\mathbf{i} + 2\mathbf{j})$ m, find the work done by the force and the speed of the particle at the end of the displacement.

The magnitude of $3\mathbf{i} + 4\mathbf{j}$ is $\sqrt{3^2 + 4^2} = 5$
Since $\mathbf{F}$ has magnitude 20, $\mathbf{F} = 4(3\mathbf{i} + 4\mathbf{j}) = 12\mathbf{i} + 16\mathbf{j}$

Work done $= \mathbf{F} \cdot \mathbf{d} = \begin{pmatrix} 12 \\ 16 \end{pmatrix} \cdot \begin{pmatrix} 1 \\ 2 \end{pmatrix} = 44\,\text{J}$

No change in PE

$\therefore$ work done = gain in KE
$\Rightarrow$ $44 = \tfrac{1}{2} \times 4 \times v^2$
$\Rightarrow$ $v = \sqrt{22}\,\text{m/s}$

Example 8   A particle moves under an acceleration $(\mathbf{i} - \mathbf{k})\,\text{m/s}^2$. If the particle has mass 2 kg and is initially at rest, find its velocity and kinetic energy after 10 s.

$$\mathbf{a} = \frac{d\mathbf{v}}{dt} = \mathbf{i} - \mathbf{k}$$

Integrating $\mathbf{v} = t\mathbf{i} - t\mathbf{k} + A\mathbf{i} + B\mathbf{j} + C\mathbf{k}$

but, when $r = 0$, particle is at rest, $\therefore\ A = B = C = 0$

$$\therefore\ \mathbf{v} = t\mathbf{i} - t\mathbf{k}$$

Kinetic Energy $= \tfrac{1}{2}m\mathbf{v} \cdot \mathbf{v} = \tfrac{1}{2} \times 2 \times \begin{pmatrix} 10 \\ -10 \end{pmatrix} \cdot \begin{pmatrix} 10 \\ -10 \end{pmatrix} = 200\,\text{J}$

10   Impulse and Momentum

Impulse = change in momentum

$$\mathbf{I} = m\mathbf{v} - m\mathbf{u}$$

11   Power

$$\text{Power} = \text{rate of work} = \mathbf{F} \cdot \mathbf{v}$$

12 *Motion Under Constant Acceleration*

Using the vector quantities $\mathbf{u} \equiv$ initial velocity
$\mathbf{v} \equiv$ final velocity
$\mathbf{a} \equiv$ acceleration
$\mathbf{s} \equiv$ displacement
and the scalar quantity $\qquad t \equiv$ time

> The equations of motion are:
>
> $$\mathbf{v} = \mathbf{u} + \mathbf{a}t \qquad\qquad \mathbf{s} = \mathbf{u}t + \tfrac{1}{2}\mathbf{a}t^2$$
>
> $$v^2 = u^2 + 2\mathbf{a}\cdot\mathbf{s} \qquad\qquad \mathbf{s} = \left(\frac{\mathbf{u}+\mathbf{v}}{2}\right)t$$

Example 9 A particle of mass 5 kg is acted on by a constant force of $(15\mathbf{i} + 20\mathbf{k})$ N. After 20 s its velocity is $(68\mathbf{i} + 5\mathbf{j} + 74\mathbf{k})$ m/s. Find its initial velocity and the rate of work of the force at the start of the motion.

$$\mathbf{F} = m\mathbf{a} \Rightarrow \qquad \begin{pmatrix} 15 \\ 0 \\ 20 \end{pmatrix} = 5\mathbf{s}$$

$$\therefore \qquad \mathbf{a} = \begin{pmatrix} 3 \\ 0 \\ 4 \end{pmatrix} \longleftarrow \boxed{\text{constant acceleration}}$$

Method 1

$$\mathbf{a} = \begin{pmatrix} 3 \\ 0 \\ 4 \end{pmatrix}$$

Integrating $\quad \mathbf{v} = \begin{pmatrix} 3t \\ 0 \\ 4t \end{pmatrix} + \begin{pmatrix} A \\ B \\ C \end{pmatrix}$

when $t = 20$, $\mathbf{v} = \begin{pmatrix} 68 \\ 5 \\ 74 \end{pmatrix} \Rightarrow \begin{pmatrix} 60 \\ 0 \\ 80 \end{pmatrix} + \begin{pmatrix} A \\ B \\ C \end{pmatrix} = \begin{pmatrix} 68 \\ 5 \\ 74 \end{pmatrix}$

$$\Rightarrow \qquad \begin{pmatrix} A \\ B \\ C \end{pmatrix} = \begin{pmatrix} 8 \\ 5 \\ -6 \end{pmatrix}$$

$$\Rightarrow \mathbf{v} = \begin{pmatrix} 3t + 8 \\ 5 \\ 4t - 6 \end{pmatrix}$$

Initial velocity is found by putting $t = 0$.

$$\therefore \; \mathbf{u} = (8\mathbf{i} + 5\mathbf{j} - 6\mathbf{k}) \text{ m/s.}$$

Method 2

$$\mathbf{a} = \begin{pmatrix} 3 \\ 0 \\ 4 \end{pmatrix} = \text{constant}$$

Using

$$\mathbf{v} = \mathbf{u} + \mathbf{a}t$$

$$\begin{pmatrix} 68 \\ 5 \\ 74 \end{pmatrix} = \mathbf{u} + 20 \begin{pmatrix} 3 \\ 0 \\ 4 \end{pmatrix}$$

$$\Rightarrow \quad \mathbf{u} = \begin{pmatrix} 68 \\ 5 \\ 74 \end{pmatrix} - \begin{pmatrix} 60 \\ 0 \\ 80 \end{pmatrix} = \begin{pmatrix} 8 \\ 5 \\ -6 \end{pmatrix}$$

$$\therefore \quad \underline{\mathbf{u} = (8\mathbf{i} + 5\mathbf{j} - 6\mathbf{k})\,\text{m/s}.}$$

Initial rate of work $= \mathbf{F} \cdot \mathbf{u} = \begin{pmatrix} 15 \\ 0 \\ 20 \end{pmatrix} \cdot \begin{pmatrix} 8 \\ 5 \\ -6 \end{pmatrix} = \underline{0}$

- Exercise 11.2   See page 332

11.3 Vector Analysis of Paths of Bodies Moving with Constant Velocity

The Vector Path of a Body

If a body is at a point A, with position vector **a**, when $t = 0$, and it is moving with constant velocity **v**, then its position vector **r**, at time t, is given by

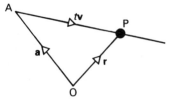

$$\boxed{\mathbf{r} = \mathbf{a} + t\mathbf{v}}$$

Collision

If the path of P_1 is given by $\mathbf{r}_1 = \mathbf{a}_1 + t\mathbf{v}_1$
and the path of P_2 is given by $\mathbf{r}_2 = \mathbf{a}_2 + t\mathbf{v}_2$
P_1 and P_2 will collide if there is a value of t for which $\mathbf{r}_1 = \mathbf{r}_2$.

Distance Apart

The distance between P_1 and P_2
$$= |\overrightarrow{P_1P_2}| = |\mathbf{r}_2 - \mathbf{r}_1|$$

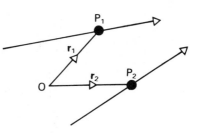

Example 10 A body is moving with a constant speed of 9 m/s in a direction $\mathbf{i} + 2\mathbf{j} - 2\mathbf{k}$. Initially, the body is at a point with position vector $5\mathbf{i} + 2\mathbf{j}$. Find the position vector of the body after: (a) t seconds   (b) 5 seconds.

The velocity $\mathbf{v}$ is in the direction $\begin{pmatrix} 1 \\ 2 \\ -3 \end{pmatrix}$

$$\Rightarrow \hat{\mathbf{v}} = \frac{1}{3} \begin{pmatrix} 1 \\ 2 \\ -2 \end{pmatrix}$$

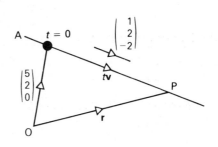

But, since $|\mathbf{v}| = 9$

$$\mathbf{v} = |\mathbf{v}|\hat{\mathbf{v}} = \begin{pmatrix} 3 \\ 6 \\ -6 \end{pmatrix}$$

At time t, $\overrightarrow{AP} = $ displacement vector $t\mathbf{v}$

$$\Rightarrow \overrightarrow{AP} = t \begin{pmatrix} 3 \\ 6 \\ -6 \end{pmatrix}$$

(a) $\mathbf{r} = \begin{pmatrix} 5 \\ 2 \\ 0 \end{pmatrix} + t \begin{pmatrix} 3 \\ 6 \\ 6 \end{pmatrix}$ or $\underline{\mathbf{r} = [5\mathbf{i} + 2\mathbf{j} + t(3\mathbf{i} + 6\mathbf{j} - 6\mathbf{k})]\, \text{m}.}$

(b) when $t = 5$, $\mathbf{r} = \begin{pmatrix} 5 \\ 2 \\ 0 \end{pmatrix} + \begin{pmatrix} 15 \\ 30 \\ -30 \end{pmatrix} \Rightarrow \underline{\mathbf{r} = (20\mathbf{i} + 32\mathbf{j} - 30\mathbf{k})\,\text{m}.}$

Example 11 When $t = 0$, two bodies, A and B, have position vectors $9\mathbf{i} + 15\mathbf{j} + 2\mathbf{k}$ and $3\mathbf{j} + 14\mathbf{k}$ respectively. Subsequently, A moves with constant velocity $\mathbf{i} - 3\mathbf{j} + 2\mathbf{k}$ and B with constant velocity $4\mathbf{i} + \mathbf{j} - 2\mathbf{k}$. Show that A and B will collide, find the time of collision and find the position vector of the point of impact.

$$\mathbf{r}_A = \begin{pmatrix} 9 \\ 15 \\ 2 \end{pmatrix} + t \begin{pmatrix} 1 \\ -3 \\ 2 \end{pmatrix} \quad \text{and} \quad \mathbf{r}_B = \begin{pmatrix} 0 \\ 3 \\ 14 \end{pmatrix} + t \begin{pmatrix} 4 \\ 1 \\ -2 \end{pmatrix}$$

If A and B collide then $\mathbf{r}_A = \mathbf{r}_B$ for one particular value of t.

$$\mathbf{r}_A = \mathbf{r}_B \quad \Rightarrow \begin{pmatrix} 9 \\ 15 \\ 2 \end{pmatrix} + t \begin{pmatrix} 1 \\ -3 \\ 2 \end{pmatrix} = \begin{pmatrix} 0 \\ 3 \\ 14 \end{pmatrix} + t \begin{pmatrix} 4 \\ 1 \\ -2 \end{pmatrix}$$

$$\Rightarrow \begin{pmatrix} 9 \\ 15 \\ 2 \end{pmatrix} - \begin{pmatrix} 0 \\ 3 \\ 14 \end{pmatrix} = t \begin{pmatrix} 4 \\ 1 \\ -2 \end{pmatrix} - t \begin{pmatrix} 1 \\ -3 \\ 2 \end{pmatrix}$$

$$\Rightarrow \begin{pmatrix} 9 \\ 12 \\ -12 \end{pmatrix} = t \begin{pmatrix} 3 \\ 4 \\ -4 \end{pmatrix}$$

and this holds for $t = 3$. $\therefore$ A and B collide when $r = 3$

when $t = 3$, $\mathbf{r}_A = \begin{pmatrix} 9 \\ 15 \\ 2 \end{pmatrix} + 3 \begin{pmatrix} 1 \\ -3 \\ 2 \end{pmatrix} = \begin{pmatrix} 12 \\ 6 \\ 8 \end{pmatrix}$

$\therefore$ position vector of point if impact $= 12\mathbf{i} + 6\mathbf{j} + 8\mathbf{k}$

Example 12. If, in the previous example, B had started moving from its initial position two seconds after A, show that A and B would not collide and find the shortest distance between them.

Taking $t = 0$ as the time when A starts to move, then at time t, B has been moving for $(t - 2)$ seconds,

$$\Rightarrow \mathbf{r}_A = \begin{pmatrix} 9 \\ 15 \\ 2 \end{pmatrix} + t \begin{pmatrix} 1 \\ -3 \\ 2 \end{pmatrix} \quad \text{and} \quad \mathbf{r}_B = \begin{pmatrix} 0 \\ 3 \\ 14 \end{pmatrix} + (t - 2) \begin{pmatrix} 4 \\ 1 \\ -2 \end{pmatrix}$$

Let $\mathbf{r}_A = \mathbf{r}_B \Rightarrow$

$$\begin{pmatrix} 9 \\ 15 \\ 2 \end{pmatrix} + t \begin{pmatrix} 1 \\ -3 \\ 2 \end{pmatrix} = \begin{pmatrix} 0 \\ 3 \\ 14 \end{pmatrix} + (t - 2) \begin{pmatrix} 4 \\ 1 \\ -2 \end{pmatrix}$$

$$\Rightarrow \begin{pmatrix} 9 \\ 15 \\ 2 \end{pmatrix} - \begin{pmatrix} 0 \\ 3 \\ 14 \end{pmatrix} + 2 \begin{pmatrix} 4 \\ 1 \\ -2 \end{pmatrix} = t \begin{pmatrix} 4 \\ 1 \\ -2 \end{pmatrix} - 4 \begin{pmatrix} 1 \\ -3 \\ 2 \end{pmatrix}$$

$$\Rightarrow \begin{pmatrix} 17 \\ 14 \\ -16 \end{pmatrix} = t \begin{pmatrix} 3 \\ 4 \\ -4 \end{pmatrix}$$

Since there is no value of t that satisfies this equation,
A and B do not collide   QED

The distance between them at time $t = |\mathbf{r}_A - \mathbf{r}_B|$

$$= \begin{pmatrix} 17 - 3t \\ 14 - 4t \\ -16 + 4t \end{pmatrix}$$

If the distance between them is D,

$$\text{then } D^2 = (17 - 3t)^2 + (14 - 4t)^2 + (4t - 16)^2$$
$$\Rightarrow D^2 = 41t^2 - 342t + 741$$

To find the least distance between A and B

Method 1 (Completing the square)

$$D^2 = 41\left[t^2 - \frac{342t}{41} + \frac{741}{41} \right]$$

$$\Rightarrow D^2 = 41\left[\left(t - \frac{171}{41} \right)^2 - \left(\frac{171}{41} \right)^2 + \frac{741}{41} \right]$$

D^2 is least when $t = \dfrac{171}{41}$ and then

$$(D_{\text{MIN}})^2 = 41\left[\frac{741}{41} - \left(\frac{171}{41} \right)^2 \right] = \frac{1140}{41}$$

$$\Rightarrow \underline{D_{\text{MIN}} = 5.27}$$

Method 2 (Using calculus)

D is least when D^2 is least that is, when $\dfrac{\mathrm{d}}{\mathrm{d}t}(D^2) = 0$

$$\frac{\mathrm{d}}{\mathrm{d}t}(D^2) = \frac{\mathrm{d}}{\mathrm{d}t}(41t^2 - 342t + 741) = 82t - 342.$$

$$\frac{\mathrm{d}}{\mathrm{d}t}(D^2) = 0 \Rightarrow t = \frac{342}{82} = \frac{171}{41}$$

$$\frac{\mathrm{d}^2}{\mathrm{d}t^2}(D^2) = 82 > 0 \quad \text{for all } t.$$

$\therefore$ when $t = \dfrac{171}{41}$, D^2 is a minimum.

$$\Rightarrow (D_{\text{MIN}})^2 = 41\left(\frac{171}{41} \right)^2 - 342\left(\frac{171}{42} \right) + 741$$

$$\Rightarrow \underline{D_{\text{MIN}} \approx 5.27}$$

Example 13. Two spheres P and Q have radii 2 and 3 units respectively. At a particular instant, the centre of P is at the point with position vector $3\mathbf{i} + 5\mathbf{j} + 5\mathbf{k}$ and the centre of Q is at the point with position vector $\mathbf{j} + \mathbf{k}$. P is moving with a constant velocity $2\mathbf{i} - \mathbf{j} + 3\mathbf{k}$ and Q with constant velocity $\mathbf{i} + \mathbf{k}$. Show that the spheres do not collide.

Assume that the spheres do collide. Then their centres will be 5 units apart at the moment of impact.

$$\mathbf{r}_P = \begin{pmatrix} 3 \\ 5 \\ 5 \end{pmatrix} + t \begin{pmatrix} 2 \\ -1 \\ 3 \end{pmatrix} \quad \text{and} \quad \mathbf{r}_Q = \begin{pmatrix} 0 \\ 1 \\ 1 \end{pmatrix} + t \begin{pmatrix} 1 \\ 0 \\ 1 \end{pmatrix}$$

$$\mathbf{r}_P - \mathbf{r}_Q = \begin{pmatrix} 3 \\ 4 \\ 4 \end{pmatrix} + t \begin{pmatrix} 1 \\ -1 \\ 2 \end{pmatrix}$$

$$D^2 = |\mathbf{r}_P - \mathbf{r}_Q|^2 = (3+t)^2 + (4-t)^2 + (4+2t)^2$$
$$= 6t^2 + 14t + 41$$

Let $D = 5 \quad \Rightarrow 6t^2 + 14t + 41 = 25$
$$\Rightarrow 6t^2 + 14t + 16 = 0$$

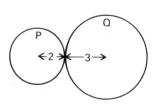

Since "$b^2 - 4ac$" $= 14^2 - 4 \times 6 \times 16 = -188 < 0$ this equation has no real solution.

∴ <u>the spheres do not collide</u>   QED

● Exercise 11.3   See page 334

11.4 Resultant Velocity: Relative Velocity I

Resultant Velocity

Velocity is a vector. The magnitude of velocity is speed

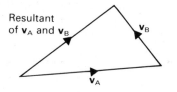

Example 14   A girl can swim in still water at 3 km/h. A river, which is 100 m wide, flows at 1 km/h. Find how long it takes her to swim across the river in each of the following cases:
(a)   the actual route she takes is directly across the river.
(b)   she tries to swim directly across the river but finally gets out downstream from where she started.

(a)

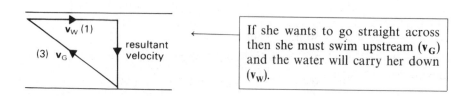

If she wants to go straight across then she must swim upstream ($\mathbf{v_G}$) and the water will carry her down ($\mathbf{v_W}$).

(Resultant velocity)$^2 = |\mathbf{v_G}|^2 - |\mathbf{v_W}|^2 = 3^2 - 1^2 = 8$

$\Rightarrow$ Resultant velocity $= \sqrt{8}$ km/h.

$$\text{time} = \frac{\text{distance}}{\text{velocity}} = \frac{0.1}{\sqrt{8}}\,\text{h}$$

$\boxed{100\,\text{m} = 0.1\,\text{km}}$

$\Rightarrow$ time $= 2.1$ minutes

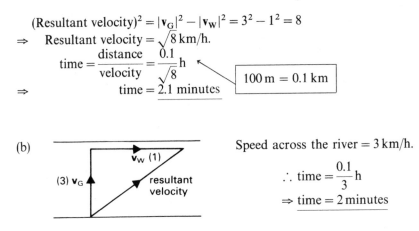

(b)

$\mathbf{v_W}$ (1)

(3) $\mathbf{v_G}$

resultant velocity

Speed across the river $= 3$ km/h.

$$\therefore \text{time} = \frac{0.1}{3}\,\text{h}$$

$\Rightarrow$ time $= 2$ minutes

Relative Velocity

When a moving object B is viewed from another moving object A, the apparent velocity of B from A is called the **velocity of B relative to A**. It is the vector difference in their velocities.

$\boxed{\text{Velocity of B relative to A} = {}_B\mathbf{v}_A = \mathbf{v}_B - \mathbf{v}_A}$

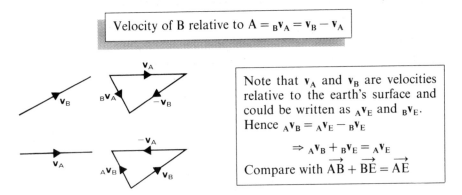

Note that $\mathbf{v}_A$ and $\mathbf{v}_B$ are velocities relative to the earth's surface and could be written as ${}_A\mathbf{v}_E$ and ${}_B\mathbf{v}_E$. Hence ${}_A\mathbf{v}_B = {}_A\mathbf{v}_E - {}_B\mathbf{v}_E$

$\Rightarrow {}_A\mathbf{v}_B + {}_B\mathbf{v}_E = {}_A\mathbf{v}_E$

Compare with $\overrightarrow{AB} + \overrightarrow{BE} = \overrightarrow{AE}$

Example 15   A car is driving due east on a straight road at 50 km/h. The driver watches a train which is travelling due south at 120 km/h. What is the apparent speed and direction of the train?

Given:

We need to find ${}_T\mathbf{v}_C = {}_T\mathbf{v}_E - {}_C\mathbf{v}_E$

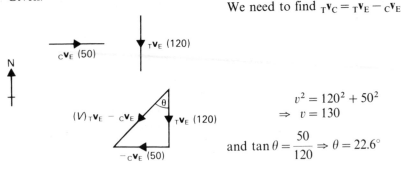

${}_C\mathbf{v}_E$ (50)

${}_T\mathbf{v}_E$ (120)

N

$(V)\,{}_T\mathbf{v}_E - {}_C\mathbf{v}_E$

θ

${}_T\mathbf{v}_E$ (120)

$-{}_C\mathbf{v}_E$ (50)

$v^2 = 120^2 + 50^2$

$\Rightarrow v = 130$

and $\tan\theta = \dfrac{50}{120} \Rightarrow \theta = 22.6°$

∴ apparent speed of the train = 130 km/h

and apparent direction = S22.6° W

Example 16   To a passenger on a train moving northwest at 90 km/h an aeroplane appears to be flying due east at 160 km/h. What is the true direction and speed of the aeroplane?

Given:

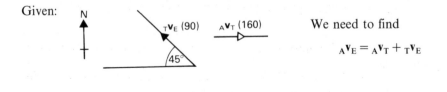

We need to find

$$_A\mathbf{V}_E = {_A}\mathbf{V}_T + {_T}\mathbf{V}_E$$

Method 1

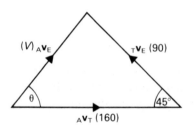

Using the cosine rule:

$$v^2 = 160^2 + 90^2 - 2 \times 160 \times 90 \cos 45°$$
$$\Rightarrow v = 115.5$$

Using the sine rule:

$$\frac{\sin \theta}{90} = \frac{\sin 45°}{115.5}$$

$$\Rightarrow \sin \theta = \frac{90}{115.5} \sin 45°$$

$$\Rightarrow \quad \theta = 33.4°$$

From the diagram θ is obviously acute.

Method 2

Taking components east and north,

	$_T\mathbf{V}_E$	$_A\mathbf{V}_T$	$_A\mathbf{V}_E = {_A}\mathbf{V}_T + {_T}\mathbf{V}_E$
→	$-90 \cos 45°$	160	96.4
↑	$90 \sin 45°$	0	63.6

Third column is sum of first two columns.

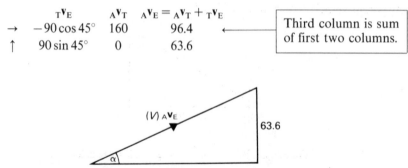

$v^2 = 96.4^2 + 63.6^2 \Rightarrow v = 115.5$

and $\tan \alpha = \dfrac{63.6}{96.4} \Rightarrow \alpha = 33.4$

$90 - \alpha$

for both methods

→ True speed of the aeroplane is 115.5 km/h and its bearing is 056.6°

> Note that great care must be taken when reading information about wind directions.
> eg (i)   a wind blowing S50°W and a wind blowing *from* N50°E are in fact blowing from the same direction.
> (ii) a north east wind blows south west

• Exercise 11.4   See page 335

11.5 Relative Velocity II

Example 17   To a yachtsman sailing due east at 18 km/h a steady breeze appears to be blowing N 30°E. When his speed drops to 15 km/h the breeze appears to blow N60°E. Find the wind speed and its true direction.

Let $_w\mathbf{V}_E = \begin{pmatrix} u \\ v \end{pmatrix}$

Yacht sailing at 18 km/h

	$_w\mathbf{V}_E$	$_y\mathbf{V}_E$	$_w\mathbf{V}_Y = {_w\mathbf{V}_E} - {_y\mathbf{V}_E}$
→	u	18	$u - 18$
↑	v	0	v

3rd column is the first column minus the second column.

Apparent velocity of wind is at N30°E

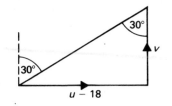

Hence $\tan 30° = \dfrac{u - 18}{v} \Rightarrow u - 18 = v \tan 30°$ (1)

Yacht sailing at 15 km/h

$$\begin{array}{ccc} {}_W\mathbf{V}_E & {}_Y\mathbf{V}_E & {}_W\mathbf{V}_Y = {}_W\mathbf{V}_E - {}_Y\mathbf{V}_E \\ \rightarrow \quad u & 15 & u-15 \\ \uparrow \quad v & 0 & v \end{array}$$

Apparent velocity of wind is at N60°E

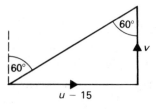

$$\text{Hence } \tan 60° = \frac{u - 15}{v} \Rightarrow u - 15 = v \tan 60° \tag{2}$$

(1) and (2) $\Rightarrow \quad v \tan 30° + 18 = v \tan 60° + 15$

$\Rightarrow \quad v[\tan 60° - \tan 30°] = 3$

$\Rightarrow v = \dfrac{3}{\tan 60° - \tan 30°} = 2.6$

and (1) $\Rightarrow u = 2.6 \tan 30° + 18 = 19.5$

Velocity of wind =

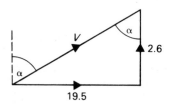

Wind Speed $= v = \sqrt{19.5^2 + 2.6^2} = 19.7 \, \text{km/h}.$

$$\tan \alpha = \frac{19.5}{2.6} \Rightarrow \alpha = 82.4°$$

$\Rightarrow$ Wind direction = N82.4°E

Example 18   On a NATO exercise, two British submarines are shadowing an American battleship. One submarine is travelling at 12 knots on a course N 30° W. The other is travelling at 20 knots on a course S 30° W. To an observer on the first submarine the battleship appears to be travelling N 50° W, whereas to an observer on the second submarine it appears to be travelling due west. What is the velocity of the battleship?

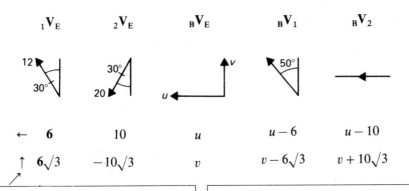

$_1V_E$	$_2V_E$	$_BV_E$	$_BV_1$	$_BV_2$
$\leftarrow$ 6	10	u	$u-6$	$u-10$
$\uparrow$ $6\sqrt{3}$	$-10\sqrt{3}$	v	$v-6\sqrt{3}$	$v+10\sqrt{3}$

Components are chosen $\leftarrow$ and $\uparrow$ rather than $\rightarrow$ and $\uparrow$ because of the predominant directions in the diagrams.	The first three columns are calculated from the diagrams. The fourth column is $_BV_E - {}_1V_E$. The fifth column is $_BV_E - {}_2V_E$.

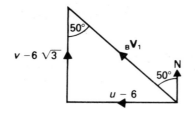

The direction of $_BV_1$ is N 50° W $\Rightarrow$

$$\Rightarrow \tan 50° = \frac{u-6}{v-6\sqrt{3}} \qquad (1)$$

The direction of $_BV_2$ is west

$$\Rightarrow v + 10\sqrt{3} = 0$$
$$\Rightarrow v = -10\sqrt{3} \qquad (2)$$

(1) and (2) $\Rightarrow \tan 50° = \dfrac{u-6}{-16\sqrt{3}}$

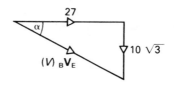

$\Rightarrow \qquad u = 6 - 16\sqrt{3}\tan 50°$

$\Rightarrow \qquad u = -27$

Hence velocity of battleship is given by

$V^2 = 27^2 + (10\sqrt{3})^2$

$\Rightarrow \qquad V = 32.1$ knots

| 1 knot = 1 nautical mile/hour |
| 1 nautical mile $\approx 1\frac{1}{8}$ miles |

$\tan \alpha = \dfrac{10\sqrt{3}}{27} \Rightarrow \alpha = 32.7$

$\therefore$ velocity of battleship is 32.1 knots at S 57.3 E

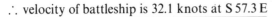

● **Exercise 11.5**   See page 336

11.6 Relative Position

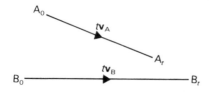

If a body leaves A_0 with constant velocity v_A then, at time t later, it will be at A_t where $\overrightarrow{A_0A_t} = tv_A$ similarly, B will be at B_t where $\overrightarrow{B_0B_t} = tv_B$.

The position of B relative to A at time t is given by $\overrightarrow{A_tB_t}$ where

$$\overrightarrow{A_tB_t} = \overrightarrow{A_tA_0} + \overrightarrow{A_0B_0} + \overrightarrow{B_0B_t}$$
$$\Rightarrow \overrightarrow{A_tB_t} = \overrightarrow{A_0B_0} + t(v_B - v_A)$$
$$\Rightarrow \overrightarrow{A_tB_t} = \overrightarrow{A_0B_0} + t_B v_A$$

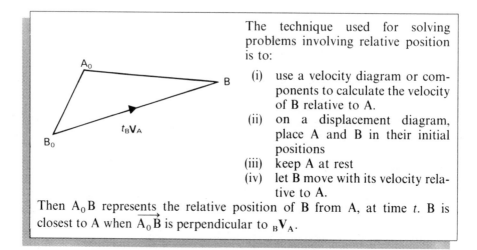

The technique used for solving problems involving relative position is to:

(i)   use a velocity diagram or components to calculate the velocity of B relative to A.
(ii) on a displacement diagram, place A and B in their initial positions
(iii) keep A at rest
(iv) let B move with its velocity relative to A.

Then A_0B represents the relative position of B from A, at time t. B is closest to A when $\overrightarrow{A_0B}$ is perpendicular to $_Bv_A$.

Example 19   At noon a ship A is 2 km north of another ship B. Their velocities are 12 km/h due west and 15 km/h NW, respectively. At what time will the ships be closest together and what will the distance between them be at that time?

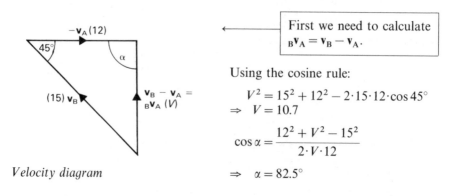

Velocity diagram

First we need to calculate $_Bv_A = v_B - v_A$.

Using the cosine rule:

$$V^2 = 15^2 + 12^2 - 2 \cdot 15 \cdot 12 \cdot \cos 45°$$
$$\Rightarrow V = 10.7$$

$$\cos \alpha = \frac{12^2 + V^2 - 15^2}{2 \cdot V \cdot 12}$$

$$\Rightarrow \alpha = 82.5°$$

Note that the sine rule would have given two possible values, 82.5° and 97.5° and that it is not obvious from the diagram to deduce for certain whether κ is acute or obtuse.

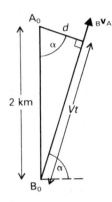

Displacement diagram

A_0, B_0 are the initial positions of A and B. A is kept at rest.
B moves with its velocity relative to A. Its relative distance travelled in time t is Vt where $V =$ relative speed.

The shortest distance between A and B is d where

$$d = 2\cos\alpha = 2\cos 82.5° = 0.261 \text{ km}.$$

If t is the time taken to reach this position, then

$$Vt = 2\sin\alpha$$

$$\Rightarrow \quad t = \frac{2\sin\alpha}{V}$$

$$= \frac{2\sin 82.5}{10.7}$$

$$= 0.185\,\text{h} \approx 11\,\text{min}.$$

∴ the ships will be closest together at 12.11 and at that time they will be 261 m apart.

● Exercise 11.6   See page 336

11.7 Choice of Course for Closest Approach

If the velocity and displacement diagrams are put together then we get:

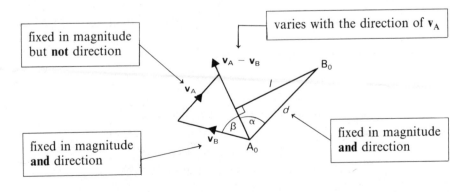

fixed in magnitude but **not** direction

varies with the direction of v_A

fixed in magnitude **and** direction

fixed in magnitude **and** direction

If the speed of A is fixed (maximum speed for closest approach) then, as the direction of v_A varies, so the direction of $v_A - v_B$ varies and hence the length of l, the shortest distance, varies. For l to be a minimum, the direction of v_A must be perpendicular to $v_A - v_B$.

> In order to approach as close as possible to B, the direction of motion of A must be perpendicular to the relative path.

Example 20   A ship is travelling due east at 50 km/h. A launch, whose top speed is 30 km/h sets off to try to intercept the ship. At the moment when the launch sets off the ship is 400 m north of the launch. How close can the launch set to the ship? What is the direction that the launch takes?

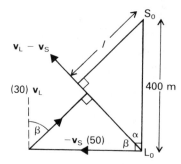

$$\sin \beta = \frac{30}{50} = \frac{3}{5}$$

$$l = 400 \sin \alpha = 400 \cos \beta$$

$$\Rightarrow \quad l = 400 \times \tfrac{4}{5} = 320$$

$$\therefore \text{ closest approach is } 320 \text{ m.}$$

The launch travels in a direction N $\beta°$ E where $\beta = \sin^{-1} \tfrac{3}{5}$.

- Exercise 11.7   See page 337

11.8 Interception

$$\overrightarrow{A_t B_t} = \overrightarrow{A_0 B_0} + t(v_B - v_A) \quad \boxed{\text{see } 11.3}$$

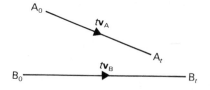

For interception, $\overrightarrow{A_t B_t} = 0$

$$\Rightarrow t(v_A - v_B) = \overrightarrow{A_0 B_0}$$

$$\Rightarrow \ _A v_B \text{ is parallel to } \overrightarrow{A_0 B_0}$$

> A will intercept B if $_A v_B$ is parallel to $\overrightarrow{A_0 B_0}$.
>
> $$\text{Time to interception} = \frac{\text{initial distance apart}}{\text{relative speed}}.$$

Example 21   A ship's engines fail and she drifts due east at a rate of 3 knots. At 11 am, a sea-tug sets off from a port 40 nautical miles south west of the ship. On what bearing should the tug travel at a speed of 10 knots and at what time will it intercept the ship?

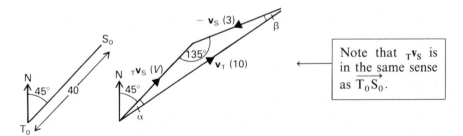

Note that $_T\mathbf{v}_S$ is in the same sense as $\overrightarrow{T_0S_0}$.

Sine rule:

$$\frac{\sin \alpha}{3} = \frac{\sin 135°}{10}$$

$$\Rightarrow \quad \alpha = 12.2°$$

∴ tug will travel on a bearing of 057.2°

$\beta = 45 - 12.2° = 32.8°$

Sine Rule:

$$\frac{V}{\sin 32.8°} = \frac{10}{\sin 135°}$$

$$\Rightarrow \quad V = 7.66 \text{ knots}$$

∴ time to interception $= \dfrac{40}{7.66}\text{h} = 5.22\,\text{h} = 5\,\text{h}\,13\,\text{min.}$

∴ interception will occur around 16.13.

Example 22   An oil tanker is travelling on a bearing of 210° at 12 knots. A ferry boat is travelling on a bearing of 315°. When the ferry is 1 nautical mile due south of the tanker, it is realised that the boats are on collision course. What is the speed of the ferry? If neither ship changes course, after what time will collision take place?

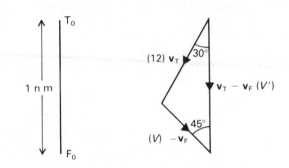

Sine Rule:

$$\frac{V}{\sin 30°} = \frac{12}{\sin 45°}$$

$$\Rightarrow \quad V = 12 \times \tfrac{1}{2} \times \sqrt{2} = 6\sqrt{2}.$$

∴ speed of ferry ≈ 8.5 knots.

Sine Rule:

$$\frac{V'}{\sin 105°} = \frac{12}{\sin 45°}$$

$$\Rightarrow \quad V' = 16.4 \text{ knots}$$

Time to collision $= \dfrac{1}{16.4}$ hours $= 3.7$ minutes.

∴ collision would be after 3.7 minutes.

- Exercise 11.8   See page 338

11.9 Vector Analysis of Relative Paths and Velocities

A_0 and B_0 are the initial positions of A and B
v_A and v_B are the velocities of A and B.
The velocity of A relative to $B = v_A - v_B$

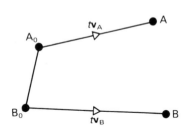

The position vector of A relative to $B = \overrightarrow{BA}$
$$= \overrightarrow{BB_0} + \overrightarrow{B_0A_0} + \overrightarrow{A_0A}$$
$$= \overrightarrow{B_0A_0} + t(v_A - v_B)$$

Example 23   At time $t = 0$, the position vectors of two particles A and B
are $i + j + k$ and $2j + k$ respectively. The particles have constant velocity vectors
$3i + j + k$ and $i - 2j + k$ respectively.
Find: (a)   the velocity of A relative to B
      (b)   the position vector of A relative to B at time t.

$$v_A = \begin{pmatrix} 3 \\ 1 \\ 1 \end{pmatrix} \quad \text{and} \quad v_B = \begin{pmatrix} 1 \\ -2 \\ 1 \end{pmatrix}$$

$$\overrightarrow{A_0B_0} = \overrightarrow{OB_0} - \overrightarrow{OA_0}$$

$$= \begin{pmatrix} 2 \\ 0 \\ 1 \end{pmatrix} - \begin{pmatrix} 1 \\ 1 \\ 1 \end{pmatrix} - \begin{pmatrix} 1 \\ -1 \\ 0 \end{pmatrix}$$

(a)   velocity of A relative to $B = \mathbf{v}_A - \mathbf{v}_B = \begin{pmatrix} 3 \\ 1 \\ 1 \end{pmatrix} - \begin{pmatrix} 1 \\ -2 \\ 1 \end{pmatrix} = \begin{pmatrix} 2 \\ 3 \\ 0 \end{pmatrix}$

$\therefore$ velocity of A relative to $B = 2\mathbf{i} + 3\mathbf{j}$.

(b)   position vector of A relative to $B = \overrightarrow{BA}$

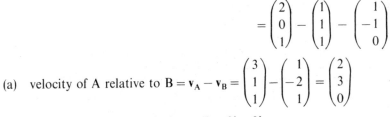

$$= \overrightarrow{B_0 A_0} + t(\mathbf{v}_A - \mathbf{v}_B)$$

$$= \begin{pmatrix} 1 \\ -1 \\ 0 \end{pmatrix} + t \begin{pmatrix} 2 \\ 3 \\ 0 \end{pmatrix}$$

$$= (2t - 1)\mathbf{i} + (3t + 1)\mathbf{j}$$

- Exercise 11.9   See page 339

11.10 + Revision of Unit: A Level Questions

- Miscellaneous Exercise 11B   See page 340

Unit 11 Exercises

Exercise 11.1

1   If the position vector of a particle at time t seconds is given by $(3t\mathbf{i} + 2t^2\mathbf{j})$ m, find its speed at $t = 1$.

2   A particle of mass m is at the point with position vector $(t^3 + t)\mathbf{i} + 2t\mathbf{i} + (5t^3 + 1)\mathbf{k}$ at time t. Find its vector acceleration at time t and the force required to produce this acceleration.

3   A body of mass 2 units is acted upon by a force which at time t is given by $(8t + 4)\mathbf{i} + 4t\mathbf{j}$. Initially the particle is at point $(0, 1)$ and moving with velocity $3\mathbf{i} + \mathbf{j}$. Find the position vector of the particle at time t.

4   A particle at time t seconds has position vector $(3t\mathbf{i} + (t^2 + 1)\mathbf{j})$ m. Find its speed and direction of motion when $t = 2$.

5   If the force $\mathbf{F} = 3\mathbf{i} + 2\mathbf{j}$ acts at the point $(2, -1)$, find the vector equation of the line of action.

6   A particle has position vector $3t^2\mathbf{i} - 4t\mathbf{j}$ at time t. Find the angle between its position vector and its direction of motion at time $t = 2$.

7   A force $\mathbf{i} - 2\mathbf{k}$ passes through the point with position vector $\mathbf{j} - \mathbf{k}$. A second force $4\mathbf{i} + 2\mathbf{j} + 3\mathbf{k}$ passes through the point with position vector $7\mathbf{i} + 3\mathbf{j} - 4\mathbf{k}$. Show that their lines of action intersect and find the point of intersection. Find the vector equation of the line of action of the resultant.

8   Show that the vectors $\mathbf{a} = 3\mathbf{i} + 4\mathbf{j}$ and $\mathbf{b} = 4\mathbf{i} - 3\mathbf{j}$ are perpendicular to each other.
    A particle moves so that its position vector at time t is given by
    $$\mathbf{r} = 2e^{(t-1)}\mathbf{i} + 3t^2\mathbf{j}$$
    Find the components of its velocity, when $t = 1$, in the directions of $\mathbf{a}$ and $\mathbf{b}$.

9   At time t seconds, where $t > 0$, the position vector of particle A with respect to a fixed origin O is $(\mathbf{i}t - \mathbf{j}\cos t + \mathbf{k}\sin t)\,\mathrm{m}$ and the position vector of particle B with respect to O is $(\mathbf{i} + \mathbf{j}e^{-t} - \mathbf{k}e^{-t})\,\mathrm{m}$. Find the least value of t for which the velocities of A and B are perpendicular and show that, at this instant, the accelerations of the particles are parallel.   [L]

10 A particle of mass m moves in such a way that at a time t its position vector is $t^2\mathbf{i} + (2t^2 - 5t)\mathbf{j} + (t^2 + 5t)\mathbf{k}$. Find the force required to maintain this motion.
    This force is the resultant of a force $m\,(14\mathbf{i} - 2\mathbf{k})$ and another force $\mathbf{R}$. Find $\mathbf{R}$ and show that it is always perpendicular to the direction of motion of the particle.   [J]

11 The position vector $\mathbf{r}$ of a moving particle at time t after the start of the motion is given by
    $$\mathbf{r} = (5 + 20\,t)\mathbf{i} + (95 + 10t - 5t^2)\mathbf{j}$$
    Find the initial velocity of the particle.
    At time $t = T$ the particle is moving at right angles to its initial direction of motion. Find the value of T and the distance of the particle from its initial position at this time.   [J]

12 The position vector $\mathbf{r}$ of a particle at time t is $\mathbf{r} = 2t^2\mathbf{i} + (t^2 - 4t)\mathbf{j} + (3t - 5)\mathbf{k}$. Find the velocity and acceleration of the particle at time t. Show that when $t = \frac{2}{5}$ the velocity and acceleration are perpendicular to each other. The velocity and acceleration are resolved into components along and perpendicular to the vector $\mathbf{i} - 3\mathbf{j} + 2\mathbf{k}$. Find the velocity and acceleration components parallel to this vector when $t = \frac{2}{5}$.   [J]

Exercise 11.2

1   Find the work done when a force $\mathbf{F} = 4\mathbf{i} + 3\mathbf{j} + 5\mathbf{k}$ moves a particle through a displacement $\mathbf{i} - 2\mathbf{j} + \mathbf{k}$.

2   Three forces of magnitudes 3 N, 18 N, 21 N act in the directions $2\mathbf{i} + 2\mathbf{j} - \mathbf{k}$, $7\mathbf{i} - 4\mathbf{j} - 4\mathbf{k}$, $2\mathbf{i} - 3\mathbf{j} + 6\mathbf{k}$, respectively. When acting together on a particle

they displace it through $(31\mathbf{i} + 22\mathbf{j} + 10\mathbf{k})$ m. Find the work done by each force. Show that the total work done by the three forces is equal to the work done by the resultant force.

3 A particle of mass 10 kg is acted on by a constant force of $(10\mathbf{i} + 20\mathbf{j} - 50\mathbf{k})$ N. Find the velocity and kinetic energy of the particle after it has been moving for 30 s, given that it starts from rest. Find the rate of work of the force at the end of that time.

4 A particle of mass 5 kg is moved from rest under the action of two forces: $\mathbf{F}_1 = (2\mathbf{i} + 5\mathbf{j})$ N and $\mathbf{F}_2 = (\mathbf{i} + 3\mathbf{j})$ N. The particle moves from the point $(1, 1)$ to $(7, 10)$. Find the total work done by the forces and deduce the speed of the particle when it reaches the second point, if the unit of time is a second and the unit of displacement is a metre, and $\mathbf{i}$ and $\mathbf{j}$ are unit vectors in a horizontal plane.

5 A particle is acted upon by a number of forces when it undergoes a displacement of $(3\mathbf{i} - 4\mathbf{j})$ m. Two of the forces are $(9\mathbf{i} + 12\mathbf{j})$ N and $(12\mathbf{i} + 9\mathbf{j})$ N. Find the work done by each of these two forces. What can you deduce about these forces from your answers?

6 A particle of mass 2 kg is moving with velocity $(4\mathbf{i} + 5\mathbf{j})$ m/s when it is struck a blow which changes its velocity to $(-3\mathbf{i} + 7\mathbf{j})$ m/s. Find the magnitude and direction of the impulse of the blow.

7 A particle of mass 500 grams is moving with a velocity of $(3\mathbf{i} + 3\mathbf{j} - \mathbf{k})$ m/s when it is given a sharp blow with an impulse of $(2\mathbf{i} - 2\mathbf{j})$ Ns. Find the velocity of the particle after the impulse.

8 Initially a particle has velocity $(3\mathbf{i} + 2\mathbf{j} + 9\mathbf{k})$ m/s and after 10 seconds of constant acceleration it has velocity $(23\mathbf{i} - 8\mathbf{j} + 39\mathbf{k})$ m/s. Find its displacement in that time and its acceleration.

9 A particle starts from the point $(0, 0, 1)$ with velocity $(3\mathbf{i} - 2\mathbf{j} + 2\mathbf{k})$ m/s and moves with a constant acceleration $(2\mathbf{i} + 2\mathbf{i} - 4\mathbf{k})$ m/s² for 5 seconds. Find the coordinates of its position at the end of the motion.

10 The only forces acting on a particle of mass 4 kg are the two constant forces $\mathbf{F}_1$ and $\mathbf{F}_2$. Initially the particle is at rest at the point $(0, 1, 0)$. Two seconds later it is at the point $(4, 7, 3)$. If $\mathbf{F}_1 = (\mathbf{i} + 2\mathbf{j} - \mathbf{k})$ N and the unit of displacement is a metre, find:
(a) the work done by $\mathbf{F}_1$ during the motion
(b) the acceleration of the particle
(c) the force $\mathbf{F}_2$
(d) the total work done.

*11 A particle of mass m starts to move when at the point with position vector $(2\mathbf{i} + \mathbf{j})$ m. It moves with constant acceleration $(3\mathbf{i} - \mathbf{j} + \mathbf{k})$ m/s² until it reaches the point with position vector $(3\mathbf{i} + 4\mathbf{j} + \mathbf{k})$ m. Find the change in kinetic energy.

Exercise 11.3

1 A body A is moving with a constant speed of $14\,\text{m/s}$ in the direction $6\mathbf{i} - 3\mathbf{j} + 2\mathbf{k}$. Find the velocity of A.
Initially, A is at a point with position vector $\mathbf{i} + \mathbf{j} + 2\mathbf{k}$. Find the position vector of the body after:
(a)　t seconds　　(b)　4 seconds.

2 A particle P has velocity $3\mathbf{i} + 2\mathbf{j} - \mathbf{k}$. At time $t = 3\,\text{s}$, P has position vector $10\mathbf{i} + 9\mathbf{i}$. Find:
(a)　the position vector of P at time $t = 0$
(b)　the position vector of P at time t.

3 A particle, travelling with constant velocity, takes 3 seconds to travel from a point with position vector $4\mathbf{i} + \mathbf{j}$ to a point with position vector $7\mathbf{i} + 7\mathbf{j} - 3\mathbf{k}$. Find its position vector 2 seconds after passing through the second point.

4 The following particles are moving with constant velocity. Determine in each case whether they collide. If they collide, find the position vector of the point of intersection. If they do not collide, find the least distance between them.
(a)　Initially A and B are at points with position vectors $\mathbf{i} + 2\mathbf{j} + 4\mathbf{k}$ and $11\mathbf{i} + 6\mathbf{j} + 14\mathbf{k}$ and have velocities $5\mathbf{i} + 3\mathbf{j} + 7\mathbf{k}$ and $\mathbf{j} + 2\mathbf{k}$ respectively.
(b)　Initially C and D are at points with position vectors $\mathbf{i}$ and $\mathbf{j} + \mathbf{k}$ and have velocities $2\mathbf{i} + 3\mathbf{j} + 2\mathbf{k}$ and $\mathbf{i} + \mathbf{j} + \mathbf{k}$ respectively.
(c)　E leaves the point with position vector $(2\mathbf{i} + 3\mathbf{j})\,\text{km}$ and travels with constant velocity $(4\mathbf{i} + \mathbf{j})\,\text{km/h}$. One hour later, F leaves the point with position vector $(\mathbf{i} - \mathbf{j})\,\text{km}$ and travels with constant velocity $(4\mathbf{i} + 8\mathbf{j})\,\text{km/h}$.

5 A sphere X, of radius 2 units, starts to move from the origin with constant velocity $5\mathbf{j}$. When X starts to move, a second sphere Y, of radius 3 units is passing through the point with position vector $11\mathbf{i} + 4\mathbf{j}$ with a constant velocity $-\mathbf{i} + 4\mathbf{j}$. Show that the spheres collide. Find the time of collision and the position vectors of their centres at the moment of impact.

6 Three space ships are each moving with constant velocity: the Klingon ship is travelling at $(\mathbf{i} - 2\mathbf{j} + 3\mathbf{k})$ metrons/second; the Tharg ship at $(2\mathbf{i} + \mathbf{k})$ metrons/second and the Enterprise at $(-3\mathbf{j} + \mathbf{k})$ metrons/second. Initially the Klingon ship is at the point with position vector $(\mathbf{i} + \mathbf{j} + 3\mathbf{k})$ metrons; two second later, the Tharg ship is at the point with position vector $(3\mathbf{i} + 7\mathbf{i} - 7\mathbf{k})$ metrons and, after three seconds, the Enterprise is at the point with position vector $(5\mathbf{i} + 12\mathbf{j} - 10\mathbf{k})$ metrons. Show that two of them will collide and find the time and position vector of the point of impact.

7 Particles A and B start at time $t = 0\,\text{s}$ from points with position vectors $(5\mathbf{i} + 13\mathbf{j})\,\text{m}$ and $(7\mathbf{i} + 5\mathbf{j})\,\text{m}$ respectively, relative to a fixed origin. The velocities of A and B are constant and equal to $(3\mathbf{i} - 5\mathbf{j})\,\text{m/s}$ and $(2\mathbf{i} - \mathbf{j})\,\text{m/s}$ respectively.

(a) Show that the particles collide and find the position vector of the point of collision.

(b) Determine the angle between the directions of motion of A and B before collision.   [A]

Exercise 11.4

1   A cyclist rides along a straight path at 10 km/h. A girl is running along a path at right angles to the first at 15 km/h. What is the apparent speed of the girl to the cyclist?

2   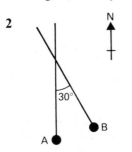 A car A is travelling due north at 80 km/h along a road towards a junction. A second car B is also approaching the junction travelling in a direction N30°W at 60 km/h. What is the velocity of B relative to A?

3   A boy can swim at 2.5 km/h in still water. A river flows at 2 km/h and is 50 m wide. Find the time taken to swim across the river if he reaches the far bank at a point directly opposite the point where he entered the water.

4   A passenger on the top deck of an open bus feels a breeze which to him appears to blow directly across the bus at 15 km/h. If the bus is travelling at 30 km/h, what is the velocity of the wind?

5   To an observer on a boat sailing N30°E at 25 km/h, a power boat appears to be travelling in a southwesterly direction at 60 km/h. What is the true velocity of the power boat?

*6   An airliner is flying due south at 450 km/h. The pilot can see another airliner which appears to be flying on a bearing of 050°. The true speed of the second airliner is 400 km/h. Draw a diagram to show that there are two possibilities for the true direction of the second airliner and find both of them.

*7   A ship is steaming north at 16 km/h, and there is a north east wind blowing at 32 km/h. Show that, to an observer on the ship, the smoke appears to blow in a direction $\tan^{-1} (2 + \sqrt{2})/2$ S of W.

8   The banks of a river are straight and parallel and at a distance a apart and the river flows with a constant uniform speed v. A and B are two points on opposite banks of the river with the line AB at right angles to the banks. Two swimmers set off at the same instant from A and B and swim with a steady speed u relative to the water, where $u > v$, so that they approach each other along the line AB. Find the time taken for the swimmers to meet.   [J]

Exercise 11.5

1   The following information is known about three moving objects:
    they each move with constant velocity
    P's velocity is 12 m/s in the direction $\overrightarrow{XY}$
    the velocity of Q relative to P is 8 m/s at 60° to $\overrightarrow{XY}$
    the velocity of R relative to Q is 20 m/s in the direction of $\overrightarrow{YX}$.
    Find the true velocity of Q and R.

2   When a boy runs due south at 8 km/h the wind appears to be blowing from
    the south west. When he turns and runs due west at the same speed, the
    wind appears to be blowing in the direction N70°E. Find the true velocity
    of the wind.

3   Two Tristar airliners are flying at the same height at a speed of 500 km/h.
    One is flying in a direction S30°E and the other is flying due west, when
    their pilots see a light aircraft which is also flying at the same height. To
    the pilot of the first airliner, the light aircraft appears to be flying due north
    but, to the pilot of the second airliner it appears to be flying on a bearing
    of 060°. On what bearing is the light aircraft actually flying.

4   An observer on a ship A travelling due north at 7 km/h sees another ship
    which appears to be travelling SE at $12\sqrt{2}$ km/h.
    Calculate:
    (i)   the speed of B
    (ii) the bearing, correct to the nearest degree, on which B is travelling.
    [J]

5   An aircraft A is flying due east at 1200 km/h. A second aircraft B is flying
    at 600 km/h on a parallel easterly course. A gun mounted on B points due
    north and fires a bullet with speed 3000 km/h relative to B.
    Determine the magnitude (correct to the nearest km/h) and the direction
    (correct to the nearest degree) of the velocity of the bullet relative to A.
    Hence, or otherwise, determine the bearing of B from A at the instant of
    firing if the bullet subsequently hits A.
    (Assume that A, B and the bullet all move in the same horizontal plane
    and that air resistance is negligible).
    [J]

Exercise 11.6

1   A cyclist A is riding along a straight level road on a bearing 030° at a
    steady speed of 5 m/s. A second cyclist B is travelling due east along another
    straight level road at a steady speed of 6 m/s. Find the exact magnitude
    and the direction, correct to the nearest degree, of the velocity of A relative
    to B.
    At a certain instant A is 500 m due south of B. Find, correct to the
    nearest second, the time that elapses before A is due west of B.   [J]

2   Two roads cross at right angles. A boy is walking along one of the roads
    towards the crossroads, at 4.5 km/h. A girl is walking along the other road

at 6 km/h. The girl reaches the crossroads when the boy is still 50 m away. Find the speed of the boy relative to the girl and show that they will be nearest together when the boy has walked 18 m further.

3   A ship is steaming in a northeasterly direction at 20 km/h when a seaman observes at 9 a.m. a liner 16 km away and southeast of the ship. The liner is travelling NNE at 40 km/h. At what time are the vessels nearest together and how far are they apart at that time?

4   A ship is travelling due east at 40 km/h. A second ship is travelling S30° E at 16 km/h. At midday the first ship is 40 km due south of the second. At what time are the ships nearest together?

5   Two aircraft A and B are flying at the same height. A is flying at 300 km/h in a southeasterly direction and B is flying at 350 km/h on a bearing of 060°. At one instant A is 1 km due north of B. How close do they get to one another?

*6

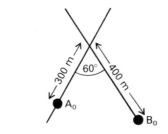

Two tracks on a piece of scrubland cross at 60°, as shown. Two boys on BMX bikes are riding towards the crossing point. A is moving at 30 km/h and B at 24 km/h. At one instant A is 300 m and B is 400 m from the crossing. What is the shortest distance between A and B?

Exercise 11.7

1   At midday two ships, A and B, are at X and Y respectively where $XY = 10$ nautical miles. A is sailing at 25 knots in a direction perpendicular to XY. B is sailing at her maximum speed of 15 knots at an angle α to YX to get as close as possible to A.
Show that:

(i) $\alpha = \sin^{-1} \frac{3}{5}$
(ii)   the closest that they get is 8 nautical miles
(iii)   they are closest at 12.18 p.m.

2   At a given instant, a ship P travelling due east at a speed of 30 km/h is 7 km due north of a second ship Q which is travelling Nθ° W at a speed of 14 km/h where $\tan \theta = \frac{3}{4}$. Show that the speed of Q relative to P is 40 km/h and find the direction of the relative velocity.
The ships continue to move with uniform velocities. Find, correct to 3 significant figures:

(i)   the distance between the ships when they are nearest together
(ii)   the time taken, in minutes, to attain this shortest distance.

If, initially, the course of Q had been altered to bring the ships as close as possible, the speed of Q and the speed and course of P being unchanged, find the direction of the new course.   [J]

3   Two ships A and B, are travelling with constant velocities of 30 km/h westwards and 40 km/h northwards respectively. Their courses meet at a point O. When B is passing through O the ship A is 15 km from O and moving towards O. Find:

  (i)   the magnitude and direction of the velocity of B relative to A
  (ii) the least distance between the ships
  (iii) the distance of the ships from O when they are nearest to one another.

  Show that, if A, while maintaining its speed, had changed course when B was passing through O in such a way as to minimise the least distance between the ships, the least distance would have been approximately 9.92 km.   [J]

4   At noon a boat is at a point A and a helicopter is at a point B, 20 km due west of A. The boat moves with a constant velocity due north with speed 60 km/h. In order to intercept the boat, the helicopter files at a constant velocity *relative to the air* of magnitude 100 km/h and direction θ East of North. Calculate the required value of $\tan \theta$ and the time at which the helicopter reaches the boat when:

  (i)   there is no wind
  (ii) the wind is 20 km/h from the north.

  In the case when the wind is 50 km/h from the north interception is impossible. Find the value of $\cos \theta$ which must be chosen if the helicopter is to approach the boat as closely as possible and show that the distance of closest approach is $20\sqrt{21}/11$ km.   [J]

Exercise 11.8

1   A galleon is sailing due south at 12 knots when it is sighted by the captain of a pirate ship. In order to intercept the galleon, the pirate ship moves off at 15 knots in a direction S 30° W. If the pirate ship overtakes the galleon 30 minutes later, find the distance and bearing of the galleon from the pirate vessel when first sighted.

2   A destroyer is travelling NE at 15 knots. At 10.30 it is sighted by an enemy submarine which is 12 nautical miles SE of the destroyer. On what bearing should it travel to intercept the destroyer if it travels at 20 knots and at what time will interception occur?

3   A flock of sheep is being driven across a field when a young ram breaks away from the flock, running in a direction S 40° W at 15 km/h. When spotted by the shepherd, it is 200 m away due west of him and his dog. In what direction should he send the dog to intercept the runaway, if the dog runs at 18 km/h?

4   A boy is drifting out to sea on an airbed. When the boy's father sees him, the airbed is 650 m away in a direction S 40° W and drifting S 20° E at $\frac{1}{4}$ m/s. If the father can swim at $\frac{1}{2}$ m/s, in what direction should he swim and how long would he take to reach his son?

5   A and B are two aircraft flying at the same height. At the moment when A is 5 km due west of B, the air traffic controller tells them that they are on collision course. A is flying at 600 km/h on a bearing of 030° and B is flying on a bearing of 340°. Find the speed of B.
Immediately A alters course to 050° and maintains his speed. What is the shortest distance between them?

6   A ship A is travelling with constant velocity 20 km/h due east and a ship B has constant velocity 15 km/h in a direction 30° east of north. At noon B is 30 km due south of A. Find the magnitude of the velocity of B relative to A, and show that the direction of this relative velocity is approximately 44° west of north.
Find, to the nearest minute, the time at which A and B are closest.
At the time of closest approach, a boat leaves A to intercept B. Find the least speed at which it must travel.   [J]

Exercise 11.9

1   A cyclist A is travelling with a constant velocity of 10 km/h due east and a cyclist B has a constant velocity of 8 km/h in a direction $\arctan(\frac{4}{3})$ east of north. At noon B is 0.6 km due south of A. Taking the position of B at noon as the origin and **i** and **j** as unit vectors due east and due north respectively, obtain expressions for the position vectors of A and B at time t hours after noon.
Hence find, at time t hours after noon:

(a)   the position vector of A relative to B
(b)   the velocity of A relative to B.

Show that the cyclists are nearest together at time 4.8 minutes after noon, and, at this time, the distance between them is 360 m.   [L]

2   Two particles A and B are free to move in the plane of the fixed unit vectors **i** and **j**. The velocity of A is $(-3\mathbf{i} + 29\mathbf{j})$ m/s whilst that of B is $v(\mathbf{i} + 7\mathbf{j})$ m/s, where v is a constant. Determine the velocity of B relative to A and find the vector $\overrightarrow{AB}$ at time t s given that, when $t = 0$, $\overrightarrow{AB} = (-56\mathbf{i} + 8\mathbf{j})$ m.
Find also that value of v such that the particles collide.
Show that, when $v = 3$, $\overrightarrow{AB}$ at time t s is given by

$$(6t - 56)\mathbf{i} + 8(1 - t)\mathbf{j}$$

and hence find t when A and B are closest together.
By evaluating a suitable scalar product show that, for your value of t and with $v = 3$, $\overrightarrow{AB}$ is perpendicular to the velocity of B relative to A.
[A]

3   An old man cycles to work each day leaving his home at a point with position vector $(-2\mathbf{i} - 8\mathbf{j})$ km relative to an origin O at the centre of the village. The unit vectors **i** and **j** point east and north respectively. One day he cycles due north at 8 km/h. At the same time, a younger man leaves his home, which has position vector $(-12\mathbf{i} - 4\mathbf{j})$ km relative to O, and cycles with velocity $(6\mathbf{i} - 6\mathbf{j})$ km/h. Show that after half an hour the cyclists are closest together and determine this shortest distance between them.

On that day as the men cycle to work there is a steady wind blowing. To the older man, cycling due north at 8 km/h, the wind appears to blow *from* the west. To the younger man, cycling with velocity $(6\mathbf{i} - 6\mathbf{j})$ km/h, the wind appears to blow *from* the south. Find the velocity of the wind as a vector.

[A]

4   At time $t = 0$ a ship A is at the point O and a ship B is at the point with position vector $10\mathbf{j}$ nautical miles referred to O. The velocities of the two ships are constant. Ship A sails at 17 knots, where 1 knot is 1 nautical mile/hour, in the direction of the vector $8\mathbf{i} + 15\mathbf{j}$, and ship B sails at 15 knots in the direction of the vector $3\mathbf{i} + 4\mathbf{j}$. Write down:

(a)   the velocity vector of each ship
(b)   the velocity of B relative to A
(c)   the position vector of B relative to A at time t hours.

Given that visibility is 5 nautical miles show that the ships are within sight of each other for 6 hours.

[L]

5   A hawk H and a sparrow S fly with constant velocities

$$p\mathbf{i} + q\mathbf{j} + r\mathbf{k} \quad \text{and} \quad -\mathbf{i} + \mathbf{j} + 2\mathbf{k},$$

respectively, and at time $t = 0$ their position vectors are $\mathbf{j}$ and $5\mathbf{i}$, respectively. Write down the velocity of H relative to S and the position vector of H relative to S at time t.

(a)   In the case when

$$p = 1, \quad q = 2, \quad r = 3.$$

find the time at which the birds are closest together and determine their least distance apart.

(b)   The hawk flies at the same speed as in (i), but the constants p, q and r are such that the hawk captures the sparrow. Show that, at the instant of capture,

$$(p^2 + q^2)t^2 = 2t^2 - 12t + 26$$

Find the value of t at this instant, giving your answer correct to two decimal places.

[J]

Miscellaneous Exercise 11 B

1   At time t seconds, the velocity, relative to a fixed origin O, of a particle P is $(9t^2\mathbf{a} + 16t\mathbf{b})$ m/s where $\mathbf{a}$ and $\mathbf{b}$ are constant vectors which are neither perpendicular nor parallel to each other. Given that, when $t = 0$, P is at the origin find:

(a) $\overrightarrow{\text{OP}}$ at subsequent time t s
(b)   the value of t for which OP is parallel to CD, where $\overrightarrow{\text{OC}} = 5\mathbf{a} + 8\mathbf{b}$, $\overrightarrow{\text{OD}} = 8\mathbf{a} + 10\mathbf{b}$
(c)   the length of OP when it is parallel to CD, given that $CD = 0.3$ m.

[A: 11.1]

2 Two forces, $2\mathbf{i} + 8\mathbf{j} + 4\mathbf{k}$ and $4\mathbf{i} + \mathbf{j} - 4\mathbf{k}$, act at points whose position vectors with respect to an origin O are $2\mathbf{i} + 3\mathbf{j} + 4\mathbf{k}$ and $5\mathbf{i} - 2\mathbf{k}$ respectively. Verify that the lines of action of these forces meet at the point whose position vector is $\mathbf{i} - \mathbf{j} + 2\mathbf{k}$. Find the magnitude of the resultant of these two forces and the vector equation of its line of action. [J: 11.1]

3 A particle is acted upon by two forces, one of magnitude 6 in the direction of the vector $2\mathbf{i} - 2\mathbf{j} + \mathbf{k}$ and the other of magnitude 5 in the direction of the vector $3\mathbf{i} - 2\mathbf{j} + 6\mathbf{k}$. Find the total work done by these forces when the particle is displaced from the point whose position vector is $\mathbf{i} + \mathbf{j} - 2\mathbf{k}$ to that whose position vector is $5\mathbf{i} - 4\mathbf{j} + 2\mathbf{k}$. [J: 11.2]

4 A force $\mathbf{F} = (9\mathbf{i} + 12\mathbf{j})\,\mathrm{N}$ acts on a particle P of mass $5\,\mathrm{kg}$ which is free to move in horizontal plane. Calculate the magnitude of the acceleration of P. Initially the particle passes through a fixed point A with speed $3\,\mathrm{m/s}$ in the same direction as $\mathbf{F}$. Show that the velocity of the particle at time t is

$$\tfrac{1}{5}(t + 1)(9\mathbf{i} + 12\mathbf{j})\,\mathrm{m/s}$$

Find the vector $\overrightarrow{AF}$ at time t.
If the position vector of A relative to O is $(30\mathbf{i} + 25\mathbf{j})\,\mathrm{m}$ find the vector $\overrightarrow{OP}$ when $t = 5$. [11.1]

5 A particle P of mass m moves so that at time t its position vector relative to a fixed origin is $\mathbf{r}$, where

$$\mathbf{r} = (a\cos\omega t)\mathbf{i} + (a\sin^2\omega t)\mathbf{j} + (a\sin\omega t\cos\omega t)\mathbf{k}$$

and a, ω are positive constants. Show that $|\mathbf{r}| = a$.
Find the velocity $\mathbf{u}$ and the acceleration $\mathbf{f}$ of P at time t and verify that $\mathbf{r}\cdot\mathbf{u} = 0$.
Find, in terms of a, m, ω and t, an expression for the kinetic energy of P. Deduce that the maximum speed of P is $a\omega\sqrt{2}$. [L: 11.1, 11.2]

6 A particle P of mass 2 units moves under the action of a force $\begin{pmatrix} 4 \\ -2 \\ 6 \end{pmatrix}$. Initially P has velocity $\begin{pmatrix} -4 \\ 1 \\ 0 \end{pmatrix}$ and is at the point with position vector $\begin{pmatrix} 1 \\ 2 \\ -3 \end{pmatrix}$. Find, at time $t = 4$:

(a) the speed of P
(b) the position vector of P. [J: 11.1]

7 A body of mass $2\,\mathrm{kg}$ moves under the action of a variable force whose value at time t seconds is

$$(3t - 2)\mathbf{i} - 8\mathbf{k}\ \text{newtons.}$$

No other force acts on the body. When $t = 0$ the velocity of the body is $(4\mathbf{i} - 2\mathbf{j} + 5\mathbf{k})\,\mathrm{m/s}$.

Find:

(a) the velocity of the body at time t
(b) the values of t for which the velocity of the body is perpendicular to its initial direction of motion
(c) the power of the force acting on the body when $t = 4$.   [J: 11.1, 11.2]

8 Two forces $\mathbf{F}_1 = \begin{pmatrix} 3 \\ 4 \end{pmatrix}$ N and $\mathbf{F}_2 = \begin{pmatrix} -1 \\ 6 \end{pmatrix}$ N act at points with position vectors $\begin{pmatrix} 2 \\ 3 \end{pmatrix}$ and $\begin{pmatrix} 6 \\ 1 \end{pmatrix}$ respectively. Find the resultant of these forces and the position vector of the point of intersection of their lines of action. Hence give a vector equation of the line of action of the resultant. A third force $\mathbf{F}_3 = \begin{pmatrix} -3 \\ -3 \end{pmatrix}$ N acting at the point with position vector $\begin{pmatrix} -3 \\ 2 \end{pmatrix}$ is now added to the system. A fourth force $\mathbf{F}_4$, acting at the origin, and a couple $\mathbf{G}$ are added to the system to produce equilibrium. Find $\mathbf{F}_4$ and $\mathbf{G}$.
[11.1]

9 At time t a particle of mass m has a position vector $\mathbf{r}$ relative to a fixed origin, where

$$\mathbf{r} = c[\mathbf{i}\cos(\lambda t^2) + \mathbf{j}\sin(\lambda t^2) + \mathbf{k}\lambda t^2]$$

and c and λ are constants. Find expressions for the velocity and the acceleration of the particle at time t.
Deduce that initially the particle is at rest and that the acceleration has magnitude $2\sqrt{2}c\lambda$.
Find, in terms of m, c, λ and t, an expression for T the kinetic energy of the particle.
Find the constant F such that $T = \mathbf{k} \cdot \mathbf{r} F$.   [L: 11.1, 11.2]

10 A particle of mass 3 units moves under the action of a force $\mathbf{F}$. At time t the velocity $\mathbf{v}$ of the particle is given by

$$\mathbf{v} = 3\mathbf{i} - \mathbf{j} + 2t\mathbf{k}$$

(i) Find the force $\mathbf{F}$.
(ii) Find the kinetic energy of the particle at time t.
(iii) Verify that the rate of change of the kinetic energy. with respect to t, is equal to the power of the force $\mathbf{F}$.
(iv) Given that the position vector of the particle when $t = 0$ is $\mathbf{i} + \mathbf{j}$, find its position vector when $t = 2$.   [J: 11.1, 11.2]

11 A particle, of mass m, which moves under the action of a force $\mathbf{F}$ has position vector $\mathbf{r}$ at time t, relative to a fixed origin, given by

$$\mathbf{r} = a\cos\omega t\,\mathbf{i} + a\sin 2\omega t\,\mathbf{j}$$

where a and ω are constants. Find:

(a) the velocity and acceleration of the particle at any time t
(b) the times in the interval $0 \leqslant t \leqslant \pi/2\omega$ when the velocity is perpendicular to the acceleration

(c) the work done by the force **F** in the interval $t = 0$ to $t = \pi/6\omega$
(d) the rate at which the force **F** is working when $t = \pi/6\omega$.

[A: 11.1, 11.2]

12 A smooth fixed straight wire passes through the origin O and the point A whose position vector is $\mathbf{i} + 3\mathbf{j} - 4\mathbf{k}$, the unit vector **k** being directed vertically upwards. A bead of mass m moves along the wire from O to A under the action of gravity, the reaction of the wire and a force **F** which is of constant magnitude, F, and acts in the direction of the vector $\mathbf{i} + 2\mathbf{j} + 2\mathbf{k}$.
  (a) Explain briefly why the work done by the reaction of the wire during the motion is zero.
  (b) Write down the work done by gravity in terms of m and g.
  (c) Find the work done by the force **F** in terms of F.
  (d) Given that the bead arrives at A with the same speed as it had at O, find F in terms of m and g. [J: 11.2]

13 [In this question the effect of gravity may be neglected.] A small ring, of unit mass, is threaded on a fixed smooth straight wire which passes through the points A $(1, -2, 3)$ and B $(4, 0, -3)$. The ring moves under the action of a force **F**, where $\mathbf{F} = 2\mathbf{i} - 8\mathbf{j} - 18\mathbf{k}$, and starts from rest at A. Show that the work done by **F** in moving the ring from A to B is 98 units, and find the speed of the ring at B.
The force **F** has two components, $\mathbf{F}_1$ in the direction of $\overrightarrow{AB}$ and $\mathbf{F}_2$ perpendicular to $\overrightarrow{AB}$. Find:
  (i) $\mathbf{F}_1$ and $\mathbf{F}_2$
  (ii) the reaction, **R**, of the wire on the ring. [J: 11.1, 11.2]

14 A particle P moves in the x–y plane in a circle of radius a with its centre at the origin. The particle is initially at the point $(a, 0)$ and moves anti-clockwise with constant angular speed ω. Write down, in terms of a, ω, t and the unit vectors **i** and **j**, the position vector **r** of P at time t. Obtain, by differentiation, an expression for $\ddot{\mathbf{r}}$ in terms of **r** and ω, and hence determine the magnitude and the direction of the acceleration of P. Two particles are connected by a light inextensible string which passes through a small smooth hole in a smooth horizontal table. One particle has mass $3m$ and moves on the table, in a circle of radius a, with constant angular speed ω; the other particle has mass m and hangs at rest. Find an expression for ω in terms of a and g. [J: 11.1, 7.1, 7.2]

15 At time t two points P and Q have position vectors **p** and **q** respectively, where

$$\mathbf{p} = 2a\mathbf{i} + (a \cos \omega t)\mathbf{j} + (a \sin \omega t)\mathbf{k},$$
$$\mathbf{q} = (a \sin \omega t)\mathbf{i} - (a \cos \omega t)\mathbf{j} + 3a\mathbf{k}$$

and a, ω are constants. Find **r**, the position vector of P relative to Q, and **v**, the velocity of P relative to Q. Find also the values of t for which **r** and **v** are perpendicular.
Determine the smallest and greatest distances between P and Q.

[L: 11.1, 11.3]

*16 A rugby player is running due north with speed $4\,\text{m/s}$. He throws the ball horizontally and the ball has an initial velocity relative to the player of $6\,\text{m/s}$ in the direction $\theta°$ west of south, i.e. on a bearing of $(180 + \theta)°$, where $\tan\theta° = \frac{4}{3}$. Find the magnitude and the direction of the initial velocity of the ball relative to a stationary spectator. Find also the bearing on which the ball appears to move initially to the referee who is running with speed $2\sqrt{2}\,\text{m/s}$ in a north-westerly direction.
(Give all results to 3 significant figures, with bearings in degrees.)
[L: 11.4, 11.5]

*17 Two ships A and B are travelling with constant speeds $2u\,\text{m/s}$ and $u\,\text{m/s}$ respectively, A on a bearing θ and B on a bearing $90° + \theta$. It is also assumed that a third ship C has a constant, but unknown, velocity which is taken to be a speed $v\,\text{m/s}$ on a bearing ϕ. To an observer on ship B the velocity of C appears to be due north. Show that

$$\frac{u}{\sin\phi} = \frac{v}{\cos\theta}$$

To an observer on ship A the velocity of C appears to be on a bearing 135°. Show that

$$2u(\cos\theta + \sin\theta) = v(\cos\phi + \sin\phi)$$

Hence find $\tan\phi$ in terms of $\tan\theta$.
Given that $\theta = 30°$ and $u = 10$, find the true velocity of C, giving your answer to 3 significant figures.
[L: 11.4, 11.5]

*18 At time $t = 0$ a ship A is at the point O and a ship B is at the point with position vector $10\mathbf{j}$ nautical miles referred to O. The velocities of the two ships are constant. Ship A sails at 17 knots, where 1 knot is 1 nautical mile per hour, in the direction of the vector $8\mathbf{i} + 15\mathbf{j}$ and ship B sails at 15 knots in the direction of the vector $3\mathbf{i} + 4\mathbf{j}$. Write down:

(a) the velocity vector of each ship
(b) the velocity of B relative to A
(c) the position vector of B relative to A at time t hours.

Given that visibility is 5 nautical miles, show that the ships are within sight of each other for $\sqrt{6}$ hours.
[L: 11.9]

19 A cyclist A is travelling with a constant velocity of $10\,\text{km/h}$ due east and a cyclist B has a constant velocity of $8\,\text{km/h}$ in a direction $\arctan\left(\frac{4}{3}\right)$ east of north. At noon, B is $0.6\,\text{km}$ due south of A. Taking the position of B at noon as the origin and $\mathbf{i}$ and $\mathbf{j}$ as unit vectors due east and due north respectively, obtain expressions for the position vectors of A and B at time t hours after noon.
Hence find, at time t hours after noon:

(a) the position vector of A relative to B
(b) the velocity of A relative to B.

Show that the cyclists are nearest together at time 4.8 minutes after noon and that, at this time, the distance between them is $360\,\text{m}$.
[L: 11.4, 11.5, 11.6, 11.9]

*20 At noon a boat A is 9 km due west of another boat B. To an observer on B the boat A always appears to be moving on a bearing of 150° (i.e. S30° E) with constant speed 2.5 m/s. Find the time at which the boats are closest together and the distance between them at this time. Find also, to the nearest minute, the length of time for which the boats are less than 8 km apart. (No credit will be given for methods using scale drawing.)

The velocities of B and of a third boat C are given to be $(3\mathbf{i} + 4\mathbf{j})$ m/s and $(5.5\mathbf{i} + 2\mathbf{j})$ m/s respectively, where $\mathbf{i}$ and $\mathbf{j}$ denote unit vectors directed east and north respectively. Express in the form $a\mathbf{i} + b\mathbf{j}$ the velocity of C relative to B. Given that, at noon, C is directly north of A and that the boats B and C are on a collision course find:

(a) the distance between A and C at noon
(b) the time at which B and C would collide.   [A: 11.4, 11.5, 11.6, 11.9]

21 A river flows at a constant speed of 5 m/s between straight parallel banks which are 240 m apart. A boat crosses the river, travelling relative to the water at a constant speed of 12 m/s. A man cycles at a constant speed of 4 m/s along the edge of one bank of the river in the direction opposite to the direction of flow of the river. At the instant when the boat leaves a point O on the opposite bank, the cyclist is 80 m downstream of O. The boat is steered relative to the water in a direction perpendicular to the banks. Taking $\mathbf{i}$ and $\mathbf{j}$ to be perpendicular horizontal unit vectors downstream and across the river from O respectively, express, in terms of $\mathbf{i}$ and $\mathbf{j}$, the velocities and the position vectors relative to O of the boat and the cyclist t seconds after the boat leaves O. Hence, or otherwise, calculate the time when the distance between the boat and the cyclist is least, giving this least distance.

If, instead, the boat were to be steered so that it crosses the river from O to a point on the other bank directly opposite to O, show that this crossing would take approximately 22 seconds.   [L: 11.4, 11.5, 11.4, 11.9]

*22 In a wind blowing from the south with constant speed w a helicopter flies horizontally with constant velocity in a direction θ east of north from a point A to a point B. The speed of the helicopter relative to the air is λw, where $\lambda > 1$. Find the speed of the helicopter along AB.

The helicopter returns from B to A with constant velocity and with the same speed λw relative to the air, and in the same wind. Show that the total time for the two journeys is

$$\frac{2c\sqrt{\lambda^2 - \sin^2\theta}}{w(\lambda^2 - 1)},$$

where $c = $ AB.   [J: 11.4, 11.5, 11.6]

*23 The map of a holiday resort shows that the point A is 1600 m north of a hotel O, B is 800 m east of O and C is 600 m north of B. The map also shows a straight horizontal road between O and C and a ski-lift carrying ski-buckets in a straight line between A, which is 150 m higher than O, to B, which is 50 m lower than O. Taking O as origin, with unit vectors $\mathbf{i}$ due east, $\mathbf{j}$ due north and $\mathbf{k}$ vertically upwards, write down the position vectors, relative to O, of A, B and C. Hence find unit vectors parallel to the directions of $\overrightarrow{\text{AB}}$ and $\overrightarrow{\text{OC}}$.

At a given instant a ski-bucket leaves A and travels towards B at a uniform speed of 9 km/h and, at the same time, a man leaves O and travels towards C at a uniform speed of 5 km/h. Write down the position vectors of the ski-bucket and the man at a subsequent time t hours. Hence find the position vector of the bucket relative to the man at time t hours.

Safety regulations require that the man and the bucket are never less than 50 m apart. By considering the distance between the man and the bucket at a time $\frac{1}{7}$ hours after their departures, or otherwise, show that the regulations are broken.   [L: 11.4, 11.5, 11.6, 11.9]

12
UNIT
FURTHER CASES OF RIGID BODY
EQUILIBRIUM: SYSTEMS OF FORCES

12.1 Light Frameworks

A framework consists of a number of rods of negligible weight that are jointed together at their ends to form a rigid construction. When external forces act on the framework, each rod will either be:

(a) **in tension**, when the rod exerts a pull at both ends

or (b) **in thrust** (or in compression), when the rod exerts a push at both ends.

In both cases, the tension or thrust acting on each end of the rod is equal and opposite.

Points to Remember when Solving Framework Problems

1   The external forces acting on the framework are in equilibrium. Find these first by resolving and/or taking moments for the whole system.
2   The forces acting at each joint are in equilibrium. Put in the forces acting on each rod. Let all of the rods be in tension, then any negative values calculated will mean that that particular rod is in the thrust.

Example 1   A framework consists of three light rods jointed together as in the diagram, with weights of 20 N hanging from corners A and B. The whole framework is suspended from C. Find the force of suspension and the forces acting in each of the rods.

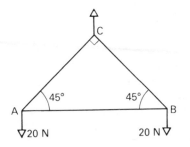

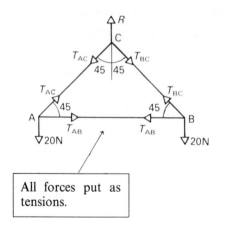

All forces put as tensions.

Resolving vertically for system:

$$R = 40$$

$\therefore$ force of suspension is $\underline{40N}$

By symmetry,

$$T_{AC} = T_{BC} \qquad (1)$$

Resolving vertically at A:

$$T_{AC} \cos 45° = 20$$

$$T_{AC} \times \frac{1}{\sqrt{2}} = 20$$

$\therefore (1) \Rightarrow \underline{T_{AC} = T_{BC} = 20\sqrt{2}N}$ (tension)

Resolving horizontally at A:

$$T_{AB} + T_{AC} \cos 45° = 0$$

$$\Rightarrow T_{AB} = -20\sqrt{2} \times \frac{1}{\sqrt{2}} = -20$$

$\therefore \underline{T_{AB}}$ = thrust of 20 N

Example 2

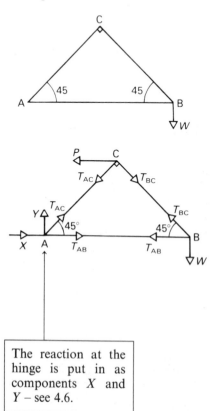

The reaction at the hinge is put in as components X and Y – see 4.6.

Three light rods form a framework ABC where $AC = BC$ and angle $ACB = 90°$. A weight W is hung from B and the framework is hinged at A and kept in equilibrium by a horizontal force applied at C. Find:

(a) the force at C
(b) the magnitude and direction of the force at the hinge
(c) the forces in each rod.

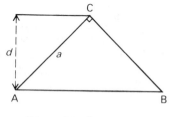

Dimension diagram

Let $AC = a$
then $AB = \sqrt{2}a$
and $d = \dfrac{a}{\sqrt{2}}$

(a) Taking moments about A for system:

$$\frac{Pa}{\sqrt{2}} = W\sqrt{2}a$$

$$\Rightarrow P = 2W$$

$$\Rightarrow \text{force at } C = 2W$$

(b) Resolving vertically for system:

$$Y = W$$

Resolving horizontally for system:

$$X = P = 2W$$

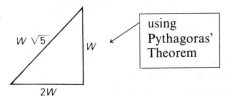

using Pythagoras' Theorem

$\therefore$ force at the hinge is $W\sqrt{5}$ at $\tan^{-1}\frac{1}{2}$ above the horizontal.

(c) Resolving vertically at A:

$$T_{AC}\cos 45° + Y = 0$$

$$\Rightarrow \quad T_{AC}\frac{1}{\sqrt{2}} + W = 0$$

$$\Rightarrow T_{AC} = -W\sqrt{2}$$

$$\therefore T_{AC} \text{ is a thrust of } W\sqrt{2}.$$

Resolving horizontally at A:

$$T_{AB} + T_{AC}\cos 45° + X = 0$$

$$\Rightarrow \quad T_{AB} + (-W) + 2W = 0$$

$$\Rightarrow T_{AB} \qquad\qquad = -W$$

$$\therefore T_{AB} \text{ is a thrust of } W.$$

Resolving vertically at C:

$$T_{BC}\cos 45° + T_{AC}\cos 45° = 0$$

$$\Rightarrow T_{BC} = -T_{AC}$$

$$\therefore T_{BC} \text{ is a tension of } W\sqrt{2}.$$

Note that a good way of checking that the tensions and thrusts are correct is to resolve horizontally and/or vertically at a point not already considered. E.g. resolving vertically at B: $T_{BC}\cos 45° = W$.

Using the value found, $W\sqrt{2} \times \dfrac{1}{\sqrt{2}} = W$ which is correct.

• Exercise 12.1 See page 358

12.2 Jointed Rods

Since we do not know the direction of forces acting at joints between rods (or at hinges), we consider the horizontal and vertical components of these forces. In order to show these components clearly in a force diagram a useful technique is to separate the rods at a joint.

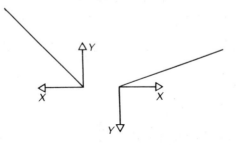

Since the forces at the joint are internal, the forces exerted by each rod on the other must be equal in magnitude and opposite in direction. Usually, the best method of tackling such problems is to resolve and take moments:

(a)   for each rod individually

and (b)   for the whole system.

Example 3   Two uniform rods AB and BC, both of length $2l$ are smoothly jointed at B and have their other ends attached to small rings which are free to slide on a horizontal wire. The weights of AB and BC are W and $2W$ respectively. The rods make angles of $60°$ to the horizontal when the system is in equilibrium. Find the magnitude of the reaction at the joint B and the normal reactions and frictional forces acting at A and C.

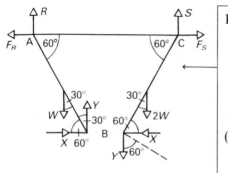

In drawing the diagram, note that:

(i) F_R and F_S must be outwards to hold the rods apart.

(ii)   Since the only other forces acting on the rods are at the joint B, they must be in the opposite directions to the frictional forces, as shown.

(iii)   It is usually difficult to predict the sense of the vertical components at the joint.

Moments are taken for each rod about the points given, in order to obtain two equation in X and Y only.

Taking moments about A for AB:

$$lW \sin 30° = 2lX \sin 60° + 2lY \sin 30°$$
$$\Rightarrow \quad W = 2\sqrt{3}X + 2Y \tag{1}$$

Taking moments about C for BC:

$$2lW \sin 30° + 2lY \cos 60° = 2lX \sin 60°$$
$$\Rightarrow \quad W + Y = X\sqrt{3} \tag{2}$$

Solving (1) and (2) simultaneously $\Rightarrow X = \dfrac{\sqrt{3}}{4}W$ and $Y = -\tfrac{1}{4}W$

Magnitude of reaction at B:

$$= [X^2 + Y^2]^{1/2}$$
$$= \frac{W}{4}[(\sqrt{3})^2 + 1^2]^{1/2} = \underline{\frac{W}{2}}$$

> Thus Y is in the opposite direction to that given in the diagram.

Resolving horizontally for AB:

$$F_R = X = \frac{\sqrt{3}}{4}W$$

Resolving horizontally for system:

$$F_S = F_R$$

Resolving vertically for AB:

$$R + Y = W \Rightarrow R = W - \left(-\frac{1}{4}W\right) = \frac{5}{4}W$$

Resolving vertically for system:

$$R + S = 3W \Rightarrow S = 3W - \frac{5}{4}W = \frac{7}{4}W.$$

$\therefore$ normal reactions at A and B are $\dfrac{5W}{4}$ and $\dfrac{7W}{4}$

and frictional forces at A and B are both $\dfrac{\sqrt{3}W}{4}$

Example 4   Repeat example 3 but with both rods of weight W.

> The system is symmetrical about a vertical line through B. This simplifies the question considerably. From this we can deduce that (i) $R = S$ (ii) $F_R = F_S$ (iii) $Y = 0$ and the diagram becomes:

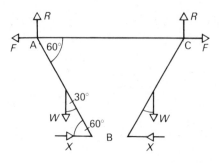

Resolving vertically for AB:
$$R = W$$

Resolving horizontally for AB:
$$F = X$$

Taking moments about A for AB:
$$lW \sin 30° = 2lX \sin 60°$$
$$\Rightarrow \qquad W = 2\sqrt{3}X$$
$$\Rightarrow \qquad X = \frac{\sqrt{3}}{6} W$$

∴ magnitude of reaction at B $= \sqrt{3}W/6$.
normal reactions at A and C $= W$.
frictional forces at A and C $= \sqrt{3}W/6$.

Example 5   A square ABCD is made up of four identical rods of weight W jointed together. The system is suspended from the joint A and a string connecting A and C keeps the framework in the form of a square. Find the tension in the string and the magnitude and direction of the reaction at B.

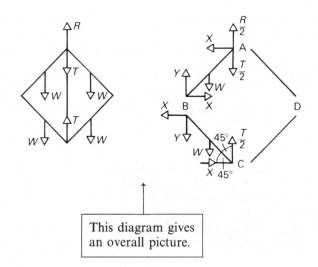

This diagram gives an overall picture.

Notes on drawing the force diagram:
  (i)   System is symmetrical, thus only half the diagram need be considered.
  (ii) Symmetry ⇒ no vertical reactions at A and C.
  (iii) Symmetry ⇒ R and T act equally on both halves of the system.
  (iv) Since the only horizontal forces are the reactions at the joints they must all be equal and act in opposite directions at each end of each rod. Be careful – this is not always so.
  (v)   Since the shape is a square, all acute amples are 45°.

Resolving vertically for system:

$$R = 4W \tag{1}$$

Resolving vertically for BC:

$$Y + W = T/2 \tag{2}$$

Taking moments about A for AB:

$$W \cos 45° + 2X \cos 45° = 2Y \sin 45°$$
$$\Rightarrow \qquad W + 2X = 2Y \tag{3}$$

Taking moments about C for BC:

$$W \cos 45° + 2Y \cos 45° + 2X \sin 45° = 0$$
$$\Rightarrow \qquad W + 2X + 2Y = 0 \tag{4}$$

$\Rightarrow$
(3) and (4) $\Rightarrow X = -W/2$ and $Y = 0$ ⟵ | This could *not* have been predicted from the symmetry.
(2) $\Rightarrow T = 2W$.

$\therefore$ Tension $= 2W$ and reaction at B is $\dfrac{W}{2}$ horizontally

---

Points to note:
(i) Equation (1) was not actually needed in this approach. However if, instead of resolving vertically for BC, we had resolved vertically for AB we would have needed to use (1) to get (2). Do not worry if you throw up equations that are not used later.
(ii) If we had resolved vertically for AB as well as for BC (and used (1)) we would have produced two identical equations. Do not worry if this happens. It is more difficult in complex questions to predict which techniques will produce independent equations.

Example 6 A ladder consists of two identical sections AB and BC each of weight W, jointed at B. A is joined to D, the midpoint of BC, by a light rope. The ladder stands on smooth horizontal ground with sections inclined at 60° to the horizontal. Find the tension in the rope.

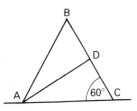

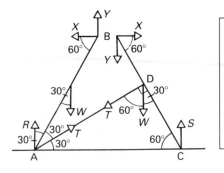

Points to note:
(i)   No symmetry.
(ii) Since the triangle is equilateral, AD is perpendicular to BC.
(iii) Since we do not require X and Y we will try to avoid using them.

Let the sections be of length $2l$.

Taking moments about B for AB:

$$lW \sin 30° + 2lT \sin 30° = 2lR \sin 30°$$
$$\Rightarrow \qquad\qquad W + 2T = 2R \qquad\qquad (1)$$

Taking moments about C for system:

$$W(l \sin 30°) + W(3l \sin 30°) = R(4l \sin 30°)$$
$$\Rightarrow \qquad\qquad\qquad R = W. \qquad\qquad (2)$$

(1) and (2) $\Rightarrow W + 2T = 2W \Rightarrow T = \frac{1}{2}W.$

- **Exercise 12.2** See page 359

12.3 More Complex Problems on Jointed Rods

- **Exercise 12.3** See page 360

12.4 Equilibrium with More than One Body

When two or more bodies are interacting it is important to be able to distinguish the forces acting on each body. The following two methods are recommended.

Either **1** Put all the forces onto one diagram using different colours to show the sets of forces acting on each body.

or **2** Draw two (or more) diagrams. On each diagram show the forces acting on *one* of the bodies.

Example 7 A rod AB, of weight W_1 and length $2a$ lies with its end A on the ground and its end B resting against a sphere of radius a and weight W_2. The contact at B is smooth but the contacts at A and C are rough. Find the normal reactions and the frictional forces at A and C. If the end A is about to slip, show that the coefficient of friction at A is $\frac{6}{41}$.

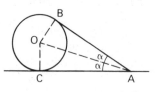

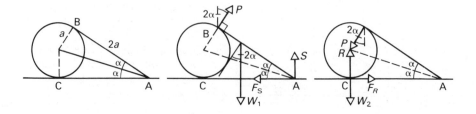

$$\tan \alpha = \frac{a}{2a} = \frac{1}{2}$$

$$\Rightarrow \tan 2\alpha = \frac{2 \tan \alpha}{1 - \tan^2 \alpha} = \frac{2 \times \frac{1}{2}}{1 - \frac{1}{4}} = \frac{4}{3}$$

$$\left. \begin{aligned} \Rightarrow \sin 2\alpha &= \frac{4}{5} \\ \cos 2\alpha &= \frac{3}{5} \end{aligned} \right\} \qquad (1)$$

Taking moments about A for rod:

$$2aP = aW_1 \cos 2\alpha$$

$$\Rightarrow \quad P = \frac{W_1}{2} \cos 2\alpha = \frac{3}{10} W_1 \qquad (2)$$

Resolving vertically for rod:

$$S + P \cos 2\alpha = W_1$$

$$\Rightarrow S = W_1 - P \cos 2\alpha = W_1 - \frac{9}{50} W_1 = \frac{41}{50} W_1 \qquad \text{using (1) and (2)}$$

Resolving horizontally for rod:

$$P \sin 2\alpha = F_S$$

$$\text{(1) and (2)} \Rightarrow F_S = \frac{3}{25} W_1$$

$\therefore$ at A, normal reaction $= \dfrac{41}{50} W_1$ and frictional force $= \dfrac{3}{25} W_1$

Resolving horizontally for sphere:

$$F_R = P \sin 2x = \frac{3}{25} W_1$$

Resolving vertically for sphere:

$$R = P \cos 2\alpha + W_2 = \frac{9}{50} W_1 + W_2$$

$\therefore$ at C, natural reaction $= \dfrac{9}{50} W_1 + W_2$ and friction force $= \dfrac{3}{25} W_1$

If end A is about to slip, $F_s = \mu S$

$$\Rightarrow \frac{3}{25} W_1 = \mu \frac{41}{50} W_1$$

$$\Rightarrow \quad \mu = \frac{6}{41} \qquad \text{QED}$$

● Exercise 12.4   See page 362

12.5 Line Action of the Resultant: Equivalent Systems of Forces

Example 8   Three forces of 1 N, 4 N, 3 N act along the sides AB, BC, CA, respectively, of an equilateral triangle ABC of side $2a$ units. Find the magnitude and direction of the resultant and the point where its line of action cuts BC produced.

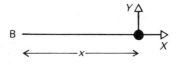

Resolving →: $X = 4 - 3\cos 60° - 1\cos 60° = 2$

Resolving ↑: $Y = 3\sin 60° - 1\sin 60° = \sqrt{3}$

To find the resultant:

$R^2 = (\sqrt{3})^2 + 2^2$

$\Rightarrow R = \sqrt{7}$

Also $\tan \alpha = \dfrac{\sqrt{3}}{2} \Rightarrow \alpha = 41°$

∴ resultant is $\sqrt{7}$ N at 41° to BC.

> Since we want to calculate a point on BC produced, we make BC go across the page.

To find the line of action of the resultant
Let the resultant act at a distance x from B, along BC. Consider the two equivalent systems:

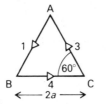 and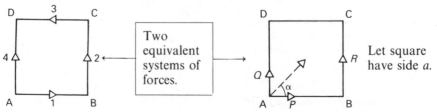

Taking moment about B for both systems:

$$2a \times 3\sin 60° = Yx$$

$$\Rightarrow \quad 3\sqrt{3}a = \sqrt{3}x$$

$$\Rightarrow \quad x = 3a$$

∴ resultant cuts BC produced at a distance a beyond C.

Example 9   Forces of 1 N, 2 N, 3 N, 4 N act along the sides AB, BC, CD, AD of a square ABCD. Find a force through A and a force along BC that are equivalent to this system of forces.

Resolving →:

$$P = 1 - 3 = -2$$

Resolving ↑:

$$Q + R = 4 + 2 = 6$$

Taking moments about A:

$$Ra = 2a + 3a \Rightarrow R = 5 \quad \text{and} \quad Q = 1$$

Force at A $= \sqrt{P^2 + Q^2} = \sqrt{4 + 1} = \sqrt{5}$

$\tan \alpha = \dfrac{Q}{P} = \dfrac{1}{-2}$

∴ Force at A is $\sqrt{5}$ N at $\tan^{-1} \frac{1}{2}$ to BA
and Force along BC is 5 N.

> $\tan^{-1}(-\frac{1}{2})$ to AB
> is the same as
> $\tan^{-1}(\frac{1}{2})$ to BA.

Example 10   If the system of forces in the previous example is equivalent to a force at B and a couple G, find the magnitude and direction of the force and the magnitude of the couple.

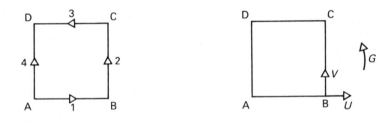

Resolving →:

$$U = 1 - 3 = -2$$

Resolving ↑:

$$V = 4 + 2 = +6$$

$\Biggr\} \Rightarrow$

$\Rightarrow R = \sqrt{36 + 4} = 2\sqrt{10}$

and $\tan \alpha = \dfrac{6}{2} = 3.$

∴ force is $2\sqrt{10}$ N at $\tan^{-1} 3$ to BA

Taking moments about A:

$5a = G + Va \Rightarrow G = -a \quad \text{or} \quad G = a$

Note that taking moments about B instead of A would have been even simpler $\Rightarrow G = -4a + 3a = -a.$

● **Exercise 12.5**   See page 363

12.6+   Revision of Unit: A Level Questions

- Miscellaneous Exercise 12B   See page 364

Unit 12 Exercises

Exercise 12.1

All the frameworks in this exercise are constructed from rods of negligible weight and each one rests in a vertical plane.

1

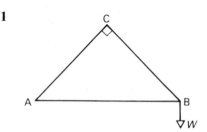

Three rods from an isosceles triangle ABC, where angle ABC is a right angle. The framework is hinged at A and a weight W hangs from B. The framework is kept with AB horizontal by a force P applied at C parallel to $\overrightarrow{BC}$.

(a)   Show that $P = \sqrt{2}W$.
(b)   Find the magnitude and direction of the force at the hinge.
(c)   Show that there is no tension or thrust in AC.
(d)   Find the forces in AB and BC, stating whether they are thrusts or tensions.

2

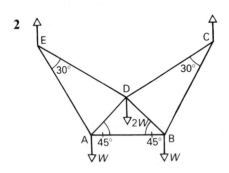

In the framework shown in the diagram, AB is horizontal, AE = DE = BC = CD and AD = BD. Equal weights W hang from A and B and $2W$ hangs from D. The framework is suspended from C and E.

Find the forces in the rods DE, AE, BD, stating whether each is a tension or a thrust. Show that the force in AB is a tension of $(\sqrt{3} + 2)W$.

3

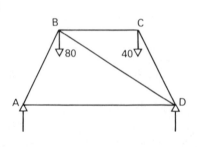

In the framework shown, AD is horizontal and 2 m long and AB = BC = CD = 1 m. The framework rests on two supports at A and D. Weights of 80 N and 40 N hang from B and C respectively.

Find:
(a)   the size of each of the angles BAD and BDA
(b)   the reactions at A and D
(c)   the forces in each of the rods.

4

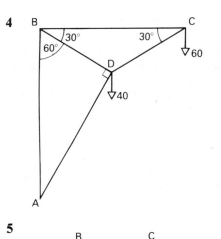

In the framework shown, if BC = 6a, show that BD = 2√3a. The framework is hinged at A and is kept with BC horizontal by a force at B which acts downwards, perpendicular to BD. Weights of 60 N and 40 N hang from C and D respectively, find:

(a)   the force at B
(b)   the magnitude and direction of the force at the hinge
(c)   the forces in each of the rods.

5

This framework is constructed from seven identical light rods. It rests on two supports at A and D.

(i)   In the case when it carries weights of 50 N and 10 N at B and C, respectively, find:
(a)   the reactions at A and D
(b)   the forces in AB, AE, BE and BC.
(ii) In the case when it carries equal weights of 50 N at both B and C, find:
(c)   the reactions at A and D
(d)   the forces in each of the rods.

Exercise 12.2

1   Two uniform identical rods AB and BC, each of weight *W*, are freely jointed at B. A and C are attached to two small light rings which can move without friction on a horizontal wire. The rods are kept at 45° to the wire, with B below A and C by means of a light strut joining the midpoints of the rods. Find the stress in the strut and the reactions at A, B and C.

2   Two uniform rods AB and BC, of weight *W* and 2*W* respectively, are freely jointed at B. They are in equilibrium standing on rough ground in a vertical plane with each rod making 45° with the horizontal. Find the frictional forces at A and C.

3   Three identical uniform rods of weight *W* are smoothly jointed to form an equilateral triangle, which is suspended from the midpoint of one of its sides. Find the magnitude and direction of the reaction at each of the joints.

4   A step ladder consists of two identical parts of weight W, jointed at the top and held together by a light rope attached to the midpoints of the two parts. When the feet of the stepladder are as far apart as possible the angle between each part of the ladder and the ground is $\tan^{-1}\frac{8}{5}$. If a man of weight $4W$ stands three-quarters of the way up the ladder find the tension in the rope and reaction at the joint given that the frictional forces between the ground and the ladder are negligible.

*5   A stepladder consists of two identical parts each of length $2a$ and weight $2W$, jointed at the top. It stands on a smooth horizontal plane and a weight W is hung from a rung of the ladder at a distance l from its lower end. A light rope of length $2b$ is attached to the feet of the two halves of the ladder. Show that the tension in the rope is

$$\frac{(l+4a)bW}{4a(4a^2-b^2)}$$

and find the pressure of each part of the ladder on the ground.

Exercise 12.3

1

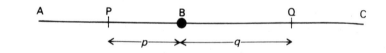

Two uniform rods, AB and BC, are smoothly jointed at B and each has length $2a$ and weight W. They rest on two knife-edge supports at P and Q, which are distances p and q from B respectively, as shown in the diagram. The system rests in equilibrium with ABC horizontal.
(a)   Find the reactions at P and Q.
(b)   Find the magnitude of the mutual reaction at the joint B.
(c)   Show that, if $p > a$ then $q < a$.

2   Two uniform rods, AB and BC, each of length $2a$, and of weights $3W$ and $2W$ respectively, are smoothly jointed at B. A and C are smoothly hinged to fixed points on the same horizontal level, and the rods hang in a vertical plane so that B is below A and C, and angle $BAC = \theta$.

(a)   Find the horizontal and vertical components of the reactions at A and C.
(b)   Show that the condition for the reaction at C to be perpendicular to the reaction at A is that $\tan^2\theta = 25/99$.
(c)   Find the horizontal and vertical components of the reaction at B.
(d)   Find the condition that the reaction at B is perpendicular to BC.

3   Two uniform ladders AB and AC of equal length $2a$ and weights $2W$ and W respectively are smoothly jointed at A and rest in a vertical plane with B and C in contact with a rough horizontal floor. The coefficient of friction

μ is assumed to be the same at both points of contact. Given that the ladders are in equilibrium with angle BAC = 2α find, in terms of W and α, the normal reactions and the friction forces at B and C. Show that, for limiting equilibrium, $\mu = 3 \tan \alpha / 5$.

A man of weight $10W$ climbs up AC. With the above value of μ find which ladder slips first and how far up AC the man then is. [A]

4 A uniform rod AB of length a and weight W is smoothly jointed at B to a uniform rod BC of length $a\sqrt{2}$ and also of weight W. The system is in equilibrium with C resting on a rough horizontal floor at a distance $2a$ from a rough vertical wall, A resting against the wall, and AB horizontal at a height a above the floor. The plane of the rods is perpendicular to the floor and the wall.

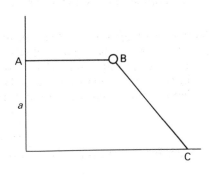

(a) Find the frictional force exerted by the wall at A.
(b) Find the vertical and horizontal components of the force exerted on BC at C.
(c) Show that equilibrium is possible only if the coefficient of friction at C is at least $\frac{2}{3}$.
(d) Find the least value of the coefficient of friction at A for equilibrium to be possible. [J]

5

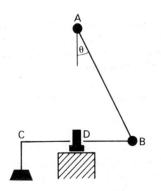

A uniform thin rod AB of length $2l$ and weight W is smoothly hinged at A to a fixed point and at B to a light rod BC of length $4l \sin \theta$. The rod BC is maintained in a horizontal position by a fixed smooth support at its mid-point D, which is vertically below A, and the rod is prevented from moving horizontally by a smooth vertical pin which is fixed to the support at D and passes through a small hole in the rod. A weight kW is suspended from C.

Find, in terms of W, θ and k, the horizontal and vertical components of:

(a) the reaction on AB at A
(b) the reaction on AB at B
(c) the reaction on BC at D

stating, or indicating clearly in a diagram, the directions you have assigned to the components.

Show that:

(d) there is no force exerted on the pin if $k = \frac{1}{2}$
(e) the reaction at A is perpendicular to AB if $k = \frac{1}{2}(1 + \cos^2 \theta)$. [J]

*Exercise 12.4

1   A uniform rod AB of weight W is
    smoothly hinged at A to a fixed
    point on a rough horizontal floor
    and is free to rotate in a vertical
    plane. A light disc is smoothly
    pivoted at its centre to the point B
    so that it is free to rotate in the
    same vertical plane. The edge of the
    disc touches the floor at C, and the
    angle ABC is θ, as shown in the
    diagram. The system is in equili-
    brium under the action of a force, of
    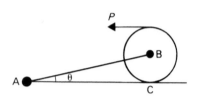
    magnitude P and direction parallel to CA, applied at the highest point
    of the disc. Find the horizontal and vertical components of the reaction at
    C and show that equilibrium is possible only if $P \leqslant \dfrac{\mu W}{2 + 4\mu \tan \theta}$, where μ
    is the coefficient of friction between the disc and the floor.
    Find the magnitude of the force exerted on the rod by the pivot A, and
    show that it acts in a direction making an angle ϕ with the horizontal,
    where $\tan \phi = \dfrac{W}{4P} + \tan \theta$.
    Find the horizontal and vertical components of the force exerted by the
    rod on the disc.   [J]

2   The diagram shows a smooth uni-
    form rod AB, of length $2a$ and mass
    m, hinged at a fixed point A on a
    smooth horizontal plane and sup-
    ported at C, where $AC = 5a/4$, by a
    fixed smooth uniform circular disc.
    The disc lies in the vertical plane
    containing AB. A particle of mass
    m is attached to the rod at B.
    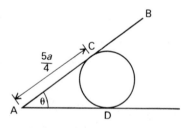

    Given that the rod and attached particle are in equilibrium with angle
    $DAB = \theta$ find:

    (a)   the reaction of the disc on the rod
    (b)   the vertical and horizontal components of the reaction at A on the rod
    (c)   the maximum value, as θ varies, of the horizontal component of the
          reaction at A on the rod
    (d)   the value of $\cos^2 \theta$ when the reaction at A is inclined at an angle θ to
          the upward vertical.

    Given that the disc is no longer fixed, but that its lowest point D is connected
    to A by a light inextensible string of length $5a/4$, find, in terms of m and θ,
    the least mass which, when placed midway along a radius of the disc, is
    sufficient to maintain equilibrium.   [A]

3

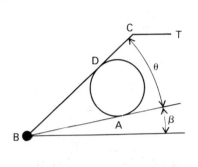

A uniform smooth rigid sphere of weight W is in contact at a point A with the upper side of a fixed smooth plane surface inclined at an angle β to the horizontal. A smooth light rod BC is smoothly hinged to the plane at a point B which is below A on the line of greatest slope through A. The mid-point of the rod rests on the sphere at D, and the plane ABC is vertical. The angle ABC is θ, and $\beta + \theta < \pi/2$. Equilibrium is maintained by a horizontal force of magnitude T applied at C.

Show that the force exerted on the rod by the hinge has magnitude T and acts in a direction making an angle $2(\beta + \theta)$ with the horizontal. Indicate the sense of this force by means of a diagram.

Show that the magnitude of the reaction at D is $W \sin \beta / \sin \theta$. Hence or otherwise find T in terms of W, β and θ.   [J]

Exercise 12.5

1 A, B, C, D lie in a straight line with $AB = BC = CD = a$. Forces of 4, 5, 2, 1N act at A, B, C and D respectively in directions making angles of $60°$, $90°$, $120°$ and $270°$ with AD. Find the magnitude and direction of the resultant and show that it cuts AD at approximately $0.6a$ from A.

2

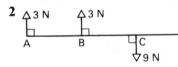

$AB = BC = CD = 1$ m

Show that this system of forces is equivalent to a force at C and a couple and find their magnitudes.

3 Forces of 4, 2, 1 N act along the sides PQ, PR, QR respectively of an equilateral triangle PQR, in the directions indicated by the order of the letters. Prove that their resultant is a force of $3\sqrt{3}$ units in a direction perpendicular to QR and find the point where the line of action meets QR.

4 Forces of 3, 4, 2, 1 N act along the sides PQ, QR, RS, SP of a square PQRS of side 1 m. Find the force at P and the couple which together are equivalent to this system.

5 A rigid square lamina ABCD of side a is subject to forces of magnitude 1, 2, 3, 1, $3\sqrt{2}$ and $\lambda\sqrt{2}$ units acting along AB, BC, CD, AD, AC and DB respectively in the directions indicated by the order of the letters. Given that the direction of the resultant force is parallel to AC, find λ.

With this value of λ, find the total moment about A of the forces acting on the lamina. Hence, or otherwise, find AE in terms of a, where E is the intersection of AB with the line of action of the resultant force.   [J]

6 A rigid rectangular lamina ABCD, with $AB = 4a$ and $BC = 3a$, is subject to forces of magnitudes $10P, P, 2P, 3P$ acting along CA, AD, DC, CB respectively in the directions indicated by the order of the letters.

(a) Find the magnitude of the resultant of the four forces.

(b) Find the tangent of the acute angle between the line of action of the resultant and the edge AB of the lamina.

(c) Find the distance from A of the point where the line of action of the resultant meets AB.

(d) Indicate clearly on a diagram the line of action and the direction of the resultant.

(e) Find the magnitude and sense of the couple G which, if added to the system, would cause the resultant force to act through E, the midpoint of CD.

(f) In the case when G is *not* applied, find forces S along AB, T along AD and U along BC which, when added to the system, would produce equilibrium.

*7 A light rigid bent rod OAB consists of two straight segments OA of length a and AB of length h; the angle between AB and OA produced is acute and equal to θ.

The rod is rigidly clamped at O and is subject to a force of magnitude F acting at B in the plane OAB and perpendicular to AB in the sense shown in the diagram. Equilibrium is maintained by the system of forces exerted by the clamp on the rod. Express this system as a force through O together with a couple, giving the magnitude of the force and the moment M_O of the couple in terms of a, h, θ and F, and showing on a diagram the direction of the force and the sense of the couple. Express likewise the system of forces exerted by the segment OA on the segment AB as a force through A and a couple. Denoting the moment of this couple by M_A, show that

$$\frac{M_O}{M_A} = 1 + \frac{a}{h}\cos\theta \qquad\qquad \text{[J]}$$

Miscellaneous Exercise 12B

1 Two uniform rods, AB and BC, each of length $2l$ and of weights W and $2W$, respectively, are smoothly hinged together at B, and AB is smoothly hinged to a fixed point at A. The points A, B and C lie in a straight line which makes an angle θ with the vertical. Equilibrium is maintained by the action of a horizontal force $5W/2$ at C and a clockwise couple of moment G applied to BC, as shown in the diagram.

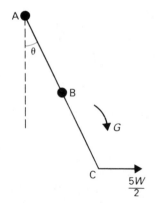

Find the horizontal and vertical components of the force exerted at B
by BC on AB, and hence find the value of θ. Find the value of G.

[J: 12.2, 12.3]

2 Two uniform rods AB and BC, each of weight W and length $2a$, are
smoothly jointed together at B. The rod AB is freely hinged at the fixed
point A to a thin smooth vertical wire. A small light ring threaded on the
wire is attached to the rod BC
at C. The rods are held at right
angles to each other, as shown
in the diagram, by means of a
light inextensible string whose
ends are attached to the midpoints D and E of AB and BC
respectively. Show that the tension in the string is of magnitude
$2W$. Find the horizontal and
vertical components of the force
exerted at B by AB on BC.

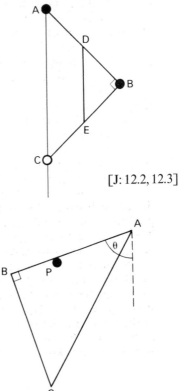

[J: 12.2, 12.3]

3 Two identical uniform rods AB
and BC, each of mass m and
length $2a$, are smoothly joined at
B. The ends A and C are joined
by a light inextensible string of
length $2\sqrt{2}a$. The diagram shows
the system in equilibrium in a
vertical plane with the road AB
supported by a rough peg at P
and the string joining A and C
taut so that the angle ABC is a
right angle. The distance AP is
denoted by $x(< 3a/2)$ and θ denotes the angle between AB and
the downward vertical.

By taking moments about P, or otherwise, express $\cot\theta$ in terms of a and x.
For the particular case when the peg is very rough so that the coefficient
of friction between the peg and the rod is 2.5 find the least value of x for
which equilibrium is possible.
Express the tension in the string in terms of m, g and θ and find the value
of x such that the reaction of the rod BC at B, on the rod AB, is parallel
to AB.

[A: 12.2, 12.3]

4 Two uniform straight rods AB and BC are smoothly hinged together at
B. They rest in equilibrium, with A, B and C in a horizontal straight line,
on two supports, one at a point D in AB and the other at a point E in
BC. The length of AB is $4a$ and its weight is $2W$; the length of BC is $6a$
and its weight is $3W$. Given that $BE = 4a$, find AD. Find also the
magnitude and direction of the force exerted on the rod BC by the rod AB.
The support formerly at D is now placed at the mid-point of AB; the

support at E is not moved. A couple is applied to the rod BC so as to maintain the rods in equilibrium in a horizontal line. Find the magnitude and sense of this couple.   [J: 12.2, 12.3]

5

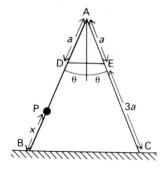

A step-ladder has two sides AB and AC, each of length $4a$ smoothly hinged at A and connected by a light taut rope DE, where $AD = AE = a$. The side AC is light, and the side AB is of weight W and is assumed to be uniform. A man of weight $4W$ stands at a point P on AB, where $BP = x$, and the system is in equilibrium with B and C on a smooth horizontal floor and the sides at an angle θ to the vertical. Find the reaction at C and the tension in the rope.

Given also that DE is an elastic rope of natural length $\frac{1}{4}a$, modulus W, show that the condition for equilibrium is

$$x = \tfrac{1}{2}a(8\cos\theta - \cot\theta - 1)$$

Show that the maximum value of x for which equilibrium is possible is just over $2a$.   [J: 12.2, 12.3, 10.1]

6

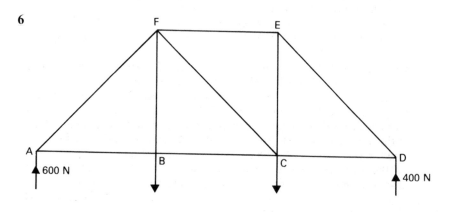

The diagram shows a framework ABCDEF consisting of nine smoothly jointed light rods AB, BC, CD, DE, EF, FA, FB, FC and EC. The rods AB, BC, CD and EF are horizontal, the rods BF and CE are vertical, whilst the other three rods are inclined at 45° to the horizontal. The framework is simply supported in a vertical plane by upwards forces at A and D and carries vertical loads at B and C. The loads at B and C are such that, when the framework is in equilibrium, the forces at A and D are of magnitude 600 N and 400 N respectively.
Find: (a) the load at C
       (b) the forces in AF, AB.
Write down the magnitudes of the forces in DE and CE.   [A: 12.1]

7 The diagram shows a framework consisting of six light rods CB, CD, DE, EA, EB and EC smoothly jointed at D, E, C and B. The framework is in a vertical plane and is smoothly hinged to fixed points A and B so that AB is horizontal.

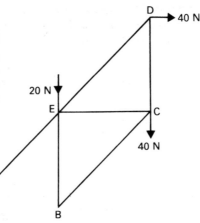

The rod EC is horizontal, EB and CD are vertical and each of the other rods makes an angle of 45° with the horizontal. Vertical loads of 20 N and 40 N are applied at E and C respectively. A horizontal force of 40 N is applied at D in the sense shown.
Find:
(a) the horizontal and vertical components of the reaction of the hinge at A
(b) the horizontal and vertical components of the reaction of the hinge at B and the magnitude of this reaction
(c) the forces in the rods EA, BC, and ED.
State which of the rods EA, BC, and ED are in tension.   [A:12.1]

*8 Three uniform straight rods, each of weight W, are smoothly jointed together to form an equilateral triangle ABC. The joint B is hinged to a fixed support and the system is free to rotate about B in a vertical plane. The system is held in equilibrium with BC horizontal and A uppermost by a horizontal light string which connects A with a fixed point. Show that the tension in the string is $W\sqrt{3}$ and find the magnitude and direction of the force exerted on the triangle by the support at B.
Find the horizontal and vertical components of all the forces which are exerted on the rod AC, and indicate them clearly on your diagram.
[J: 12.2, 12.3]

*9 A light structure ABC consists of a vertical post AB of height h fixed in the ground at A, joined rigidly to a horizontal cross bar BC of length b. A force of magnitude F is applied at C, in the plane ABC, at an angle θ to the downward vertical, as shown in the diagram. Express the system of forces exerted by the ground on the post as a force through A together with a couple, giving the magnitude of the force and the moment M_A of the couple in

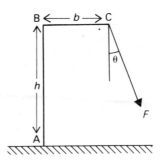

terms of h, b, θ and F. Show on a diagram the direction of the force and the sense of the couple. Express likewise the system of forces exerted by the post on the crossbar as a force through B and a couple M_B.

The angle θ is varied while the magnitude F of the force applied at C remains constant. State the directions of the force applied at C for which M_A and M_B, respectively, are maximum.   [J: 12.5]

*10

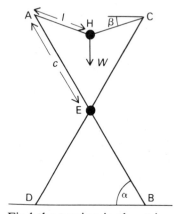

The diagram shows two light rods, AB and CD, each of length $2c$, joined at their midpoints by a smooth pivot E. The ends A and C are connected by a light inextensible string, of length $2l$, where $l < c$, which carries a particle of weight W at its midpoint H. The whole system is confined to a vertical plane, and is in equilibrium with B and D resting on a rough horizontal floor, the rods at an angle α to the horizontal, and the strings AH and CH at an angle β to the horizontal.

Find the tension in the string AH in terms of W and β. Show that the force of friction exerted on AB at B is $W(\frac{1}{2}\cot\beta - \cot\alpha)$ in the direction DB.

Show that the reaction exerted at E on the rod AB by the rod CD is horizontal and of magnitude $W(\cot\beta - \cot\alpha)$, and indicate its direction on a diagram.

Show that $l\cos\beta = c\cos\alpha$.

Given that $8l = 5c$, find the least value of the coefficient of friction between the rods and the floor for which the system can rest in equilibrium with $\alpha = 60°$.   [J: 12.2, 12.3]

ANSWERS

Unit 1

Exercise 1.1

1 (a) $3i+j$ (b) $-3i+j$ (c) $3i$ (d) $-3i-j$ (e) $3i-j$ (f) $-3j$

2 (a) $\begin{pmatrix} -2 \\ -4 \end{pmatrix}$ (b) $\begin{pmatrix} 4 \\ -2 \end{pmatrix}$ (c) $\begin{pmatrix} -5 \\ 0 \end{pmatrix}$

3 (a) 5 (b) 25 (c) $4i+2j$ (d) $i+j$ (e) $\sqrt{5}$ (f) 3
   (g) $2i-4j$ (h) $1.5i+2j$ (i) $i-5j$ (j) $16j$ (k) $\sqrt{2}$

4 (a) $5.2i+3j$ (b) $7.07i+7.07j$ (c) $-3j$ (d) $2.83i-2.83j$
   (e) $-4i-6.93j$ (f) $-7i$

5 (a) 13 at $67.4°$ above the positive x-axis
   (b) $\sqrt{2}$ at $45°$ above the positive x-axis
   (c) $\sqrt{2}$ at $135°$ above the positive x-axis
   (d) $\sqrt{13}$ at $56.4°$ below the positive x-axis.

6 $9.6i+2.8j$

7 $-3.6i+4.8j$

8 (a) $\sqrt{34}$ (b) $\sqrt{2}$ (c) $\begin{pmatrix} 4 \\ 4 \end{pmatrix}$ (d) $\begin{pmatrix} 3 \\ 2 \end{pmatrix}$ (e) $\sqrt{13}$ (f) 2.83

9 (a) $\begin{pmatrix} 3 \\ 4 \end{pmatrix}$ (b) $\begin{pmatrix} -2 \\ 5 \end{pmatrix}$ (c) $\begin{pmatrix} 5 \\ -1 \end{pmatrix}$ (d) $\begin{pmatrix} -5 \\ 1 \end{pmatrix}$

10 (a) $1.92i-4.62j$ (b) $-7.8i+10.4j$ (c) $0.54i-1.31j$ (d) $3.54i+3.54j$

Exercise 1.2

1 $20\,\text{m/s}$ 2 $9.72\,\text{m/s}$ 3 $234\,\text{km/h}$
4 $36\,\text{km/h}$ 5 $50\,\text{m/s}$ 6 $6.25\,\text{m/s}$
7 Either $41.7\,\text{m/s}$ **or** $150\,\text{km/h}$
8 $1116\,\text{m}$ 9 $310\,\text{m}$ 10 $20\,\text{s}$
11 $50.4\,\text{s}$ 12 $1360\,\text{m}$ 13 $8\,\text{min}\,20\,\text{s}$
14 (a) $9.85\,\text{m/s}$ (b) $35.47\,\text{km/h}$
15 $3\,\text{min}\,54.78\,\text{s}$
16 (a) $2.08\,\text{m/s}$ (b) $1.25\,\text{m/s}$
17 (a) $12.5\,\text{min}$ (b) $36\,\text{min}$ (c) $9.9\,\text{km/h}$ (d) $8.2\,\text{km/h}$
18 $4.8\,\text{km}$
19 (a) $124.5\,\text{km/h}$ (b) $98\,\text{km/h}$ (c) $147\,\text{km/h}$ (d) $108.6\,\text{km/h}$
    (e) $49.8\,\text{km/h}$
20 $0.6\,\text{m/s}^2; 2.25\,\text{km}$

Exercise 1.3

1 $26\,\text{m/s}$ 2 $24\,\text{m}$ 3 $20\,\text{m/s}$
4 $200\,\text{m}$ 5 $10\,\text{m/s}$ 6 $3\,\text{s}$
7 $10\,\text{m/s}^2$ 8 $-\frac{3}{2}\,\text{m/s}^2$ 9 $\frac{10}{3}\,\text{s}$
10 $5\,\text{m/s}$ 11 $140\,\text{m}; 1.2\,\text{m/s}^2$ 12 $1.5\,\text{m/s}; 75\,\text{m}$
13 $3.75\,\text{m/s}^2; 14.6\,\text{m/s}$ 14 $69\,\text{cm}; \frac{7}{6}\,\text{cm/s}^2$ 15 $126\,\text{km/h}$
16 $1\,\text{m/s}^2; 8.5\,\text{m}$

Exercise 1.4
1 (a) 490 m   (b) 4.5 s   (c) 44.3 m/s
2 164.1 m   3 30.6 m; 1 s   4 14 m/s
5 78.4 m   6 1.22 s, 1.84 m, 6 m/s   7 2 s
8 0.61 m   9 10.5 m   10 4 s
11 51 m   12 122.5 m

Exercise 1.5
1 $816\frac{2}{3}$ m   2 4 m/s²; 12 m/s
3 4.77 s after first throw; 16.75 m/s↓, 2.15 m/s↓

Exercise 1.6
1 16.032 km   2 15 m/s   3 0.1 m/s²; 0.2 m/s²   4 200 s

Miscellaneous Exercise 1A
1 (a) 2   (b) 5   (c) $5\mathbf{i} + 2\mathbf{j}$ (d) $\sqrt{5}$; N26.6°E   (e) $3\mathbf{i} + 6\mathbf{j}$

2 $\begin{pmatrix} 1 \\ 3 \end{pmatrix}$ 3 B   4 6   5 (a) $2\mathbf{i} + \mathbf{j}$ (b) $7\mathbf{j}$ 6 $3, -1$

7 2; 5 h   8 $83\frac{1}{3}$ s, $2083\frac{1}{3}$ m

Miscellaneous Exercise 1B
1 (a) 4 s   (b) 20 m   2 40 m/s² from O to B, 10 m/s from B to O
3 7 min 40 s   4 (a) $\frac{1}{6}$ (b) 800 m   5 $1\frac{3}{7}$ s
6 20 m/s   7 $T = \frac{1}{3}V/f$
8 (a) 900, 1800 (b) 4 min (c) 1 km (d) 46 min
9 126 m; 30 m/s; PQ $= |80 - 20t - 2t^2|$; 2 s and 4 s
10 20 s

Unit 2

Exercise 2.2
1 (a) 0.5 m/s² (b) 500 m/s²
2 5 grams   3 0.16 N   4 $(0.25\mathbf{i} + 2\mathbf{j})$ m/s²
5 $(30\mathbf{i} - 10\mathbf{j})$ N   6 2 N   7 8 s
8 (a) 80 N (b) 40 N
9 2.5 m/s²   10 23 N
11 (a) 25.6 N (b) 13.6 N (c) 19.6 N
12 2060 N   13 1.53×10^5 N   14 102 s

Exercise 2.3
1 $2g/5$ m/s²; $21g/5$ N   2 $3g/7$ m/s²; $g/70$ N
3 $g/6$ m/s²; $25g/3$ N; 31.3 m/s   4 380 N
5 $6g/29$ m/s²; $230g/29$ N and $140g/29$ N

6 $\dfrac{(m_1 - m_2)g}{m_1 + m_2}$

7 $1183\frac{1}{3}$ N; $\frac{7}{6}$ m/s²; $\frac{5}{12}$ m/s²; a thrust of $\frac{250}{3}$ N.

Exercise 2.4
1 (a) 2.83→, 2.83↑ (b) 3→, 0↑   (c) 8.66→, − 5↑
    (d) − 1→, 1.73↑ (e) − 2→, − 3.46↑ (f) 6→, $2\sqrt{3}$↑
2 (a) $-4\mathbf{i} + 6.93\mathbf{j}$ (b) $4\mathbf{i} - 4\mathbf{j}$ (c) $-4\mathbf{i} - 4\mathbf{j}$
    (d) $R\cos\alpha\mathbf{i} + R\sin\alpha\mathbf{j}$ (e) $-S\cos\beta\mathbf{i} + S\sin\beta\mathbf{j}$ (f) $-T\sin\gamma\mathbf{i} - T\cos\gamma\mathbf{j}$
3 (a) 84.87 N (b) 49 N   4 $3g$ N   5 (a) 9.57 (b) 1.01

6 (a) 6.2 (b) 8.66   **7** (a) -2.16 (b) 5.73
8 (a) $R\cos\alpha - T\sin\beta$ (b) $R\sin\alpha + T\cos\beta - S$
9 (a) 14.66 N (b) 0   **10** (a) 5.66 N (b) 1.8 N
11 (a) 0 (b) 11.14 N   **12** (a) -15.31 N (b) -2 N

Exercise 2.5
1 $a = -3,\ b = -2$
2 (a) 5; 53.1° to direction of **i** (b) 5; $-53.1°$ to direction of **i**
  (c) 5; $-126.9°$ to direction of **i**
3 2.83 N at 45° to AB
4 $X = 5.56$ N, $Y = 8.55$ N; R is 10.19 N at 57° above the positive x-axis
5 $X = -1.14$ N, $Y = -13.26$ N; R is 13.31 N at 85° below the negative x-axis
6 $X = 5.57$ N, $Y = -9.5$ N; R is 11.01 N at 59.6° below the positive x-axis
7 $X = -14.65$ N, $Y = 4.14$ N; R is 15.22 N at 15.8° above the negative x-axis
8 3 N at 19.5° to PQ   **9** 8.23 N at 44.5° to AB.
10 $R = 25$ N

Exercise 2.6
1 $R = 6$ N, $S = 2\sqrt{2}$ N   **2** $P = 4$ N, $Q = 2\sqrt{2}$ N   **3** $X = 8$ N, $Y = 4$ N
4 $\theta = 26.6°$, $Z = 6.7$ N   **5** $P = 3$ N, $R = 3\sqrt{3}$ N   **6** $3\sqrt{2}$ N, $R = 5\sqrt{2}$ N
7 (a) $-6, -3$ (b) $1, 2$ (c) $2, -1$ **8** $0.79g$ N, $1.48g$ N
9 (a) $4\sqrt{3}g$ N (b) $8\sqrt{3}g$ N   **10** 38.4 N, 21.4°

Exercise 2.7
3 $\theta = 41.8°$, $T = 2g$ N, $R = 2.24g$ N   **4** 16.1°
5 52.9 N, 22.2°; 63.2 N, 52.2°   **6** $5\sqrt{2}g$ N, $5\sqrt{2}g$ N, $10g$ N
7 $mg[1 + k^2 - 2k\cos\alpha]^{1/2}$

Exercise 2.8
1 (a) 690 N (b) 290 N   **2** (a) 19.6 N (b) 25.6 N
3 0.2 m/s²   **4** $\frac{10}{13}$ m/s²   **5** 1.36 km
6 0.0575   **7** $m = 9$, $a = \frac{10}{9}$ m/s²   **8** 44.6 m

Exercise 2.9
1 $g/16$ m/s², $45g/16$ N   **2** $g/14$ in the opposite direction
3 2.02 s, 1.21 s   **4** (a) 1.6 s (b) 1.01 s

5 $a = \dfrac{mg - Mg\sin\alpha}{m + M}$; $\sin\alpha < m/M$

Exercise 2.10
1 $10g/13$ m/s²; $18g/13$ N   **2** $2Mmg/(4M + m)$
3 $4g/11$ m/s²; $6g/11$ m/s²; $12g/11$ N   **4** $g/9\downarrow, g/9\downarrow, g/3\uparrow$; $8g/3$ N

5 $g\sqrt{3}/15$; $14\sqrt{3}g/15$ N; $8.4g$ N   **6.** $\dfrac{3MM'}{M + 4M'}g$

7 $mg\sin\alpha\cos\alpha$ **8** $8g/3$ N

Miscellaneous Exercise 2A
1 $2\mathbf{i} + \mathbf{j}$ **2** 19 N   **3** $\frac{25}{18}$ m/s²; $\frac{10000}{9}$ N
4 (a) $R\cos\alpha + F\sin\alpha - W$ (b) $F\cos\alpha - R\sin\alpha$ (c) $F - W\sin\alpha$ (d) $R - W\cos\alpha$
5 (a) (i) -9 N (ii) 1 N (b) 9.06 N at 6.3° to AB
6 $g/5$ m/s²; 30 cm   **7** 9.47 N at 30° to PQ
8 $460g$ N; 11.8 m/s   **9** $2g$ N and $2\sqrt{3}g$ N
10 7 m/s²

Miscellaneous Exercise 2B

1 $\sqrt{\dfrac{2gl}{3}}$ **2** $\frac{1}{2}\,\text{m/s}^2; 3\,\text{m/s}; \frac{3}{4}\,\text{m/s}^2; 6\,\text{m}$

3 $2mg - T = 2ma;\ S - 6mg = 6ma;\ T - S = 8ma;\ g/4;\ 2.5mg$ and $4.5mg;\ 5x/4$

4 $m(g+f)$ **5**   840 N

6 $\frac{13}{10}mg$ (b) $\frac{20}{11}Mg$

7 $R + \dfrac{m\sin\alpha}{h}(gh + 2u^2)$

8 $\left(\dfrac{k-4}{k+4}\right)g;$ (a) $\dfrac{6mgk}{k+8}$ (b) $\dfrac{3mgk}{k+8}$

Unit 3

Exercise 3.1

1 (a) $10g$ N (b) $20\sqrt{2}g/3$ N   **2** $\frac{1}{4}$ **3** 29.4 N   **4** 259.8 N; no.

5 (a) $1.1\,\text{m/s}^2$ (b) $1.325\,\text{m/s}^2$ (c) $2.6\,\text{m/s}^2$ **7** 0.064

8 (i) (a) 7.84 (b) 7.35 (c) 11.8 (ii) 13.5

Exercise 3.2

1 (a) 17 N; at rest (b) 33.5 N; at rest; (c) 42.4 N; not at rest   **2** $1/\sqrt{3}$

3 $\frac{1}{8}$ **4** $2\sqrt{3}g$ N   **5** (a) $8.8g$ N (b) $15.2g$ N   **6** $a = 114, b = 2085$

Exercise 3.3

1 3.91 s; 7.67 m/s   **2** (a) $6\,\text{m/s}^2; 22$ N (b) $1.1\,\text{m/s}^2; 22$ N

3 The cake will slip   **4** 34.3 m; 9.8 m/s   **6** $2.24\,\text{m/s}^2$

7 (a) $g/6$ (b) 1.36 s (c) 1.6 m/s

Exercise 3.4

1 $m_1 > m_2 + \mu M;$ $T_1 = \dfrac{m_1 g[2m_2 + M(1+\mu)]}{m_1 + m_2 + M}$

4 (a) $a = \dfrac{(m - M\sin\alpha - \mu M\cos\alpha)}{M+m}g;$ $T = \dfrac{mMg}{M+m}(1 + \sin\alpha + \mu\cos\alpha)$

(b) $a = \dfrac{(M\sin\alpha - \mu M\cos\alpha - m)}{M+m}g;$ $T = \dfrac{mMg}{m+M}(1 + \sin\alpha - \mu\cos\alpha)$

Exercise 3.5

1 $25°$ **2** $26.2°$ **3** 47.5 N at $14.1°$ above the horizontal

4 721 N at $\tan^{-1}\frac{3}{4}$ to the plane

Miscellaneous Exercise 3A

1 (a) 300 N; does not move (b) 343 N; moves; $1.14\,\text{m/s}^2$

2 286   **3** 156 N, 39 N, $3.25\,\text{m/s}^2$ **5** $\tan\alpha$ **6** 0.06

7 0.29   **8** 14.5 N   **9** (a) 26 N (b) 1107 N

10 $7Mmg/4(M+m;\ (4M-3m)g/4(M+m)$

Miscellaneous Exercise 3B

1 0.077   **2** 0.304   **3** $\sqrt{3}-1$ **4** 25 m

5 (a) $\frac{9}{25}g$; $\dfrac{48mg}{25}$ (b) $\dfrac{7g}{25}$; $\dfrac{54mg}{25}$

6 $4.5\,\text{m/s}^2$; $42\,\text{N}$ and $24\,\text{N}$ (b) $2\,\text{m/s}^2$; $42\,\text{N}$ and $24\,\text{N}$

7 (c) $mg\cot\theta$ (d) $3mg\cot\theta$ **8** $\dfrac{9mg}{10}$ **9** $\dfrac{W(\tan\alpha-\mu)}{1+\mu\tan\alpha}$

10 $W\cos\alpha\sec(\alpha-\lambda)$

Unit 4

Exercise 4.1
1 (a) $5\,\text{N},36.9°$ (b) $19.4\,\text{N},10.4°$ (c) $7.6\,\text{N},28.5°$
(d) $10.5\,\text{N},10.8°$ (e) $13.5\,\text{N},45.7°$ **2** $1.6\,\text{N},38.3°$
3 $130.5°$ **4** $P\sqrt{7},40.9°$ **5** $9.4\,\text{N}$, at $20°$ to both forces
6 P at $60°$ to both **7** $5.2\,\text{N},\text{N}33°\text{E}$
8 (a) $26.5\,\text{N},139.2°$ (b) $5.4\,\text{N},146.8°$ (c) $4\,\text{N},6.9\,\text{N}$
9 3 and $2\sqrt{3}$ **10** 5, east, $18.03\,\text{N}$, $\text{S}43.9°\text{W}$.

Exercise 4.2
1 $15\,\text{Nm}\circlearrowright$ **2** $15\,\text{Nm}\circlearrowleft$ **3** $18\,\text{Nm}\circlearrowright$ **4** $1\,\text{Nm}\circlearrowleft$ **5** 0
6 $6\,\text{Nm}\circlearrowright$ **7** $7\,\text{Nm}\circlearrowleft$ **8** $6\,\text{Nm}\circlearrowleft$ **9** $7\,\text{Nm}\circlearrowleft$ **10** $3\,\text{Nm}\circlearrowright$
11 $16\,\text{Nm}\circlearrowleft$ **12** $6\,\text{Nm}\circlearrowright$ **13** $10\,\text{Nm}\circlearrowleft$
14 (a) $12\,\text{Nm}\circlearrowright$ (b) $4\,\text{Nm}\circlearrowright$ (c) $4\,\text{Nm}\circlearrowleft$ (d) $8\,\text{Nm}\circlearrowleft$ (e) $20\,\text{Nm}\circlearrowleft$
15 $6\sqrt{3}\,\text{Nm}\circlearrowleft$ **16** $10\,\text{Nm}\circlearrowright$ **17** $2\,\text{Nm}\circlearrowright$ **18** (a) $2\,\text{Nm}\circlearrowleft$ (b) $1\,\text{Nm}\circlearrowright$
19 $20\,\text{Nm}\circlearrowleft$ **20** $12\sqrt{3}\,\text{Nm}\circlearrowright$ **21** 0 **22** $13\,\text{Nm}\circlearrowright$

Exercise 4.3
1 (a) $110\,\text{N}$ in direction of forces, $14\,\text{cm}$
(b) $30\,\text{N}$ in direction of $70\,\text{N}$, $51\frac{1}{3}\,\text{cm}$
2 (a) $6\,\text{Nm}\circlearrowleft$ (b) No (c) $33\,\text{Nm}\circlearrowleft$ (d) $8\,\text{Nm}\circlearrowleft$ (e) No
(f) No (g) $6\,\text{Nm}\circlearrowright$ (h) $7\,\text{Nm}\circlearrowright$
3 $P=12$; $10\,\text{m}$ **4** $25\,\text{N}$ and $75\,\text{N}$ **5** $1\frac{1}{3}\,\text{Nm}$ **6** $12\,\text{Nm}\circlearrowright$
7 $3\,\text{Nm}\circlearrowright$ **8** $X=8,Y=12$ **9** $Y=40,Z=32$
10 $4\,\text{N},\uparrow$, on BA produced, $5\,\text{m}$ from A

Exercise 4.4
1 $12g\,\text{N},14g\,\text{N}$ **2** $20g\,\text{N},40g\,\text{N}$ **3** $20g\,\text{N},30g\,\text{N}$ **4** $20\,\text{kg}$
5 $86.5\,\text{cm}$ from the end from which measurements were taken
6 $20\,\text{kg}$ **7** $2.18\,\text{m}$ from A
8 Within $50\,\text{cm}$ either side of centre
9 $\frac{50}{3}\,\text{kg}$ **10** $(25+5m_1-m_2)g/4$; $\frac{25}{6}$ and $\frac{35}{6}$

Exercise 4.5
1 (a) $0.3g\,\text{N}$ (b) $5g/13\,\text{N}$ (c) $g\sqrt{3}/3\,\text{N}$ (d) $g\sin\theta\,\text{N}$
2 (a) $3\sqrt{3}g/4\,\text{N}$ (b) $7\sqrt{3}g/2\,\text{N}$ (c) $3\sqrt{3}g/2\,\text{N}$ (d) $3g\cos\alpha/(2\sin\theta)$
3 $11\sqrt{3}g/2\,\text{N}$ **4** $10\sqrt{3}Mg/3\,\text{N}$ **6** $50\sqrt{3}g/3\,\text{N}$

Exercise 4.6
1 (a) $3\sqrt{3}g\,\text{N},3g\,\text{N},6g\,\text{N}$ (b) $3g\,\text{N},3g\,\text{N},3\sqrt{2}g\,\text{N}$
2 (a) $4g\,\text{N},4g\,\text{N}$ (b) $2\sqrt{2}g\,\text{N},2\sqrt{10}g\,\text{N}$
3 $4\sqrt{3}g\,\text{N},2\sqrt{3}g\,\text{N},2g\,\text{N}$ **4** (a) $10g\,\text{N}$ (b) $10\sqrt{3}g/3\,\text{N}$
5 $5\sqrt{3}g/3\,\text{N},10g\,\text{N},5\sqrt{3}g/3\,\text{N},\sqrt{3}/6$ **6** (a) $W/3$ (b) $W/2$

Exercise 4.7

1 (a) $5\sqrt{2}g$ N; $5\sqrt{2}g$ N; $45°$ (b) $5\sqrt{3}/3g$ N; $10.4g$ N; $16.1°$
2 60 cm; 8.72 N; $36.6°$ to the vertical   **3** $54.5°$
4 $36.1°$ **7** 0.269   **8** 7 m; $\frac{6}{19}$

Exercise 4.8

2 $2.5g$ N   **3** $6.5g$ N   **5** $W/6$; $W\sqrt{37}/6$ at $\tan^{-1}6$
6 $5g$ N and $5\sqrt{3}g$ N; $30°$
7 $77.1g$ N; $163.3g$ N at $63.7°$ to the horizontal.

Exercise 4.9

1 $\frac{3}{4}W, \frac{1}{4}W; \mu \geqslant \frac{1}{3}$ **2** $\dfrac{mg\sqrt{7}}{3}$

3 $mg\left[1 + \dfrac{\cos\alpha\sin\theta}{2\sin(\alpha-\theta)}\right], \dfrac{mg\cos\alpha\cos\theta}{2\sin(\alpha-\theta)}$

4 $\dfrac{W}{2\cos\alpha}[2-\sin^2\alpha]; \dfrac{W}{2}\sin\alpha; \dfrac{W}{2}\tan\alpha$

5 $2W\operatorname{cosec}\theta; 2W\cot\theta; 3W; \frac{12}{5}a$

6 $x \geqslant \dfrac{b-a}{\mu}$

Miscellaneous Exercise 4A
1 9.5 N, $17.3°$ **2** $148.5°$ **3** 5.6; $129.4°$ **4** 1 Nm
5 (a) 60 N in the direction of both forces, at 40 cm from the 20 N force
        (b) 20 N in the direction of the 40 N force, at 120 cm from the 20 N force
6 20 N and 30 N   **7** $40g/3$ N, $20g/3$ N; 20 N   **8** $2\sqrt{3}g/3$ N; $4\sqrt{3}g/3$ N
9 $9\sqrt{3}g/2$ N; $9\sqrt{3}g/4$ N and $g/4$ N
11 $16g$ N and $18.3g$ N; $40.9°$ **12** 12.5 cm

Miscellaneous Exercise 4B
1 $3\frac{3}{4}$ N, $1\frac{1}{4}$ N; 20 Nm   **2** $2\sqrt{2}$ at $\pi/4$ below x-axis; 7
4 $\frac{1}{4}; \frac{3}{2}l$ **5** $3W\cos\theta - W\sec\theta; 3W\cos\theta - 2W\sec\theta$
6 (a) $\frac{5}{6}mg$ (b) $\frac{1}{6}mg\sqrt{13}; \frac{2}{3}; \frac{7}{10}mg$
8 (a) $\frac{1}{7}(M+m)g, \frac{2}{7}(M+m)g$ (b) $(8m-M)a/3m$
9 $\dfrac{\mu P\sin 2\theta}{2(\sin\theta + \mu\cos\theta)}; \dfrac{P\mu\cos^2\theta}{\sin\theta + \mu\cos\theta}$

10 $w - W\cos\theta$; Force at H is $W(b+c)\sin\theta/b$ horizontally; Force at K is $Wc\sin\theta/b$ horizontally

Unit 5

Exercise 5.1
1 147 J   **2** 58.8 J   **3** 5880 J   **4** 1176 J
5 117.6 J   **6** 36 J   **7** 78.4 J   **8** 294 J
9 (a) 45 J (b) 45 J (c) 1350 kJ (d) 0.5 J
10 (a) 1323 J (b) 35.28 J   **11** (a) 1568 J (b) 7644 J
12 (a) 75 kJ (b) 22.5 J   **13** 700 kJ   **14** 6 m/s   **15** 112.5 kJ
16 $\frac{1}{10}$ **17** 3 kJ   **18** 12.5 kJ   **19** (a) 1176 J (b) 448 J

Exercise 5.2
1 (a) 225 J   (b) 225 J   (c) 22.5 m   2 (a) 275 J   (b) 275 J   (c) 11 N
3 (a) 600 J   (b) 600 J   (c) $5\sqrt{5}$ m/s
4 (a) 48 J   (b) 58.8 J (c) 106.8 J   (d) 26.7 N
5 636 J   6 (a) 900 J   (b) 900 J   (c) $\frac{1}{4}$ 7 8.82 J; 4.2 m/s
8 3.7 m/s   9 2.53 N   10 960 J; 4800 N   11 32.4 cm
12 7.67 m/s   13 324 J; 0.14   14 44.7 m/s   15 $\mu mgd \cos \alpha$

Exercise 5.3
1 98 W   2 250 W   3 $65\frac{1}{3}$ W   4 0.15 W   5 1380 W
6 20 m/s   7 500 N; 12.5 kW   8 $\frac{1}{6}$ m/s²; $33\frac{1}{3}$ m/s²
9 310 N   10 0.505 m/s²; 0.995 m/s²   11 2400 N; 0.48 m/s²
12 3.2 m/s   13 517 W   14 (a) 10 m/s   (b) 4.46 kW   (c) 5.575 kW

15   $\dfrac{3P}{2MV} - \dfrac{g}{2n}$

Exercise 5.4
1 (a) 22500 (b) 12.5 (c) 525 (d) 1 (e) 160 000 (f) 0.001 (g) 17500
2 (a) 8 Ns in original direction   (b) 8 Ns in opposite direction
   (c) 32 Ns in opposite direction   (d) 40 Ns in opposite direction
3 60 N   4 (a) 200 Ns (b) 200 Ns (c) 40 m/s
5 55 m/s   6 2.56 Ns
7 (a) 27 Ns away from the wall   (b) 27 Ns in the opposite direction
8 32 Ns parallel to the positive direction of the x-axis
9 6i m/s   10 -6i Ns   11 28
12 (4i − 3j) m/s; 5 m/s   13 (22i + 53j) m/s   14 1 and 3
15 4500 N   16 0.05 N   17 34.56 N

Exercise 5.5
1 2 m/s   2 5.43 m/s   3 30 000 Ns; 8.57 m/s   4 10 m/s
5 21g 6 6 m/s; 0.75 J   7 1 m/s in the original direction
8 9.836 m/s 8.5 mm   9 2 m/s; 25 000 N   11 11.1 m/s; 15.7 m

Exercise 5.6
1 6 m/s and 7.5 m/s in original direction
2 53/13 m/s and 66/13 m/s in original direction
3 (a) 8/9 m/s and 64/9 m/s both in reversed directions
   (b) 8/9 m/s in same direction and 44/9 m/s in reversed direction
4 $\frac{3}{4}$
7 $\dfrac{(m_1 - em_2)v_1}{m_1 + m_2}; \dfrac{m_1 v_1 (m_2 - em_3)(1 + e)}{(m_2 + m_3)(m_1 + m_2)}; \dfrac{m_1 m_2 v_1 (1 + e)(1 + e')}{(m_1 + m_2)(m_2 + m_3)}$

Exercise 5.7
1 10.9 m/s and 2.6 m/s both reversed; 85.05 J
2 0.6 m/s and 1.5 m/s; 1350 J

Exercise 5.8
1 $\frac{3}{4}$ 2 0.6   3 9.06 m/s at 38.7° to the cushion
4 0.079 J   5 6.3 m/s at 9.4° to the plane; 4.625 J
6 (a) 4.5 m   (b) 2.53 m; 1.9 s

Exercise 5.9

1 (a) (i) $\frac{1}{2}u$ (ii) $\frac{1}{3}$ (iii) $\frac{5}{6}mu^2$ (b) (i) $\frac{3}{8}v$ (ii) $\frac{15}{8}mv$ (iii) $\frac{15}{16}mv^2$

2 $\dfrac{u}{3}(7 - 8e), \dfrac{u}{3}(7 + 16e); \frac{1}{8}; \dfrac{3u}{1 + k}; k > \frac{1}{2}$

3 $2\,\text{m/s}; \frac{11}{3}\,\text{m/s}; \frac{17}{18}\,\text{m}; \frac{29}{12}\,\text{J}$

4 $\frac{1}{2}(1 - e)u, \frac{1}{4}(1 + e)^2 u, \frac{1}{4}(e^2 + 4e - 1)u; \frac{1}{4}mu(e + 1)(e + 3)$ in the reverse direction.

5 $\dfrac{u(1 + e)(1 - e\lambda)}{(1 + \lambda)^2}, \dfrac{u(1 + e)^2}{(1 + \lambda)^2}$

6 u (a) $2mu$ (b) $mu^2; u^2/8g$

Miscellaneous Exercise 5B

2 $J/3$ **3** $25\,\text{m/s}$ **4** $\frac{1}{3}u$ **5** $\sqrt{(E/6m)}$ and $\sqrt{(3E/2m)}; \sqrt{(8E/3m)}$

6 $4mg - T = 4ma; T - 3mg = 3ma; a = g/7; T = 24mg/7$
(a) $7V/9$ (b) $8mV/9$ (c) $49V^2/18g$

7 (a) $u, 2u$ (b) $3mu$ **8** $10500\,\text{N}; 84\,\text{kW}; 15\,\text{m/s}$

9 $\frac{10}{3}mga$ **10** $m\sqrt{\dfrac{gl}{2}}; \frac{3}{4}mg$ **11** $79.5\,\text{kJ}$

12 $\frac{1}{2}(1 - e)u, \frac{1}{4}(1 - e^2)u, \frac{1}{4}(1 + e)^2 u$ **13** $400; \frac{1}{2}\,\text{m/s}^2; 16.8\,\text{kJ}$

14 $4.5 \times 10^4\,\text{N}; 450\,\text{kW}; 24\,\text{m/s}$

15 (a) $(3c - 1)u; (4 - 2c)u$ (b) $c - 1$ (c) $1 \leqslant c \leqslant 2$ (d) $\frac{5}{4}$

16 $\sqrt{3}:1$ **17** $\dfrac{u}{4}(3 - 2e - e^2), \dfrac{u}{4}(3 + e^2), \frac{1}{2}u(1 + e); e = 1$

18 $\sqrt{\dfrac{2gl}{3}}; m\sqrt{6gl}$ **19** $\frac{3}{5}$ **20** $17.55\,\text{kW}; 720\,\text{N}$ **21** $\dfrac{1}{\sqrt{e}}$ **22** $0.015\,\text{m/s}^2$

23 (a) $1.5\,\text{m/s}^2$ (b) $1050\,\text{N}; P = 4850; 72.75\,\text{kW}; 110\,\text{m}$

24 $\frac{1}{4}\pi; \frac{4}{3}mu\sqrt{e}$ **25** $\frac{4}{5}(u - v); \frac{2}{5}mu^2; \frac{1}{8}$

Unit 6

Exercise 6.1

1 $(2\frac{1}{2}, \frac{5}{8})$ **2** $(\frac{2}{3}, \frac{5}{8})$ **3** $(-0.72, 1)$ **4** $5.6\,\text{cm}, 4\,\text{cm}$
5 $1.5\mathbf{i} + 2.7\mathbf{j}$ **6** $(2, 1\frac{1}{2})$ **7** $(2\frac{1}{2}, 2)$ **8** $(3, 3)$
9 $(4, 1)$ **10** $(3, 1)$ **11** $(5, 1)$ **12** $(4, 3)$
13 $(7\frac{2}{3}, 3\frac{1}{3})$ **14** $1\,\text{cm}, 4.2\,\text{cm}$ **15** $7, 14$ **16** $(2\frac{1}{2}, -7\frac{1}{2})$

Exercise 6.2

1 $(2, 2.6)$ **2** $(2.6, 1.8)$ **3** $(6, 2\frac{2}{3})$ **4** $(4, 3.38)$
5 $28.3\,\text{cm}$ **8** $5\,\text{cm}$ **9** $18\frac{4}{7}\,\text{cm}$ **10** $3\,\text{cm}$ from each
11 $28a/9\pi$ **12** $(-0.71, 0.36)$ **13** $2\sqrt{3}a/9$ **14** $33/19\,\text{m}$

Exercise 6.3

1 $\left(\dfrac{3h}{5}, 0\right)$ **2** $\left(\dfrac{16}{5}, \dfrac{128}{7}\right)$ **3** $(2, 1.6)$ **4** $\left(\dfrac{30}{7}, 0\right)$

5 $(2.875, 5.96)$ **6** $(1.5, 3.6)$ **7** $(0.45, 0.45)$

Exercise 6.4

1 $\left(\dfrac{16}{3},0\right)$ 2 $\left(\dfrac{16}{5},0\right)$ 5 $33h/76$

Exercise 6.5

1 (b) $2r/\pi$ 8 $2h/3$

Exercise 6.6

2 $\frac{4}{3}$

4 $W\cos\alpha+\frac{1}{3}W\sin\alpha$

6 $2h/3$ or $2h/5$; 7.5; $\dfrac{W}{\sqrt{5}}\left(\dfrac{4}{15}\sin\theta-2\cos\theta\right)$ 8 $\dfrac{45}{112}$.

Miscellaneous Exercise 6B

1 $3a/4$ 2 $0.5\,\text{cm}$; $8.05\,\text{cm}$ 3 $22°$ 4 $(7a/18,4a/9)$
5 $\frac{11}{15}h$ 6 $2a/\pi$ 7 $28a/9\pi$ 8 $a\sqrt{3}/9$, $4a\sqrt{3}/9$;
(a) $\sqrt{3}/9$ (b) $10mg/3$ 10 $(a^2+ab+b^2)/(a+b)$

11 $3W/16$; $3W/16\leftarrow$, $11W/16\downarrow$ 12 $-1+\sqrt{\dfrac{1+\sin\phi}{\sin\theta}}$

13 $51a/8$ 14 $\dfrac{6l^2+4hl+h^2}{12l+4h}$ from base of cylinder

15 $57a/52$; $21\,Wa\sin\theta/52$ 17 $(H+k):(3H-k)$
18 (a) $\frac{1}{2}W\operatorname{cosec}\theta$ (b) $\mu W/(\sin\theta-\mu\cos\theta)$; $\mu<1/(2+\cot\theta)$; $\frac{1}{2}W(\sin\theta-\cos\theta)$
19 (a) $4.4a$ (b) $3.6a$; $T_A=0.4W$, $T_D=0.6W$
20 (a) $W/\sqrt{15}$; $4W/\sqrt{15}$ (b) 6

Unit 7

Exercise 7.1

1 (a) $40\pi\,\text{rad/min}$ (b) $\pi\,\text{rad/s}$ (c) $750/\pi\,\text{rev/min}$ ($\approx239\,\text{rev/min}$)
2 (a) $12\,\text{rev/min}$ (b) $2\pi/5\,\text{rad/s}$ 3 (a) $3.5\,\text{rad/s}$ (b) $52\,\text{cm/s}$
4 (a) $4.5\,\text{m/s}^2$ towards the centre   (b) $18\,\text{m/s}^2$ towards the centre
5 $0.4\,\text{m/s}^2$; $8\,\text{m/s}^2$ 6 $5.76\,\text{m/s}^2$ 7 $2.4\,\text{rev/min}$
8 $1.26\,\text{m/s}$; 3π seconds   9 $0.92\,\text{m/s}$ 10 $1.88\,\text{m/s}$, $1.47\,\text{m/s}$
11 $4.19\,\text{mm/s}$ 12 (a) $177\,\text{m/s}$ (b) $637.5\,\text{m/s}$; $398\,\text{mph}$
13 $1.023\,\text{km/s}$ 14 $0.2618\,\text{rad/h}$; $1668\,\text{km/h}$, $1444\,\text{km/h}$, $933\,\text{km/h}$
15 $25\,\text{m/s}^2$, $3750\,\text{N}$, 2.5

Exercise 7.2

1 $12\,\text{N}$ 2 $3.5\,\text{rad/s}$ 3 $75000\,\text{N}$ 4 $210/\pi$
5 $800\,\text{N}$ 6 $15.7\,\text{cm}$ 7 $38.6\,\text{rev/min}$ 8 0.66
9 $0.98\,\text{N}$ upwards and $2.25\,\text{N}$ towards the centre.
10 $26.56\,\text{m/s}$ 11 $41.8\,\text{cm}$ 12 $2mg$ 13 $(lm\omega^2-Mg)/m\omega^2$

Exercise 7.3

1 $4.76\,\text{rad/s}$ 2 $212\,\text{N}$; $82°$ 3 $3.84\,\text{N}$; $1.03\,\text{cm}$
4 g/ω^2 5 $10.8\,\text{N}$; $12.4\,\text{N}$; 24.5 6 $77.2\,\text{rev/min}$ 8 $60°$

Exercise 7.4
1 $6.125\,\text{N}, 1.725\,\text{N}$ 2 $12\,\text{rad/s}; g\,\text{N}$ 3 $ml\omega^2; m(g-h\omega^2); g/h$

5 $5:3$ 6 $5mg/4; \pi\sqrt{(3l/2g)}$ 7 $5m\left[\dfrac{g}{8}+\dfrac{u^2}{9a}\right]; 5m\left[\dfrac{u^2}{9a}-\dfrac{g}{8}\right]$

Exercise 7.5
1 $14\,\text{m/s}$ 2 $17.7\,\text{cm}$ 3 $66.5°$ 4 $36.4g\,\text{kN}$
5 $31\,\text{m/s}$ 6 0.42 7 $7\,\text{m/s}; 21\,\text{m/s}$ 8 $0.4; 21.8°$
9 $7.3°; 5/4g$ 10 $22.2°$

Exercise 7.6
1 $22.2\,\text{N}; 18\,\text{cm}$ 2   (a) $4.2\,\text{m/s}$ (b) $\sqrt{44.1}\approx 6.64\,\text{m/s}$ 3 $3g/8\,\text{N}$

4 $m\left(\dfrac{u^2}{l}+3g\sin\theta\right)$ 6 $7:1$ 7 $2l/3$

Exercise 7.7
1 $22.56\,\text{kN}$ 2 $12.5\,\text{m/s}$ 3 $r/3$ 4 $2mg+3mg\cos\theta$
5 $u\geqslant 2\sqrt{ga};$ (a) outward   (b) inward   7 $2:5$ 9   (ii) $7ag/2; \sqrt{ag/2}$

Exercise 7.8
1 $\frac{3}{2}g\,\text{N}; \frac{7}{2}g\,\text{N}$; complete revolutions   2 $\frac{2}{3}l$
4 $\frac{3}{2}mg[1-\cos\theta+\sin\theta]; \frac{1}{4}g(3\sin\theta+\cos\theta); \frac{3}{4}mg[\cos\theta-\sin\theta]$
5 $\frac{2}{3}mg(1+\cos\theta); \frac{1}{3}mg[5\sin\theta-4\theta]$

Exercise 7.9
1 $u\geqslant 2(ga)^{1/2}; (mu^2/a)-mg(2+3\cos\theta)$
2 (a) $\sqrt{2ga\sin\phi}$ (b) $2g\sin\phi, g\cos\varphi$ (c) $3mg\sin\varphi; \frac{1}{2}$

3 $\dfrac{d^2\theta}{dt^2}=-\dfrac{g\sin\theta}{a}; R=\dfrac{mu^2}{a}+mg(3\cos\theta-2); \sqrt{e^2u^2+2ga(1-e^2)}$

4 $\sqrt{ga(2\cos\theta-1)}; 2mg(3\cos\theta-1); 8a/25$

Miscellaneous Exercise 7B

1 (a) $2, 160$ (b) $10\,\text{m/s}$ or $30\,\text{m/s}$ (c) $K\leqslant 40$ 2 $T=2\pi\sqrt{\dfrac{ma}{F}}; 1.60\,\text{h}$

3 $f=\dfrac{a^2g}{r^2}; \dfrac{2\pi(a+h)^{3/2}}{a\sqrt{g}}$ 4 $m(g-h\omega^2)$ 6 $\frac{2}{3}\sqrt{3mg}$ and $\dfrac{mu^2}{l}-\dfrac{\sqrt{3mg}}{3}$

7 $\sqrt{g/h}; 5:4$ 8   (a) $mp\omega^2$ (b) $Mq\omega^2-mp\omega^2$
9 $5u/12$ and $u/6; 2\pi/3, 0.405\,\text{J}; 0.0075\,\text{s}$ 10 $\sqrt{3ga/2}$

11 $\frac{1}{2}m\left(\dfrac{g}{\sin\alpha}+l\omega^2\right)$ 12 $m\left(\dfrac{ag}{b}+\dfrac{1}{2}a\omega^2\right); m\left(\dfrac{1}{2}a\omega^2-\dfrac{ag}{b}\right)$ 13 $\frac{3}{4}$.

14 $mg(2+3\cos\theta); 2\sqrt{\dfrac{g}{a}}\cos\theta/2$ 15 $\dfrac{mu^2}{2a}-3mg-3mg\sin\phi$

16 $\frac{1}{4}mg(17+12\cos\theta); \frac{3}{2}m\sqrt{ga}$
17 $(mu^2/a)-2mg+3mg\cos\theta; mg$
18 $m[g\sin\alpha+v^2/l\cot\alpha]; m[(v^2/l)-g\cos\alpha]$

Unit 8

Exercise 8.1
1	4.95 s; 99 m	**2**	303 m	**3**	72 m; 44.1 m; 84.4 m
4	24.8 m	**5**	0.45 s; 3.32 m/s	**6**	23.6 cm
7	100 m/s; 49 m/s.	**8**	20.4 m/s; 78.9° to wall	**9**	904 m
10	15.7 m/s; 5 m	**11**	83.2°, 14.1 m/s	**12**	21.5 m/s; yes
13	$2l$				

Exercise 8.2
1 (a) 25.8 m (b) 4.6 s (c) 179 m   **2** 4.43 km   **3** 7.2 m.
5 25.1 m/s; 23.7° below horizontal; 6   **6** 16.2 m
7 23 m/s at 51.4° above horizontal; 23 m/s at 51.4 below horizontal; 3.05 s
8 3.6 m; 1.76 m.   **9** 81° or 40°
10 $\tan^{-1}\frac{4}{3}$; 63.9 m/s; 10.4 s

Exercise 8.3
1 22.9 m   **2** 18.75 m/s and 20 m/s   **3** 28.1 m/s at 21.4° above horizontal; 5.4 m
4 1.38 m; 58.7 m   **5** 20.4 m/s at 35° above the horizontal
6 (a) 25 m/s; 16.3° (b) 4 s (c) 96 m (d) 40.8 m/s at 54° below horizontal
7 15.6° or 67.6°   **9** 16.1 m/s; 65.6°

Exercise 8.4
1 6.25 km   **2** 1105 m/s   **3** 4.70 m
5 36 km; 16.9° and 73.1°   **6** (a) $\dfrac{U^2\sqrt{3}}{2g}$ (b) $\dfrac{2\sqrt{2}U^2}{3g}$
7 156 m   **8** 30 m/s; 32.2 m/s; 4.2 s; 3.2 s; 5.6 s.
9 (a) 19° or 71° (b) $\dfrac{u\sin\alpha - gr}{u\cos\alpha}; \frac{2}{5}$

Exercise 8.5

2 $\tan^{-1}\frac{7}{4}$; 86.4 m   **3** $\sqrt{\dfrac{13ga}{3}}$ **4** $r^2\tan^2\alpha - 2hr\tan\alpha + (r^2 - 2h^2) = 0$

5 $\dfrac{2b}{a}; \left(2b + \dfrac{a^2}{2b}\right)g; p = \dfrac{b}{a^2}, q = a, r = b; \dfrac{1}{\sqrt{2}}$

Exercise 8.6
1 $4\sqrt{3}$ s; 240 m   **2** 137 m   **3** (a) 19.9 m (b) 74.3 m
5 47.8° and 72.2°

Exercise 8.7
1 $0, -g$; $x = V\cos\alpha t$, $y = V\sin\alpha t - \frac{1}{2}gt^2$

2 $u, (v^2 - 2gh)^{1/2}$; $v = \frac{1}{2}\sqrt{g(8h + d)}$ **3** $g\dfrac{b}{V}$ **4** $\sqrt{\dfrac{10gh}{3}}$

Miscellaneous Exercise 8B

1 $\sqrt{320}$ m/s   **2** (a) $\dfrac{v^2\sin^2\alpha}{2g}$ (b) $\dfrac{v^2\sin 2\alpha}{g}$

3 20, 30 (a) 33.75 m

    (b) $\frac{3}{4}$, 25 m/s (c) 90 m   **4** $4a/T, \frac{1}{2}gT$; $\dfrac{(256a^2 + g^2T^4)^{1/2}}{4T}$

5 (a) $21\frac{2}{3}$ N (b) 15.6 m; 4   **6** $\tan^{-1}\frac{3}{2}$, 20.6 m/s; 10 m, $\tan^{-1}3$

7 $u\sin\alpha/g$; $\frac{1}{2}u\cos\alpha$; $u^2\sin2\alpha/4g$

8 $x = u\cos\alpha t$, $y = u\sin\alpha t - \frac{1}{2}gt^2$; u^2/g (ii) $mg(\frac{1}{2}R - y)$

9 20 m/s, 60 m/s; $\frac{40}{3}$ m/s; 40 m.

10 $\sqrt{h/g}$; $\sqrt{2h/g}$; $\sqrt{2gh}$ **11** $\dfrac{u\sin\alpha}{\sin\beta}$; $\dfrac{u^2}{2\sin\beta}\sin(\beta-\alpha)$; $\dfrac{u^2}{4g}(1+\sqrt{3})$

12 (a) 5 m (b) $30\sqrt{3}$ m (c) -20 m/s (d) 100 N; 2.25 kW

13 $2\tan\beta$; $\dfrac{2V}{g\cos\beta}$ **14** ($V\cos\alpha t$, $Vt\sin\alpha - \frac{1}{2}gt^2$); $\dfrac{V^2}{2g}\sin^2\alpha$; $\dfrac{4V^2\tan\alpha}{g(4+\tan^2\alpha)}$; $1/\sqrt{3}$

15 $V^2\cos^2\alpha(\tan^2\alpha - \tan^2\beta)/2g$

16 $\left[\dfrac{2gh}{2\tan\alpha + 1}\right]^{1/2}\Big/\cos\alpha$; $\left[\dfrac{2h}{g}(2\tan\alpha + 1)\right]^{1/2}$; $\frac{1}{3}$ and 1.

17 $\left[\dfrac{u^2\sin^2\alpha}{2g}, \dfrac{u^2\sin^2\alpha}{2g}\right]$; $90° - 1^{st}$ value of α; $b^2/4h$

Unit 9

Exercise 9.1

1 $6\cos2t - 8\sin2t$, $-12\sin2t - 16\cos2t$; 2

2 $\frac{3}{5}t^2 + 2t + 3$; $\frac{1}{5}t^3 + t^2 + 3t$ **(3)** $t^3 - t + 4$, 10 m/s   **4** $\frac{1}{3}$ s, 3 s; 15 m

5 $\sqrt{10t + 4}$; 3.76 m/s   **(6)** $3\sqrt{10}$ m/s   **(7)** 2 m   **(8)** $u/(1 - 20u^2)^{1/2}$; $1/4u$

Exercise 9.2

1 5 m/s   **5** 3   **6** 20 m   **7** $u^2[1 - 2/a]/b$

Exercise 9.3

1 3.7 s   **2** $(b/u) + kb^4/12m$ **4** $u\sqrt{3}$ **5** (i) $\dfrac{S}{mv} - g\sin\alpha - \dfrac{kv}{m}$

6 $mg(kt + me^{-kt/m} - m)/k^2$; $\dfrac{m^2g^2}{k^2}(tk + me^{-kt/m} - m) - \dfrac{m^3g^2}{2k^2}(1 - e^{-kt/m})^2$

Exercise 9.4

1 (a) $\pi/3$ s (b) π s (c) 1 s   **2** $\pi\sqrt{2}$ s   **(3)** 16 m/s^2 **4** 5 m/s^2

5 (a) at mean position (b) at farthest points from mean position

6 (a) 1.6 m/s (b) 3.2 m/s^2 (c) 1.25 m/s

7 (a) $\frac{3}{2}$ m/s (b) $\sqrt{5}/2$ m/s (c) $\sqrt{5}$ m

8 (a) 1.3 m (b) π s (c) 5.2 m/s^2

9 3 m/s (b) 60 cm (c) $2\pi/5$ s (d) 9 J

10 (a) 5 m (b) 8 m/s   **12** (a) 12 s (b) 1 m (c) 2 m

13 (a) $\frac{4}{3}$ s (b) $2\frac{2}{3}$ s (c) $5\frac{1}{3}$ s (d) $6\frac{2}{3}$ s

14 (a) $4\pi/15$ m/s (b) 0.42 m/s (c) 0

Exercise 9.5

1 96 cm   **2** 1 s   **3** 0.695 m/s, 1.66 m   **4** 0.253; 2.1 s

6 0.244 s   **7** $\frac{5}{2}$; $x - 2 = 2.5\sin2t$ **8** $5\pi^2/18$; 2 s.

9 (a) 5 m (b) $4\sqrt{21}$ m/s (c) 0.103; $200\sqrt{3}$ W; 400 W

Miscellaneous Exercise 9B

2 1.5 s   **3** 20 cm   **4** 5 m/s; 18 s   **5** 5 m

6 $u/9$ **7** $u^2(3 + 2u)/6k$ **8** $\frac{5}{6}$ s; $8\pi^2/25$ N

9 $v = [2k(1 - e^{-\alpha x})/m\alpha - 2\lambda x]^{1/2}$ **10** 2.3 m (a) 4 m/s, 2 m/s (b) 0.6 kg m/s

11 $\dfrac{c}{g}\tan^{-1}(V/c)$ **12** $x = (u + k/\omega)t - k\sin\omega t/\omega^2$ (ii) $u + k/\omega = 0$

13 $m(g + a\omega^2\cos\omega t); 2\pi/3\omega$ **14** $\dfrac{\mu g}{k^2}(1 - e^{-kt}) + \dfrac{V}{k}(1 - e^{-kt}) - \dfrac{\mu g t}{k}$

15 17.31 m; 17.45 m   **16** 20 m/s; 4/15 m/s^2

17 126 m, 30 m/s, PQ $= |80 - 20t - 2t^2|$; 2 s and 4 s, 10 s

19 $x = \dfrac{c^2}{2g}\ln\left(\dfrac{c^2 + b^2}{c^2 + v^2}\right); \dfrac{dw}{dy} = \dfrac{g}{w}\left(1 - \dfrac{w^2}{c^2}\right); y = \dfrac{c^2}{2g}\ln\left(\dfrac{c^2}{c^2 - w^2}\right)$

20 0.15175 m; $v = \dfrac{g}{k}(1 - e^{-kt})$; 0.15293 m

22 $2T\theta - 2mg + 3mg\cos\theta; \dfrac{d^2\theta}{dt^2} = \dfrac{T}{ma} - \dfrac{g\sin\theta}{a}; 2\pi\left(\dfrac{2a}{\sqrt{3g}}\right)^{1/2}$

Unit 10

Exercise 10.1
1 (a) 1.2 N (b) 2.25 m   **2** 3 N   **3** 25 N   **4** 18 cm   **5** 1.6 m
6 8g N   **7** 104 cm   **8** 4.9 m/s^2 **9** 1.79 m/s   **10** 8g N, 2g N
12 (a) 2 m/s^2 (b) $\frac{11}{30}$ m/s^2

Exercise 10.2
1 1 J   **2** 0.185 J   **3** 3.2 J   **4** 1.6 J   **5** 0.24 J
6 86 cm/s   **7** 12.65 m/s   **9** 12.5 cm   **11** $\sqrt{8gl/3}$

Exercise 10.3
1 24g N; 8g N; 16g N; 2.42 m/s   **3** $\sqrt{13}mg$, $3\sqrt{2}mg$; 56.3°, 45°
4 $[ma(a + b)/\lambda b]^{1/2}\pi/2$ **6** $\lambda(2 - \cos\theta)/\cos\theta; \lambda(2 - \cos\theta) + w; \lambda(2 - \cos\theta)\tan\theta$
7 0.32 m; $3\pi/125$; $[0.36\cos(250t/3), 0.36\sin(250t/3)]$; $3\pi/35$

Exercise 10.4
1 0.4 m/s   **2** 1.29 m/s   **3** 1.4 m/s

Exercise 10.5
1 45 cm; 1.35 s, 15 cm; $e \leqslant 45$ cm   **2** 78.9 N
3 500 grams, 79 N   **4** $mga/\lambda; 2\pi\sqrt{am/\lambda}$;   **5** $a/4$

Exercise 10.6
1 $l/4$ **2** $2a; 3\sqrt{ga}; \frac{2}{3}\pi\sqrt{(a/3g)}$ **3** $mg/2; 3\sqrt{2ag}/4$.
4 $a/2$ **5** 6g N; 0.78 s; 21 cm; 0.49 s, 53 cm below P

Miscellaneous Exercise 10B
1 99.4 cm   **2** 7/5000   **3** loses 44 s   **5** 2 m/s

6 $4mg; x = \dfrac{3a}{2}\sin 2\sqrt{\dfrac{g}{a}}t; \dfrac{3a}{2}, \dfrac{\pi}{12}\sqrt{\dfrac{a}{g}}$ **7** $\sqrt{2a\lambda/m}$ **8** $\frac{3}{4}$

9 $mg/2; mg/2$ (i) $3l/2$ (ii) $\pi\sqrt{(2l/g)}$ **10** $A = -2g\cos\alpha/l,\, B = -2mg\cos\alpha;$

$\dfrac{d^2\theta}{dt^2} = -g\sin\theta/l;\ \theta = \alpha\cos\sqrt{\dfrac{g}{l}}\cdot t$ **11** $T = ml\omega^2;\ R = m(g - h\omega^2);\ g/h;$

$\dfrac{g}{2a} \leqslant \omega^2 < g/a$ **12** π/ω **13** $2l$ **14** $7a;\ \dfrac{2\pi}{3}\sqrt{\dfrac{2a}{g}};\ \sqrt{3ga/2}$

15 5; 0.5 m; 4/3; 0.5 m, 2.5 m/s   **16** $\frac{1}{2}l$ **17** (a) $\pi(2l/g)^{1/2}, l/4$

18 $3g/n^2;\ 2\pi/n$ **20** $8u, u$ (a) $9mu$ (b) $18mu^2$

21 $\sqrt{2gx(5 - 2x/l)};\ 2.5\sqrt{gl}$ **22** (i) $(mg/h)(a\cos\omega t + h)$
(ii) $(mg/h)(a\cos wt + 2h)$ **23** $5a/2$ **24** (a) 2.5 m/s   (b) $\pi/60\,$s

(c) 4.5 N   (d) 1 m   **25** $\sqrt{gl};\ \dfrac{d^2x}{dt^2} = -\dfrac{g}{l}(3x + l);\ H = l/3,\ k = l/\sqrt{3};$

$2\sqrt{\dfrac{l}{g}}\left[\sqrt{3} - 1 + \dfrac{\pi}{3\sqrt{3}}\right]$ **26** $\sqrt{2gb};\ \dfrac{\pi}{4}\sqrt{\dfrac{b}{g}}$

Unit 11

Exercise 11.1
1 5 m/s   **2** $6ti + 30tk;\ 6mt(i + 5k)$
3 $(\frac{2}{3}t^3 + t^2 + 3t)i + (\frac{1}{3}t^3 + t + 1)j$ **4** 5 m/s at $\tan^{-1}\frac{4}{3}$ to the **i** direction.
5 $(2 + 3t)i + (2t - 1)j$ **6** 15.3°   **7** $(3, 1, -7);\ (3 + 5t)i + (1 + 2t)j + (t - 7)k$
8 6 and -2 **9** $\pi/4$
10 $2m(i + 2j + k);\ m(-12i + 4j + 4k)$
11 $20i + 10j$; 5; 125 units
12 $4ti + (2t - 4)j + 3k;\ 4i + 2j;\ 43\sqrt{14}/35,\ -2\sqrt{14}/14$

Exercise 11.2
1 3 units   **2** 96 J, 178 J, 168 J   **3** $(30i + 60j - 150k)$ m/s; 135 kJ; 9 kW
4 90 J; 6 m/s   **5** -21 J; 0 J; first force resists the motion of the particle; second force
is perpendicular to the displacement.
6 14.6 Ns at 164° to **i** direction   **7** $7i - j - k$
8 $(130i - 30j + 240k)$ m; $(2i - j + 3k)$ m/s^2
9 $(40, 15, -39)$ **10** (a) 13 J   (b) $(2i + 3j + 1.5k)$ m/s^2.
(c) $(7i + 10j + 7k)$ N   (d) 122 J   **11** m J

Exercise 11.3
1 $12i - 6j + 4k$ (a) $(1 + 12t)i + (1 - 6t)j + (2 + 4t)k$ (b) $49i - 23j + 18k$
2 (a) $i + 3j + 3k$ (b) $(1 + 3t)i + (3 + 2t)j + (3 - t)k$ **3** $9i + 11j - 5k$
4 (a) collide at point with position vector $11i + 8j + 18k$ (b) no collision; $\sqrt{7/3}$
(c) no collision; 5 km.
5 $t = 7;\ 35j,\ 4i + 32j$ **6** $t = 4;\ (5i + 9j - 9k)$ metrons.
7 (a) $11i + 3j$ (b) 85.6°

Exercise 11.4
1 18 km/h   **2** 41 km/h at S46.9°W
3 2 min.   **4** 33.5 km/h at 26.6° to direction of bus
5 36.4 km/h at S55.2°W   **6** Bearing of 109.5° or 170.5°

8 $\dfrac{a}{2\sqrt{u^2 - v^2}}$

Exercise 11.5

1 17.4 m/s at 23.4° to $\overrightarrow{XY}$, 8 m/s at 60° to $\overrightarrow{XY}$
2 19.4 km/h at N61.9°E   3   030°
4 (i) 13 km/h   (ii) 113°   5   3059 km/h, 349°; 169°

Exercise 11.6

1 $\sqrt{31}$ m/s, 321°, 116 s   2   7.5 km/h   3   9.28; 11.86 km
4 12.27   5   228 m   6   151 m

Exercise 11.7

2 N73.7°W   (i) 6.72 km   (ii) 2.94 min; $\arcsin\frac{7}{15}$ E of N
3 (i) 50 km/h at N36.9°E   (ii) 12 km   (iii) 7.2 km and 9.6 km;
4 (i) $\frac{4}{3}$; 12.15   (ii) $\frac{3}{4}$; 12.20: $\frac{10}{11}$

Exercise 11.8

1 3.78 nmi, 262°   2   003.6°; 11.25   3   S50.3°W
4 S14.3°W; 8 min 20 s   5   553 km/h; 1015 m.
6 18.03 km/h; 13.12; 4.13 km/h.

Exercise 11.9

1 $(10t\mathbf{i} + 0.6\mathbf{j})$ km, $(6.4t\mathbf{i} + 4.8t\mathbf{j})$ (a) $(3.6t\mathbf{i} + (0.6 - 4.8t)\mathbf{j})$ km
   (b) $(3.6\mathbf{i} - 4.8\mathbf{j})$ km/h.
2 $(v + 3)\mathbf{i} + (7v - 29)\mathbf{j}$; $\overrightarrow{AB} = [-56 + t(v+3)]\mathbf{i} + [8 + t(7v - 29)]\mathbf{j}$; $v = 4$; $t = 4$
3 $\sqrt{58}$ km; $(6\mathbf{i} + 8\mathbf{j})$ km/h.
4 (a) $(8\mathbf{i} + 15\mathbf{j})$ knots, $(9\mathbf{i} + 12\mathbf{j})$ knots   (b) $(\mathbf{i} - 3\mathbf{j})$ knots   (c) $t\mathbf{i} + (10 - 3t)\mathbf{j}$
5 $(p + 1)\mathbf{i} + (q - 1)\mathbf{j} + (r - 2)\mathbf{k}$; $[(p + 1)t - 5]\mathbf{i} + [(q - 1)t + 1]\mathbf{j} + (r - 2)t\mathbf{k}$
   (a) $t = 1\frac{1}{2}$; $5/\sqrt{2}$ (b) 1.20.

Miscellaneous Exercise 11B

1 (a) $\overrightarrow{OP} = 3t^3\mathbf{a} + 8t^2\mathbf{b}$ (b) 4   (c) 19.2 m
2 $\sqrt{117}$; $\mathbf{r} = (\mathbf{i} - \mathbf{j} + 2\mathbf{k}) + t(6\mathbf{i} + 9\mathbf{j})$ 3 $76\frac{6}{7}$ J.
4 3 m/s; $\overrightarrow{AP} = (1/5)(t^2 + t)(9\mathbf{i} + 12\mathbf{j})$; $84\mathbf{i} + 97\mathbf{j}$
5 $\mathbf{u} = -a\omega \sin\omega t\mathbf{i} + a\omega \sin 2\omega t\mathbf{j} + a\omega \cos 2\omega t\mathbf{k}$; $\frac{1}{2}ma^2\omega^2[1 + \sin^2\omega t]$

6 (a) 13   (b) $\begin{pmatrix} 1 \\ -2 \\ 21 \end{pmatrix}$ 7   (a) $\mathbf{v} = \left(\frac{3t^2}{4} - t + 4\right)\mathbf{i} - 2\mathbf{j} + (5 - 4t)\mathbf{k}$ (b) 3, 5

   (c) 208 W   8 $\begin{pmatrix} 2 \\ 10 \end{pmatrix}$; $\begin{pmatrix} 5 \\ 7 \end{pmatrix}$; $\mathbf{r} = \begin{pmatrix} 5 \\ 7 \end{pmatrix} + \lambda\begin{pmatrix} 2 \\ 10 \end{pmatrix}$; $\begin{pmatrix} 1 \\ -7 \end{pmatrix}$; 39λ

9 $\mathbf{v} = 2c\lambda t[\sin\lambda t^2\mathbf{i} - \cos\lambda t^2\mathbf{j} + \mathbf{k}]$,
   $\mathbf{a} = 2c\lambda[(\sin\lambda t^2 + 2\lambda t^2 \cos\lambda t^2)\mathbf{i} - (\cos\lambda t^2 + 2\lambda t^2 \sin\lambda t^2)\mathbf{j} + \mathbf{k}]$;
   $T = 4mc^2\lambda^2 t^2$; $F = 4m\lambda c$
10 (i) $6\mathbf{k}$ (ii) $15 + 6t^2$ (iv) $7\mathbf{i} - \mathbf{j} + 4\mathbf{k}$
   (ii) (a) $\mathbf{v} = -a\omega \sin\omega t\mathbf{i} + 2a\omega \cos 2\omega t\mathbf{j}$, $\mathbf{a} = -a\omega^2 \cos\omega t\mathbf{i} - 4a\omega^2 \sin 2\omega t\mathbf{j}$
   (b) $0, 0.66/\omega, \pi/2\omega$ (c) $\frac{11}{8}ma^2\omega^2$ (d) $-7\sqrt{3}a^2\omega^3/4$
12 (b) $4mg$ (c) $-1/3$ F   (d) $24mg$
13 14 units   (i) $6\mathbf{i} + 4\mathbf{j} - 12\mathbf{k}$, $-4\mathbf{i} - 12\mathbf{j} - 6\mathbf{k}$ (ii) $4\mathbf{i} + 12\mathbf{j} + 6\mathbf{k}$
14 $\mathbf{r} = a\cos\omega t\mathbf{i} + a\sin\omega t\mathbf{j}$, $\ddot{\mathbf{r}} = -\omega^2\mathbf{r}$; $a\omega^2$ along PO; $\sqrt{g/3a}$
15 $\mathbf{r} = \mathbf{a}(2 - \sin\omega t)\mathbf{i} + 2a\cos\omega t\mathbf{j} + a(\sin\omega t - 3)\mathbf{k}$,
   $\mathbf{v} = -a\omega\cos\omega t\mathbf{i} - 2a\omega\sin\omega t\mathbf{j} + a\omega\cos\omega t\mathbf{k}$; $(2n + 1)\pi/2\omega, n = 0, 1, 2, \ldots$;
   $\sqrt{5}a, 5a$ 16   4.82 m/s on bearing 275°, 240°

17 $\tan\phi = \dfrac{1}{1 + 2\tan\theta}$; $\phi = 24.9°$, $v = 20.6$

18   (a) $8\mathbf{i} + 15\mathbf{j}, 9\mathbf{i} + 12\mathbf{j}$ (b) $\mathbf{i} - 3\mathbf{j}$ (c) $t\mathbf{i} + (10 - 3t)\mathbf{j}$

19 $\mathbf{a} = (10t\mathbf{i} + 0.6\mathbf{j})\,\text{km}$, $\mathbf{b} = (6.4t\mathbf{i} + 4.8t\mathbf{j})\,\text{km}$
   (a) $[3.6t\mathbf{i} + (0.6 - 4.8t)\mathbf{j}]\,\text{km}$ (b) $(3.6\mathbf{i} + 4.8\mathbf{j})\,\text{km/h}$.

20   12.30 pm; 7.79 km; 24 min; $(2.5\mathbf{i} - 2\mathbf{j})\,\text{m/s}$ (a) $11\frac{1}{4}\,\text{km}$ (b) 13.15

21 $5\mathbf{i} + 12\mathbf{j}, -4\mathbf{i}; 5t\mathbf{i} + 12\mathbf{j}; (80 - 4t)\mathbf{i}; 16\,\text{s}, 80\,\text{m}$

22 $w(\cos\theta + \sqrt{\lambda^2 - \sin^2\theta})$

23 $(1600\mathbf{j} + 150\mathbf{k})\,\text{m}, (800\mathbf{i} - 50\mathbf{k})\,\text{m}, (800\mathbf{i} + 600\mathbf{j})\,\text{m}; \frac{1}{9}(4\mathbf{i} - 8\mathbf{j} - \mathbf{k}), \frac{1}{5}(4\mathbf{i} + 3\mathbf{j})$;
   $(1100t - 1600)\mathbf{j} + (1000t - 150)\mathbf{k}$

Unit 12

Exercise 12.1

1   (b) W horizontally   (d) $T_{AB} = W$ (tension), $T_{BC} = \sqrt{2}W$ (tension)

2 $T_{DE} =$ thrust of $2W$, $T_{AE} =$ tension of $2\sqrt{3}W$, $T_{BD} =$ thrust of $2\sqrt{2}W$

3   (a) $60°, 30°$,   (b) 50 N, 70 N;   (c) $T_{AB} =$ thrust of $140/\sqrt{3}$, $T_{AD} =$ tension of $70/\sqrt{3}$,
   $T_{BD} =$ thrust of 20, $T_{AC} =$ thrust of $80/\sqrt{3}$, $T_{BC} =$ thrust of $140/\sqrt{3}$

4   (a) $80\sqrt{3}\,\text{N}$ (b) 231 N at 72.5° to the horiz.   (c) $T_{AB} =$ thrust of 100 N,
   $T_{AD} =$ thrust of $80\sqrt{3}\,\text{N}$, $T_{CD} =$ thrust of 120 N, $T_{BC} =$ tension of $60\sqrt{3}\,\text{N}$,
   $T_{BD} =$ thrust of 40 N

5   (a) 40 N, 20 N   (b) $T_{AB} =$ thrust of $80/\sqrt{3}\,\text{N}$, $T_{AE} =$ tension of $40/\sqrt{3}\,\text{N}$,
   $T_{BE} =$ thrust of $20/\sqrt{3}\,\text{N}$, $T_{BC} =$ thrust of $30/\sqrt{3}\,\text{N}$.   (c) 50 N, 50 N
   (d) $T_{AB} = T_{CO} = 100/\sqrt{3}\,\text{N}$, $T_{AE} = T_{DE} =$ tension of $50/\sqrt{3}\,\text{N}$, $T_{BE} = T_{CE} = 0$
   $T_{BC} =$ thrust of $50/\sqrt{3}\,\text{N}$

Exercise 12.2

1 W; At A and C, reaction is W, vertically; At B reaction is W, horizontally

2 $\frac{3}{4}W, \frac{3}{4}W$ **3** $W\sqrt{\frac{3}{6}}$, horizontally at A; $W\sqrt{39}/6$ at $\tan^{-1}2\sqrt{3}$ to the horizontal,
   at B and C.

4 $5W/2; \sqrt{(17/2)}W$ at $\tan^{-1}\frac{3}{5}$ to the horizontal

5 $\left(\dfrac{l}{4a} + 2\right)W, \left(3 - \dfrac{l}{4a}\right)W$

Exercise 12.3

1   (a) $\dfrac{2pW}{p+q}, \dfrac{2qW}{p+q}$ (b) $\dfrac{W(p-q)}{p+q}$

2   (a) At A, $\dfrac{5W}{4}\cot\theta\leftarrow, \dfrac{11W}{4}\uparrow$, At C, $\dfrac{5W}{4}\cot\theta\rightarrow, \dfrac{9W}{4}\uparrow$

   (c) $\dfrac{5W}{4}\cot\theta, \dfrac{W}{4}$ (d) $\tan^2\theta = 5$

3   At B, $\dfrac{7W}{4}$ and $\dfrac{3W}{4}\tan\alpha$; At C, $\dfrac{5W}{4}$ and $\dfrac{3W}{4}\tan\alpha$; B steps, $x = 3a/10$

4   (a) $W/2$ (b) $3W/2, W$ (c) $\frac{1}{2}$

5   (a) $(k - \frac{1}{2})W\tan\theta, (1 - k)W$ (b) $(k - \frac{1}{2})W\tan\theta, kW$ (c) $(k - \frac{1}{2})W\tan\theta, 2k\omega$

Exercise 12.4

1 $P, W/2, -2P\tan\theta; \sqrt{4P^2\sec^2\theta + 2PW\tan\theta + W^2/4}; 2P, W/2 - 2P\tan\theta.$
2 (a) $12mg\cos\theta/5$ (b) $12mg\sin\theta\cos\theta/5$ (c) $1.2mg$ (d) $5/12; 4.8m\sin\theta\cos\theta$

3 $T = \dfrac{W\sin\beta}{2\sin(\theta + \beta)\sin\theta}$

Exercise 12.5

1 $9.25\,\text{N}$ at $83.8°$ to AD   2 $1\,\text{N}, 5\,\text{N}\mu\circlearrowright$ 3 $\frac{1}{3}$ along QR from Q
4 $4\sqrt{10}\,\text{N}$ at B, $3\sqrt{10}\,\text{N}$ at C   5 $5/2; 5a/2; \text{AE} = 5a/7$
6 (a) $10P$ (b) $\frac{4}{3}$ (c) $9a/4$ (e) $20Pa$ (f) $S = 6P, T = 7P/2, U = 9P/2$
7 Force $= F$ at $(90 - \theta)$ to OA, $M_0 = F(a\cos\theta + h)\circlearrowright$; F perpendicular to BA, $M_A = Fh$

Exercise 12B

1 $5W/2 \rightarrow, 2W\downarrow; \theta = \tan^{-1}\frac{1}{2}; G = 8Wl/\sqrt{5}$ 2 $\frac{1}{2}W, W$
3 $\cot\theta = (3a - 2x)/a; a/4; (1/\sqrt{2})mg\cos\theta; x = a/2$

4 $\frac{3}{4}W\uparrow; 3Wa\circlearrowright$ 5 $\left(\dfrac{a + 2x}{4a}\right)W; \left(\dfrac{a + 2x}{a}\right)W\tan\theta$

6 (a) $200\,\text{N}$ (b) thrust of $600\sqrt{2}\,\text{N}$, tension of $600\,\text{N}$; $400\sqrt{2}\,\text{N}, 400\,\text{N}$
7 (a) $120\,\text{N}, 120\,\text{N}$ (b) $180\,\text{N}, 80\,\text{N}; 197\,\text{N}$ (c) $120\sqrt{2}\,\text{N}, 80\sqrt{2}\,\text{N}, 40\sqrt{2}\,\text{N}$;
    AE, DE are in tension

8 $2\sqrt{3}W$ at $60°$ to BC; horizontal, $\dfrac{W}{2\sqrt{3}}, \dfrac{W}{2\sqrt{3}}$; vertical $0, W$

9 Force at $A = F, M_A = F(b\cos\theta + h\sin\theta)$; Force at $B = F$,
    $M_B = bF\cos\theta; \theta = \tan^{-1}(h/b); \theta = 0$
10 $W\operatorname{cosec}\beta/2; 2(2 - \sqrt{3})/3.$

INDEX